Lecture Notes in Mechanical Engineering

Lecture Notes in Mechanical Engineering (LNME) publishes the latest developments in Mechanical Engineering—quickly, informally and with high quality. Original research reported in proceedings and post-proceedings represents the core of LNME. Volumes published in LNME embrace all aspects, subfields and new challenges of mechanical engineering. Topics in the series include:

- Engineering Design
- Machinery and Machine Elements
- Mechanical Structures and Stress Analysis
- Automotive Engineering
- Engine Technology
- Aerospace Technology and Astronautics
- Nanotechnology and Microengineering
- Control, Robotics, Mechatronics
- MEMS
- Theoretical and Applied Mechanics
- Dynamical Systems, Control
- Fluid Mechanics
- Engineering Thermodynamics, Heat and Mass Transfer
- Manufacturing
- Precision Engineering, Instrumentation, Measurement
- Materials Engineering
- Tribology and Surface Technology

To submit a proposal or request further information, please contact the Springer Editor of your location:

China: Ms. Ella Zhang at ella.zhang@springer.com
India: Priya Vyas at priya.vyas@springer.com
Rest of Asia, Australia, New Zealand: Swati Meherishi at swati.meherishi@springer.com
All other countries: Dr. Leontina Di Cecco at Leontina.dicecco@springer.com

To submit a proposal for a monograph, please check our Springer Tracts in Mechanical Engineering at https://link.springer.com/bookseries/11693 or contact Leontina.dicecco@springer.com

Indexed by SCOPUS. All books published in the series are submitted for consideration in Web of Science.

Debi Prasad Mishra · Ashok Kumar Dewangan ·
Achhaibar Singh
Editors

Recent Trends in Thermal and Fluid Sciences

Select Proceedings of INCOME 2021

Springer

Editors
Debi Prasad Mishra
NITTTR, Kolkata
Kolkata, India

Ashok Kumar Dewangan
Department of Mechanical Engineering
National Institute of Technology Delhi
Delhi, India

Achhaibar Singh
Department of Mechanical Engineering
Netaji Subhas University of Technology
New Delhi, India

ISSN 2195-4356 ISSN 2195-4364 (electronic)
Lecture Notes in Mechanical Engineering
ISBN 978-981-19-3497-1 ISBN 978-981-19-3498-8 (eBook)
https://doi.org/10.1007/978-981-19-3498-8

This Springer imprint is published by the registered company Springer Nature Singapore Pte Ltd.
The registered company address is: 152 Beach Road, #21-01/04 Gateway East, Singapore 189721, Singapore

Preface

This book presents the selected research papers on recent trends in thermal and fluid sciences from International Conference on Mechanical Engineering (INCOME-2021), which was organized by Netaji Subhas University of Technology, New Delhi, India, from November 25 to 26, 2021.

The conference was aimed at providing a platform to the researchers and engineers to share and discuss various aspects of mechanical engineering and its applications. The conference provided an excellent opportunity for the presentation of new inventions, discoveries, implementations, improvements, product innovations and manufacturing techniques. The conference was a right place for the exchange of new ideas and transfer of knowledge. The INCOME-2021 was a good platform for interaction between industry and academic professionals.

The conference was initially planned in offline mode. But it was held in mixed mode due to COVID-19 restrictions at many places. A few papers were presented at the conference venue, and a few papers were presented online.

About 60 research papers were presented during 2 days conference. The researchers from industry and academics shared the most recent studies, held discussions and exchanged ideas on several practical topics. The conference provided an opportunity for collaboration among academicians and practitioners. The recent developments and innovations in various fields of mechanical engineering were also shared through keynote lectures. The book covers the thermal and fluid topics in mechanical engineering areas such as combustion and biofuels, heat and mass transfer, refrigeration and air conditioning, fluid–structure interaction, flow control, aerodynamic, aeroacoustics, turbulence and instabilities, rheology and complex fluids, turbomachinery and thermodynamics. This book connects together various topics of thermal and fluid sciences and will be beneficial for industry as well as for academic professionals working in this field.

The editors acknowledge all the authors for their contribution to this volume. We sincerely express our profound gratitude to the candid support extended by, Netaji Subhas University of Technology, New Delhi. The editors also thank faculty and staff of the department of mechanical engineering for their enthusiastic support to accomplish this goal. We would also like to thank all those who directly or indirectly assisted us in this work. We express our sincere gratitude to the honourable Vice Chancellor, Dr. J. P. Saini for his encouragements and blessings. Last but not the least; we are grateful to the publishers of this book for publishing the selected proceedings of INCOME-2021. The editors welcome the readers' suggestions for improvement in this book.

Kolkata, India
Delhi, India
New Delhi, India
January 2022

Debi Prasad Mishra
Ashok Kumar Dewangan
Achhaibar Singh

Contents

About the Editors

Dr. Debi Prasad Mishra a professor in the Department of Aerospace Engineering and Design Program, Indian Institute of Technology Kanpur (IITK), Kanpur, India, is currently working as the Director of National Institute of Technical Teachers Training and Research (NITTTR), Kolkata. His major areas of research interests are propulsion, combustion, and atomization. He had held Indian Oil Golden Jubilee Professional Chair in IIT Kanpur. He has been conferred with several awards and recognitions. He has authored 5 textbooks and 6 edited books in the field of combustion and propulsion. He has published more than 234 research papers in international/national journals and six patents to his credit. He has undertaken several research and consultancy projects from various agencies, research labs and industries. He has developed 6 online courses under Swayam Platform, a national initiative of MHRD.

Dr. Ashok Kumar Dewangan is an Assistant Professor in the Department of Mechanical Engineering, National Institute of Technology, Delhi, India. He received his undergraduate degree from National Institute of Technology, Raipur, Chhattisgarh; postgraduate degree from Indian Institute of Technology Kharagpur, West Bengal, and Ph.D. from Indian Institute of Technology Roorkee, Uttarakhand. His areas of research are refrigeration and air- conditioning, cryogenics, two-phase flow & heat transfer, instrumentation and measurements. He has published 17 research papers in the reputed peer-reviewed international journals and conferences. Currently he is working in the areas of boiling and condensation of refrigerants, thermal modelling of selective laser melting and single/twin wire gas metal arc welding.

Dr. Achhaibar Singh is a visiting faculty in the Department of Mechanical Engineering, Netaji Subhas University of Technology, New Delhi, India. He obtained graduate degree in engineering from Madan Mohan Malviya University of Technology Gorakhpur; postgraduate degree from Indian Institute of Technology Bombay, and Ph.D. from Indian Institute of Technology Bombay. He has worked 18 years in industry as engineering manager and R&D head. His areas of research are refrigeration and air- conditioning, rotating flows in turbo machines, and computational fluid dynamics. He has published 7 research papers in the reputed peer-reviewed international journals and several papers in conferences. Currently, he is working in the areas of flow in rotor stator system and CFD of rotating flow using finite element method.

Effect of Ambient Temperature on Thermodynamic Performance of a Split Type Air Conditioner

Sunil Kumar Gupta, B. B. Arora, and Akhilesh Arora

Abstract This study investigates the effect of ambient temperature (T_0) on the thermodynamic performance of a split type air conditioner. The analysis was carried out with three different refrigerants R410A, R32, and R454B by varying T_0 from 25 °C to 40 °C. The simulation was performed to obtain volumetric cooling effect (VCE), COP and exergy destruction at different T_0. The refrigerant R32 produced 8 to 13.6% and 11.9 to 14.2% higher VCE than that of R410A and R454B respectively. The COP of R410A reduced from 5.9 to 3.17 respectively. R32 and R454B showed almost the same reduction in COP from 6.06 to 3.3. For both R32 and R454B, the exergy efficiency (η_{ex}) increased by 6% with T_0 rising from 25 to 40 °C, while an increase of 5% was found with R410A. The refrigerants R32 and R410A showed the lowest (18.66%) and greatest (22.27%) falls respectively in II law efficiency.

Keywords Ambient temperature · Exergy destruction · Exergy efficiency · COP

1 Introduction

The refrigeration and air conditioning systems provide controlled environmental conditions to increase the efficiency both of humans and machines. The heating, ventilating, air conditioning, and refrigeration (HVACR) equipment are employed in all walks of life such as industry, hotel, mall, hospital, home, and transportation, etc. For domestic use, the demand for small air conditioners has increased enormously. The HVAC designers have the challenge to improve the system's thermodynamic performance as well as reduce global warming. An air conditioning system works in diverse climate conditions at different atmospheric temperatures. So it becomes imperative to investigate the effect of ambient temperature on the system performance.

According to Oruç & Devecioǥlu [1], the ambient temperature has a considerable effect on the thermodynamic performance of the refrigeration system. Jani et al. [2]

S. K. Gupta (✉) · B. B. Arora · A. Arora
Department of Mechanical Engineering, Delhi Technological University, New Delhi 110042, India
e-mail: sunilgupta_me@yahoo.co.in

D. P. Mishra et al. (eds.), *Recent Trends in Thermal and Fluid Sciences*, Lecture Notes in Mechanical Engineering, https://doi.org/10.1007/978-981-19-3498-8_1

showed the high influence of variation in the ambient state on the system performance. Bilgili et al. demonstrated that the rise in ambient temperature has significant effects on the performance of the system and its components [3].

Arora et al. [4] reported that the exergetic efficiency is affected by the varying dead-state temperature. Exergy efficiency and exergy destruction depend on the reference temperature. It was concluded that exergy efficiency increases at elevated environmental temperatures [5]. Arora & Kaushik [6] found that the ambient temperature has a positive effect on exergetic efficiency and EDR reduction. Islam & Dincer [7] found that heat exchangers counter lesser exergy destruction with an increase in dead-state temperature.

Ribeiro [8] obtained lower COP values for the vapor-compression unit at higher ambient temperatures. COP decreased by increasing the dry-bulb temperature of ambient air [9]. Oruç et al. obtained higher COP values at elevated ambient temperatures in hot climate regions [10]. The refrigeration capacity and COP reduce against the increased compression ratio and the input power with the rise in the ambient temperature [11].

Medium pressure refrigerants were shown to be more efficient than high-pressure refrigerants at elevated ambient temperatures [12].

Going through the literature, a need was realized to explore the effect of ambient temperature on thermodynamic performance parameters for the assessment of a small split type air conditioning system. With consideration to global warming intermediate GWP refrigerants, R32 and R454B were included besides the presently used R410A.

2 Materials and Method

2.1 *The System*

A 1.5 Ton (5.25 kW) split air conditioner working on a vapor compression refrigeration (VCR) cycle was selected for the study. A liquid vapor heat exchanger (LVHE) was installed to ensure a bit of superheated vapor at the suction to the compressor and a simultaneous increase in cooling effect by subcooling the liquid leaving the condenser. Some of the important properties of the refrigerants used are presented in Table 1.

Figure 1a, b represent the schematic and p–H diagrams of the VCR cycle used in the air conditioning system.

2.2 *Analysis*

The energy effectiveness of a refrigeration system is usually represented by the coefficient of performance (COP).

Table 1 Properties of refrigerants (Genetron properties software v14.1)

Property	Chemical composition	Mol.wt (kg/kmol)	Cr. temp. (°C)	Cr. pr. (MPa)	GWP	Safety class	Temp glide (K)
R410A	R32/R125 (w/w = 50/50)	72.585	70.17	4.77	2088	A1	1.5
R32	Single component	52.024	78.4	5.186	675	A2L	–
R454B	R32/R1234yf (w/w = 68.9/31.1)	62.2	77	5.014	466	A2L	0.05

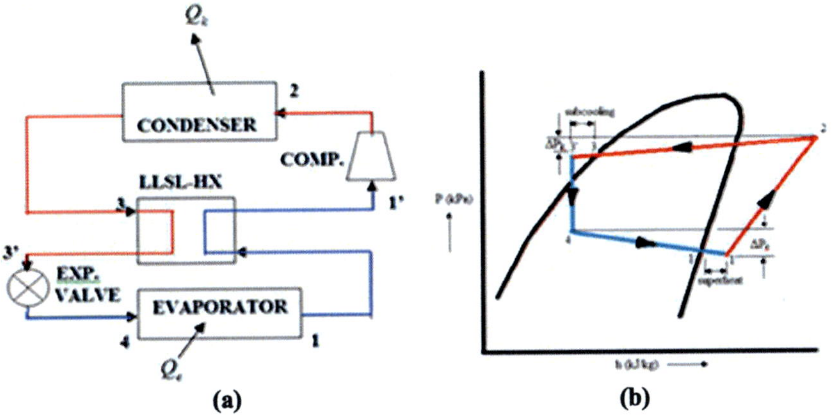

Fig. 1 VCR cycle with LVHE (**a**) schematic (**b**) p–h diagram

$$COP = \frac{\dot{Q}_e}{\dot{W}_c} \tag{1}$$

where $\dot{Q}_e$ is the thermal load and $\dot{W}_c$ is the compression work.

The volumetric cooling effect (VCE) is a measure of the cooling effect (Δh) produced in a unit volume (ν) at the evaporator exit, i.e.

$$VCE = \frac{\Delta h}{\nu} \tag{2}$$

A realistic performance can be measured by II law efficiency which is expressed as the ratio of actual COP to the maximum possible (reversible) COP subject to the same conditions.

Thus,

$$\eta_{II} = \frac{COP}{COP_R} \tag{3}$$

The exergy analysis accounts for the losses occurred during a cycle. The specific exergy is represented by:

$$\dot{E} = (h - h_0) - T_0(s - s_0) \tag{4}$$

where h_0 and s_0 denotes enthalpy and entropy of environment.

The exergy destructions in the main components of the refrigeration system can be found by the following relations:

$$\textbf{Evaporator}: \dot{E}_{D,e} = \dot{m}_r\{(h_4 - h_1) - T_0(s_4 - s_1)\} + \dot{Q}_e\left[1 - \left(\frac{T_0}{T_R}\right)\right] + W_{e,fan} \tag{5}$$

$$\textbf{Compressor}: \dot{E}_{D,c} = \mathrm{W_c} \underset{\mathrm{r}}{+} \mathrm{m}\{(\mathrm{h}_{1'} - \mathrm{h}_2) - \mathrm{T}_0(\mathrm{s}_{1'} - \mathrm{s}_2)\} = \dot{\mathrm{m}}_\mathrm{r}\mathrm{T}_0(\mathrm{s}_2 - \mathrm{s}_{1'}) \tag{6}$$

$$\textbf{Condenser}: \dot{E}_{D,k} = \dot{m}_r\{(h_2 - h_3) - T_0(s_2 - s_3)\} - \dot{Q}_k\left[1 - \left(\frac{T_0}{T_k}\right)\right] + W_{k,fan} \tag{7}$$

$$\textbf{Expansion valve}: \dot{E}_{D,v} = \dot{m}_r T_0(s_4 - s_3) \tag{8}$$

$$\textbf{LVHE}: \dot{E}_{D,HX} = \dot{m}_r\{(h_1 - h_{1'}) - T_0(s_1 - s_{1'}) + (h_3 - h_{3'}) - T_0(s_3 - s_{3'})\} \tag{9}$$

$$\textbf{System}: \dot{E}_{D,tot} = \dot{E}_{D,e} + \dot{E}_{D,c} + \dot{E}_{D,k} + \dot{E}_{D,v} + \dot{E}_{D,HX} \tag{10}$$

where $W_{e,fan}$ and $W_{k,fan}$ are the evaporator and condenser fan power taken as 68 W each.

The second law efficiency in terms of exergy is expressed as the useful fraction of exergy input [13].

$$\eta_{ex} = \frac{exergy output}{exergy input}$$

$$\text{Or } \eta_{ex} = 1 - \frac{\dot{E}_{D,tot}}{\dot{E}_{in}} \tag{11}$$

The exergy input in a VCR cycle is the sum of compressor motor power, evaporator fan, and condenser fan power. Thus

$$\dot{E}_{in} = \dot{W}_c + \dot{W}_{e,fan} + \dot{W}_{k,fan} \tag{12}$$

$$\dot{E}_{out} = \dot{E}_{in} - \dot{E}_{D,tot} \tag{13}$$

It was shown by Bobbo et al. [14] that the variation in compressor isentropic efficiency can lead to the non-realistic performance of the VCR system. Lee et al. [15] calculated the isentropic efficiency as

$$\eta_{is}=0.85-0.0467*PR \tag{14}$$

Where PR is the compression ratio.

2.3 Input Parameters

Cold room temperature, $T_R = 18°C$ to 27 °C (normal human comfort range)
Difference between cold room and evaporator temperature, $(T_R - T_e) = 15K$ [4]
Condenser temperature, $T_k = 10K$ higher than ambient temperature, [15]
$T_0 = 25$–40 °C in steps of 5 °C (as per climatic conditions)
Superheat = 5K [15]
Pressure drop in the evaporator = 20 kPa [4]
Pressure drop in the condenser = 10 kPa [4].

A model was developed to compute the performance parameters using the Genetron software (Honeywell) for different refrigerants' properties.

3 Results and Discussion

The effect of ambient temperature on the performance parameters of a split type air conditioning system working on a vapor compression cycle was explored in this work. Three different refrigerants R410A, R32, and R454B were used. The results have been predicted at extreme conditions of ambient temperature keeping the T_e constant (7 °C).

At elevated ambient temperatures, the VCE reduces considerably (Fig. 2). With the increase in pressure ratio, the p–H curve converges and so the cooling effect (Δh) reduces. R32 shows the highest VCE due to its large latent heat value and R454B has the lowest VCE. R410A shows intermediary behaviour.

Similar trends are reflected for COP and second law efficiency (η_{II}). With the increase in ambient temperature, both the COP and η_{II} decrease as shown in Fig. 3. An increase in compression ratio results in higher compressor work and so the COP reduces. Both the COP and η_{II} of R32 are slightly higher than R454B and these parameters are the least for R410A.

Figure 4 depicts the behaviour of the $\dot{E}_{D,tot}$ and the η_{ex}. As can be seen, both parameters increase with the rise in the ambient temperature. R410A shows a higher

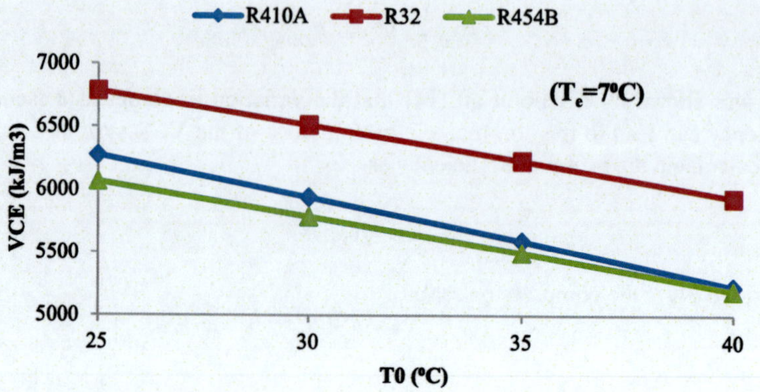

Fig. 2 Effect of T_0 on VCE at T_e=7°C

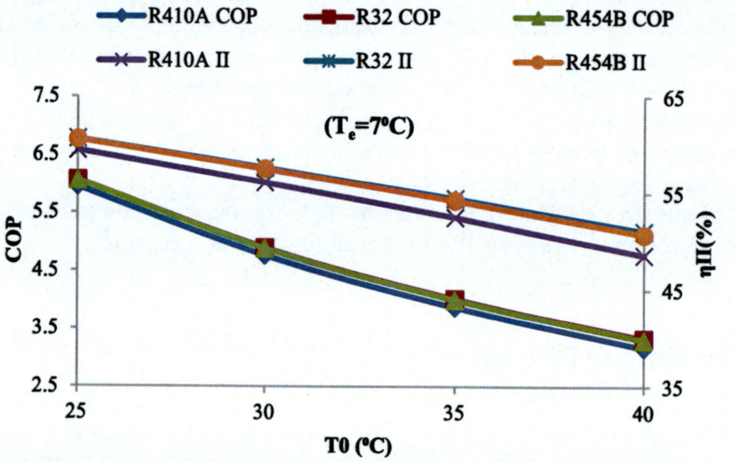

Fig. 3 Effect of T_0 on COP and η_{II}

$\dot{E}_{D,tot}$ than R32 and R454B. The increase in compressor work is more than the increase in $\dot{E}_{D,tot}$, and therefore, the η_{ex} increases. The η_{ex} for all the refrigerants is almost the same at lower ambient temperatures. However, R410A shows a decline in η_{ex} at elevated temperatures.

At lower ambient temperature (25 °C), the $\dot{E}_D$ in the evaporator is the highest followed by the compressor. At elevated ambient temperature, the compressor produces the largest $\dot{E}_D$ followed by the evaporator. The LVHE shows the least $\dot{E}_D$ in all situations (Fig. 5).

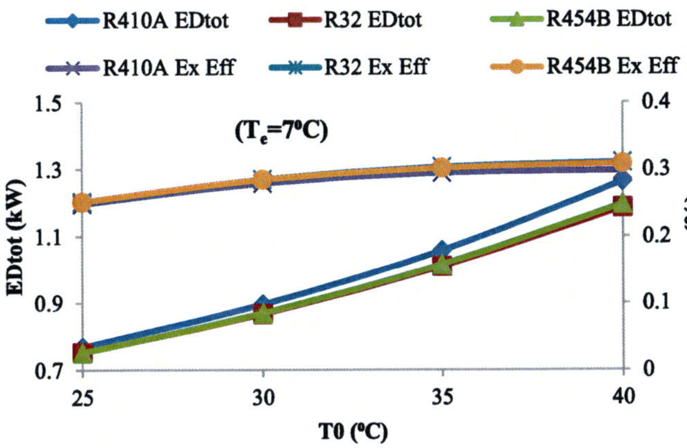

Fig. 4 Effect of T_0 on $\dot{E}_{D,tot}$ and η_{ex}

4 Conclusions

The following conclusions were derived from the study:

1. The VCE reduced at elevated T_0 for all refrigerants considered. The reason might be the increase in the specific volume of the vapor leaving the evaporator at a higher temperature. VCE of R32 was found higher than that of R410A and R454B respectively.
2. Both COP and η_{II} reduced with the increase in T_0. The increase in ambient temperature would mean greater exergy destruction in the compressor leading to higher energy input. For R410A, the COP reduced from 5.9 at 25–3.17 at 40 °C. R32 and R454B showed almost the same COP reduction i.e. 6.06 to 3.3 at the corresponding T_0.
3. The $\dot{E}_{D,tot}$ for R410A increased from 0.77 at $T_0 = 25$ °C to 1.26 at $T_0 = 40$ °C. The corresponding values for both R32 and R454B were 0.75 and 1.19.
4. The η_{ex} varied from 25% ($T_0 = 25$ °C) to 31% ($T_0 = 40$ °C) for both R32 and R454B. The η_{ex} for R410A, however, deviated at higher T_0 by 29.8%.
5. The compressor showed the highest $\dot{E}_D$ followed by the evaporator at a higher T_0. The LVHE produced the least $\dot{E}_D$ in all conditions.

The above results conclude that the ambient temperature has a significant effect on the thermodynamic performance of the refrigeration system. Thus, the designer should consider the climatic conditions to ascertain variations in ambient temperature throughout the year and obtain the optimum performance of an air conditioner.

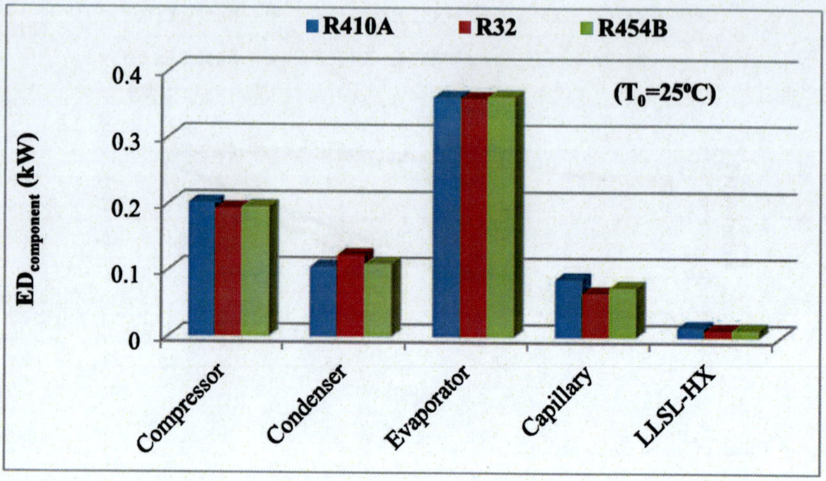

(a)

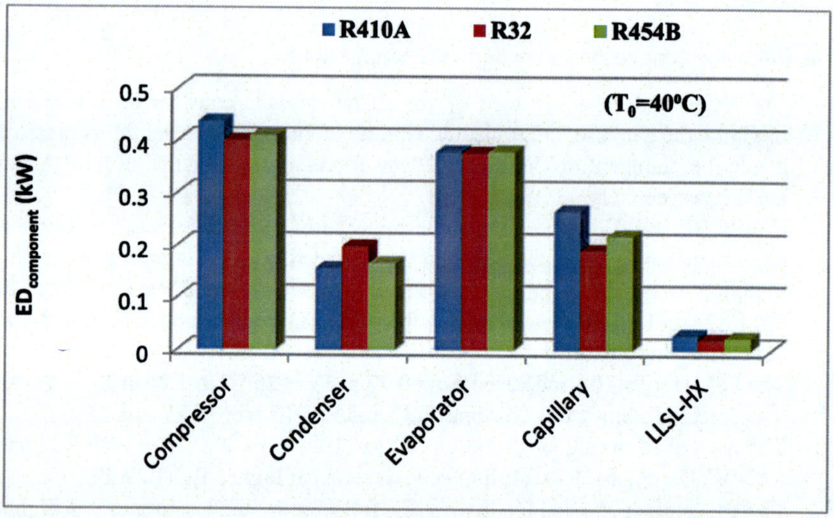

(b)

Fig. 5 Effect of T_0 on component $\dot{E}_D$ at (**a**) $T_0 = 25$ °C and (**b**) $T_0 = 40$ °C

References

1. Oruç V, Devecioąlu AG (2015) Thermodynamic performance of air conditioners working with R417A and R424A as alternatives to R22. Int J Refrig 55:120–128
2. Jani DB, Mishra M, Sahoo PK (2016) Experimental investigation on solid desiccant-vapor compression hybrid air-conditioning system in hot and humid weather. Appl Therm Eng 104:556–564

3. Bilgili M, Ozbek A, Yasar A, Simsek E, Sahin B (2016) Effect of atmospheric temperature on exergy efficiency and destruction of a typical residential split air conditioning system". Int J Exergy 20(1):66–84
4. Arora A, Arora BB, Pathak BD, Sachdev HL, Arora A (2007) Exergy analysis of Vapour Compression Refrigeration system with R-22, R-407C and R-410A. Int J Exergy 4(4):441–454
5. Hepbasli A (2007) Thermoeconomic analysis of household refrigerators. Int J ENERGY Res 31:947–959
6. Arora A, Kaushik SC (2008) Theoretical analysis of a vapour compression refrigeration system with R502, R404A and R507A. Int J Refrig 31(6):998–1005
7. Islam S, Dincer I (2016) Comparative performance study of an integrated air-cycle refrigeration and power generation system. Appl Therm Eng 106:493–505
8. Ribeiro GB (2016) Une nouvelle boucle frigorifique à compression de vapeur pour de hautes températures ambiantes. Int J Refrig 70:103–107
9. Liu H, Zhou Q, Liu Y, Wang P, Wang D (2015) Experimental study on cooling performance of air conditioning system with dual independent evaporative condenser. Int J Refrig 55:85–92
10. Oruç V, Devecioղlu AG, Berk U, Vural I (2016) Experimental comparison of the energy parameters of HFCs used as alternatives to HCFC-22 in split type air conditioners. Int J Refrig 63:125–132
11. Kumar A.: A Performance of a Window Air Conditioner Using Alternative Refrigerants R22 AND R410A ., Int. J. Eng. Sci. Res. Technol., vol. 2, no. 7, pp. 1842–1848, 2013.
12. Sethi A, Becerra EV, Yana MS, F., Spatz M. W., (2015) Low GWP R22 replacement for air conditioning in high ambient conditions. Int J Refrig 57:26–34
13. Y. A. Ç., Boles M. A.: Thermodynamics: An Engineering Approach, 5th ed. McGraw-Hill, 2006.
14. Bobbo S (2019) Energetic and Exergetic Analysis of Low Global Warming Potential Refrigerants as Substitutes for R410A in Ground Source Heat Pumps. Energies 12:3538
15. Lee H, Hwang Y, Radermacher R, Chun HH (2015) Performance investigation of multi-stage saturation cycle with natural working fluids and low GWP working fluids. Int J Refrig 51:103–111

Effects of Freestream Mach Numbers on Flow Field Features of a Sonic Jet in a Supersonic Crossflow

Imran Rasheed and Debi Prasad Mishra

Abstract A transverse air jet discharging into supersonic crossflow is numerically studied using 3D Reynolds-averaged Navier–Stokes equation (RANS), with Menter's Shear Stress Transport (SST) and k-ω turbulence model. The jet to crossflow momentum flux ratio (J) is fixed as 1.73, and a set of three crossflow Mach numbers are studied ($M_\infty = 1.6$, 2, and 4). The velocity and pressure measurements are compared with experimental data to validate the numerical setup. The aim of the current study is to investigate the effect of crossflow Mach number on the jet penetration and mixing of the injected jet. The jet penetration height seems to increase with an increase in freestream Mach number. The maximum jet mass fraction decay shows a decrease in mixing of injected jet and freestream for larger Mach number. The mixing layer between the crossflow and jet increases with an increase in freestream Mach number.

Keywords Turbulence · Compressible · Mixing

Notations

M	Mach number
Y	Mass fraction
J	Jet to crossflow momentum flux ratio
D	Jet injector diameter

I. Rasheed (✉)
Indian Institute of Technology, Uttar Pradesh, Kanpur 208016, India
e-mail: imranr@iitk.ac.in

D. P. Mishra
National Institute of Technical Teachers Training and Research (NITTTR), Kolkata, India

D. P. Mishra et al. (eds.), *Recent Trends in Thermal and Fluid Sciences*, Lecture Notes in Mechanical Engineering, https://doi.org/10.1007/978-981-19-3498-8_2

Subscripts

∞ Crossflow or freestream conditions
j Injected jet conditions

1 Introduction

For the past few decades, significant research has been done on hypersonic propulsion. One of the major challenges associated with high-speed propulsion is the combustion of fuel at very high speeds. With a residence time of a few milliseconds, fuel needs to mix well with the crossflow air to achieve efficient combustion. A significant amount of literature investigated mixing at supersonic speeds [1–7] and the subsequent combustion of fuel [8–10]. A transverse injection is a simple and effective method for injecting fuel in a supersonic flow field and offers excellent penetration of fuel into crossflow and better air–fuel mixing [11]. Although extensive studies on transverse jets have been done in the past, the effect of incoming Mach number on a transverse jet penetration and subsequent air–fuel mixing is not addressed in detail. Papamoschou and Hubbard [12] and Portz and Segal [13] in their experimental works, found evidence of an increase in jet penetration with freestream Mach number. At the same time, Takahashi et al. [14] reported an increase in eddies size because of Rayleigh–Taylor (R-T) instability with $M\infty$, although no significant improvement in penetration was reported. Watanabe et al. [15] observed a wider windward mixing layer with the increase in freestream Mach number. As for mixing studies, Watanabe et al. [15] reported a decrease in mixing efficiency with freestream Mach number. Most of the studies reported were for Mach numbers less than or equal to 3. There is still a scarcity of literature to address the effect of incoming freestream or crossflow Mach number on the mixing and flow characteristics of a jet in supersonic crossflow.

The present study consists of a transverse jet discharging into supersonic crossflow, where the details of the computational domain are from the experimental reports of Santiago and Dutton[16], VanLerberghe et al. [17] and Everett [18]. The jet to crossflow momentum flux ratio (J) is given by:

$$J = \frac{\gamma_j P_j M_j^2}{\gamma_\infty P_\infty M_\infty^2} \tag{1}$$

where subscript '∞' denotes freestream or crossflow conditions and subscript 'j' denotes the injected jet conditions. γ is the ratio of specific heats, p is the static pressure, and M represents the Mach number. Air is taken as the working fluid for both crossflow and jet. For the present study, J is fixed as 1.73 and the air jet is injected

Table 1 Flow conditions selected for current study

	Mach number	Stagnation temperature (K)	Stagnation pressure (kPa)
Freestream	1.6	295	241
	2	295	284
	4	295	1373.925
Air jet	1	295	476

into the crossflow at sonic conditions. The freestream Mach number is varied as $M_\infty = 1.6$, 2 and 4.

2 Computational Details

2.1 *Physical Model and Computational Grid*

A commercial solver, ANSYS Fluent V18.0 is employed for the present numerical investigation. The computational domain consists of a rectangular box and the sonic jet (air) is introduced into the supersonic freestream (air). The injector is a circular port, with a diameter, D of 4 mm. For capturing the complex three-dimensional flow features, a mesh with a higher density is employed near the injector region. Three different meshes are explored (i) a coarse mesh having 836,500 cells, (ii) medium having 1,269,100 cells, and (iii) fine with 1,501,367 cells, where the pressure-based grid adaptation is made to capture the shock features properly. All the results discussed in this work are based on fine mesh. The height of the first cell in the fine mesh is 1.6×10^{-6} m from the plate surface, maintaining a y^+ value close to 1. The origin of the simulation domain is set at the center of the injector. The flow conditions selected for the study are shown in Table 1. The boundary layer thickness (δ) is 3.1 mm at $X/D = 5$. Apart from that the turbulent intensity, I and viscosity ratio are taken as 5% and 10 respectively. To generate the required boundary layer thickness at the injection location, an additional three-dimensional simulation of the flat plate without injection is done prior to the main simulation. The properties at the plane corresponding to the required boundary layer thickness are then mapped onto the inlet of the main simulation with injection (Fig. 1).

2.2 *Governing Equations*

The three-dimensional Favre averaged or density averaged governing equations for a steady state flow are given by

Mass conservation

$$\frac{\partial\left(\overline{\rho}\,\tilde{u}_i\right)}{\partial x_i} = 0 \tag{2}$$

Momentum conservation

$$\frac{\partial}{\partial x_j}\left(\overline{\rho}\,\tilde{u}_j\tilde{u}_i\right) = -\frac{\partial \overline{p}}{\partial x_i} + \frac{\partial}{\partial x_j}\left(\tilde{\tau}_{ij} - \overline{\rho}\overline{u_i'' u_j''}\right) \tag{3}$$

Energy conservation

$$\frac{\partial}{\partial x_j}\left(\tilde{u}_j\,\overline{\rho}\tilde{H}\right) = \frac{\partial}{\partial x_j}\left(k_{th}\frac{\partial \tilde{T}}{\partial x_j} - \overline{\rho u_j'' h''} + \overline{u_i'' \tau_{ij}} - \overline{\rho u_j'' k}\right) + \frac{\partial}{\partial x_j}\left[\tilde{u}_i\left(\tilde{\tau}_{ij} - \overline{\rho}\overline{u_i'' u_j''}\right)\right] \tag{4}$$

where k is the turbulent kinetic energy and H is the total enthalpy.

Also,

$$\overline{\rho}\tilde{H} = \overline{\rho}\tilde{h} + \overline{\rho}\tilde{k} + \frac{1}{2}\overline{\rho}\,\tilde{u}_i\tilde{u}_i \tag{5}$$

$$\overline{\rho}\tilde{k} = \frac{\overline{\rho}\,\widetilde{u_i'' u_i''}}{2} \tag{6}$$

where specific enthalpy,

$$\tilde{h} = C_p\tilde{T} \tag{7}$$

An overbar (—) represents the time-averaged values and a tilde (~) represents mass (or Favre) averaged values. k_{th} is the air thermal conductivity. $\overline{\rho}$ and $\tilde{u}_i$ denotes the mean density and velocity. $\overline{p}$ is the pressure, T is the static temperature. $\tilde{\tau}_{ij} - \overline{\rho}\overline{u_i'' u_j''}$ gives the mean total stress. $\tilde{\tau}_{ij}$ is the laminar component.

$$\tilde{\tau}_{ij} = \mu\left(\frac{\partial\,\tilde{u}_i}{\partial x_j} + \frac{\partial\,\tilde{u}_j}{\partial x_i} - \frac{2}{3}\frac{\partial\,\tilde{u}_k}{\partial x_k}\delta_{ij}\right) \tag{8}$$

where μ is the dynamic viscosity.

The final term from Eq. (4) is the Reynolds stress tensor $({\tau_{ij}}^T)$

$${\tau_{ij}}^T = -\overline{\rho}\overline{u_i'' u_j''} \tag{9}$$

Viscosity is computed using Sutherland's transport law and the SST k-ω turbulence model [19] is chosen to model the Reynolds stress terms. The SST k-ω model shows good performance in flows with strong pressure gradients and has been employed to model transverse injection in supersonic crossflow [20, 21] in the past.

3 Results and Discussion

3.1 Validation

The validation of the present numerical setup has been carried out in our previous study[4], where the velocity and wall pressure data at freestream Mach number 1.6 are compared to experimental data [16, 18]. For further clarity, the validation of wall pressure data at various Z/D planes is shown in Fig. 2. The numerical analysis shows excellent agreement with the experimental data. Three levels of meshes are compared and only minor variations are observed in the wall pressure plot. The pressure is under-predicted near the injected region in all cases, with a maximum variation of 23% at Z/D = 1. This region corresponds to the recirculation region in front of the injector, where because of the complex flow structures, accurate measurement of pressure data is difficult.

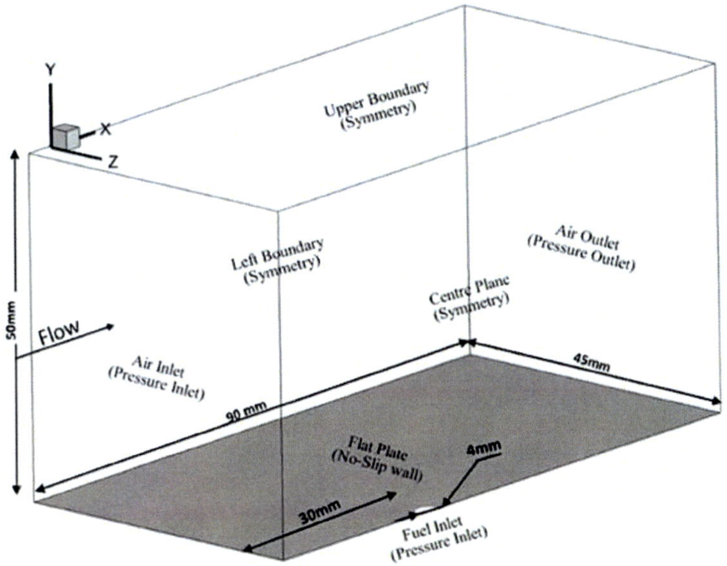

Fig. 1 Boundary conditions for present study

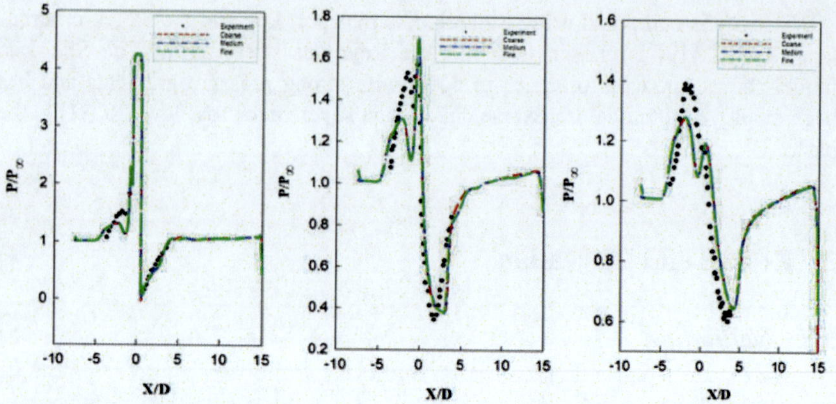

Fig. 2 Comparison of numerically predicted wall pressure with experimental data at different Z/D planes, Z/D = 0, 1 and 2 for freestream Mach number 1.6 and three levels of mesh refinement [4]

3.2 Mean Flow Features

The mean velocity contour for $M_\infty = 2$ and 4 is shown in Fig. 3. For simplicity, only the important flow features are marked in the figure. The common flow features include the recirculation regions and shock systems including a barrel shock, a bow shock, and a Mach disk. Comparing $M_\infty = 2$ and 4, the upstream recirculation zone is wider in the case of $M_\infty = 4$. The inclination of the bow shock and barrel shock and the size of barrel shock increases with M_∞ due to high flow momentum associated with larger Mach numbers.

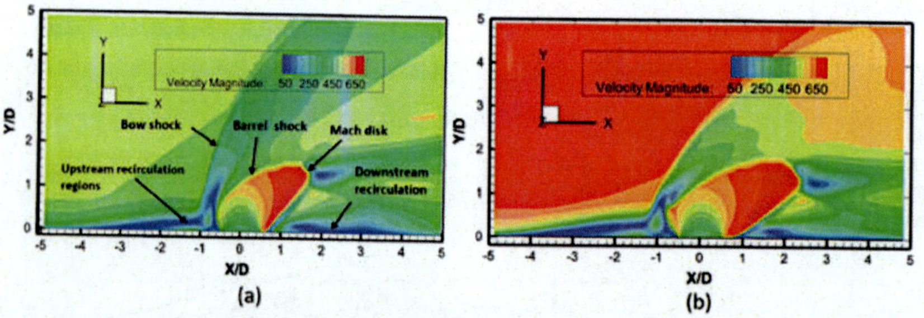

Fig. 3 Velocity magnitude contour showing the flow features near-field of injector for M_∞ of **a** 2 and **b** 4

3.3 Jet Penetration

The jet penetration into the crossflow is defined in different ways in literature, which includes jet outer edge based on schlieren visualization [12, 13], maximum concentration trajectory [14], the injectant boundary with mass fraction, Y_{jet} of 0.01 [15, 22]. For the present study, the jet penetration height is taken as the height from the flat plate surface where the injected jet mass fraction reaches 0.01. This assumption provides a valid comparison to that of experiments where the jet boundary is usually taken as the visible outer edge based on flow visualization techniques such as schlieren.

Figure 4 shows the jet penetration height for the three Mach numbers plotted along the jet centerline plane. It is clear that for larger Mach numbers, the jet penetration height is higher. For $M_\infty = 1.6$ and 2, the difference is not substantial, with a maximum difference of only 0.2D.

As the freestream Mach number reaches 4, the difference in penetration heights becomes significant. The maximum injectant mass fraction ($Y_{jet,max}$) trajectory is also plotted in Fig. 4. The $Y_{jet,max}$ trajectory for $M_\infty = 1.6$ and 2 is almost similar, but for the case of $M_\infty = 4$, the trajectory is lower than that of smaller freestream Mach numbers. A lower trajectory for jet maximum mass fraction is expected for $M_\infty = 4$ case due to the jet being more inclined with the crossflow. Comparison between the jet outer edge and maximum mass fraction trajectories suggests a wider windward mixing layer for higher freestream Mach numbers, which is also evident from Fig. 6. This wider mixing layer explains the better jet penetration in the case of high freestream Mach numbers.

It is observed that some quantity of injected air moves upstream of the injector into the upstream recirculation zone. For the case of $M_\infty = 1.6$ and 2, only a minuscule of the injected jet is present at the recirculation zone, reaching a distance of 0.5D–1D from the injector front edge, as shown in Fig. 4. Contrary to the low freestream Mach

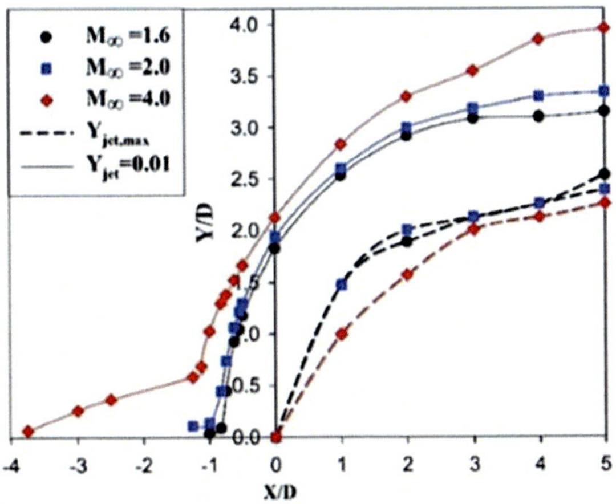

Fig. 4 Jet penetration height and maximum jet mass fraction trajectory for various freestream Mach numbers at centerline plane

number cases, $M_\infty = 4$ case shows a significant amount of injected air jet in the recirculation zones. For reactive flows, these recirculation zones act as the hotspots that stabilize the combustion [9, 10]. It is clear from the present study that freestream Mach number affects the quantity of injected jet in the upper recirculation zone. Thus, the Mach number will significantly influence the combustion characteristics of a reactive flow.

3.4 Mixing of Jet with Crossflow

The difference in the flow field structure and penetration characteristics with M_∞ will affect the mixing characteristics also. Recently Cody et al. [23] compared several mixing metrics for high-speed flows, including mixing efficiency, decay of fuel maximum mass fraction, spatial unmixedness parameter, etc. The maximum fuel mass fraction decay is employed in many studies to compare the relative performance between injectors or configurations [24, 25]. In the current study, the concept is applied to compare the mixing performance for different freestream Mach numbers. Figure 5 shows the plot of one minus maximum jet mass fraction decay along the axial location at the jet centerline plane. As the freestream Mach number increases, the slope of decay of the maximum mass fraction curve decreases. For $M_\infty = 4$, the value of $1\text{-}Y_{max}$ is 0.043 at $X/D = 10$, which means that the core of the air jet is still unmixed with the crossflow.

Figure 6 shows the mass fraction contour of the air jet at different cross-sections along the axial direction. The core of the jet remains intact as the freestream Mach number increases. However, for a higher Mach number, $M_\infty = 4$, the jet plume is more spread than the lower Mach number cases. Comparing Fig. 6d–l, a higher mass fraction of fuel can be seen at the wake of the jet and the jet spreads more at jet peripheries in both lateral and transverse directions for higher freestream Mach

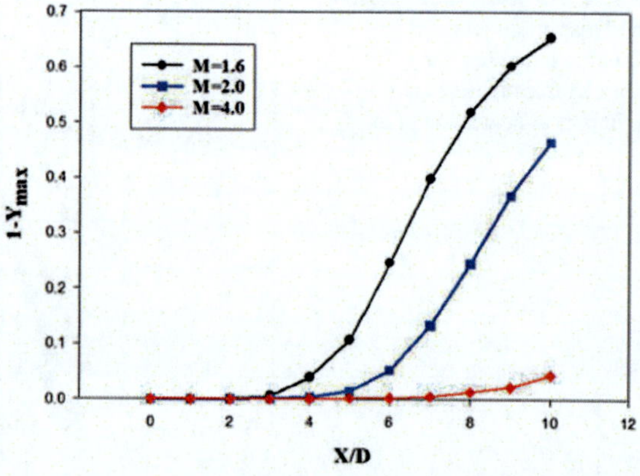

Fig. 5 One minus maximum jet mass fraction along the jet centerline plane for various freestream Mach numbers

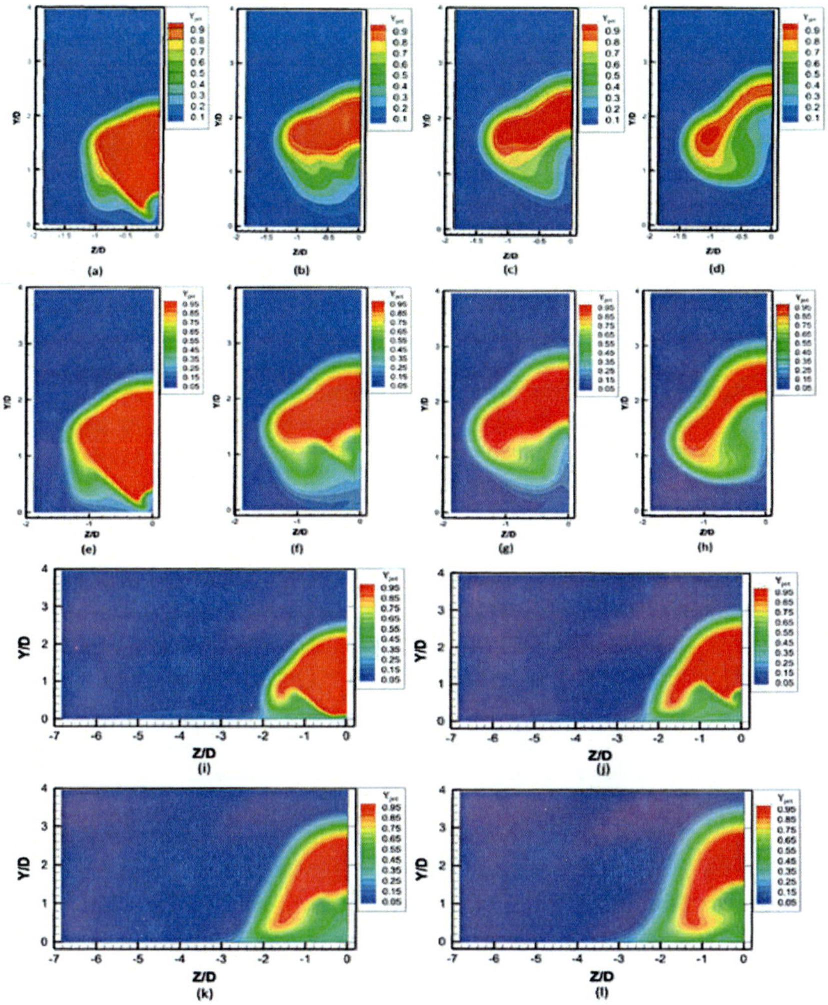

Fig. 6 Jet mass fraction at different cross-sections at **a–d** X/D = 1, 2, 3, 5 for $M_\infty = 1.6$, **e–h** X/D = 1, 2, 3, 5 for $M_\infty = 2$, **i–l** X/D = 1, 2, 3, 5 for $M_\infty = 4$

numbers. The spread can be attributed to the broader mixing layer at the windward side between the air jet and crossflow for higher M_∞.

4 Conclusion

A three-dimensional numerical study of a sonic air jet discharging into a supersonic flow is carried out. The jet to crossflow momentum flux ratio (J) is fixed as 1.73 and three different crossflow or freestream Mach numbers, $M_\infty = 1.6$, 2, and 4, are employed. The jet penetration, mixing performance, and mean flow field are compared for the different Mach number conditions.

1. The mean flow field remains similar for all the Mach numbers with bow shock, barrel shock, and separation regions. The size of the upstream recirculation region increases with the freestream Mach number. The bow and barrel shock becomes more inclined with the flow direction due to increased crossflow momentum at higher freestream Mach numbers.
2. The jet penetration height increases with freestream Mach number. The trajectory of maximum mass fraction for $M_\infty = 4$ case is lower than $M_\infty = 1.6$ and 2, suggesting a wider windward mixing layer for higher M_∞.
3. The decay of maximum jet mass fraction is employed for comparison of mixing performance at different freestream Mach numbers. The slope of decay of jet maximum mass fraction becomes smaller with an increase in M_∞ due to increased inclination of jet with crossflow. At the same time, the jet seems to spread more in the peripheries and wake of the jet as a result of a wider windward mixing layer between jet and freestream.

Acknowledgements The authors acknowledge the HPC facility at IIT Kanpur, funded by DST and IITK, for providing the necessary resources for conducting the study.

References

1. Peterson DM (2011) Simulations of injection, mixing, and combustion in supersonic flow using a hybrid rans/les approach
2. Viti V, Neel R, Schetz JA (2009) Detailed flow physics of the supersonic jet interaction flow field. Phys Fluids 21:046101
3. Rana ZA, Thornber B, Drikakis D (2011) Transverse jet injection into a supersonic turbulent cross-flow. Phys Fluids 23:046103
4. Rasheed I, Mishra DP (2020) Numerical study of a sonic jet in a supersonic crossflow over a flat plate. Phys Fluids 32:126113
5. Lee S-H (2006) Characteristics of dual transverse injection in scramjet combustor, part 1: Mixing. J Propul Power 22:1012–1019
6. Gruber MR, Nejad AS, Chen TH, Dutton JC (1995) Mixing and penetration studies of Sonic Jets in a Mach 2 freestream. J Propul Power 11:315–323
7. Gruber M, Nejad A, Chen T, Dutton J (1994) Mixing and penetration studies of Sonic Jets in a Mach 2 freestream. In: 32nd aerospace sciences meeting and exhibit
8. Ben-Yakar A, Kamel M, Morris C, Hanson R (1998) Hypersonic combustion and mixing studies using simultaneous OH-PLIF and schlieren imaging. In: 36th AIAA aerospace sciences meeting and exhibit

9. Ben-Yakar A, Mungal MG, Hanson RK (2006) Time evolution and mixing characteristics of hydrogen and ethylene transverse jets in supersonic crossflows. Phys Fluids 18:026101
10. Gamba M, Mungal MG (2015) Ignition, flame structure and near-wall burning in transverse hydrogen jets in supersonic crossflow. J Fluid Mech 780:226–273
11. Drummond JP, Cutler AD, Diskin GS (2002) Fuel-air mixing and combustion in scramjets. AIAA, Reston
12. Papamoschou D, Hubbard DG (1993) Visual observations of supersonic transverse jets. Exp Fluids 14:468–476
13. Portz R, Segal C (2006) Penetration of gaseous jets in supersonic flows. In: 44th AIAA aerospace sciences meeting and exhibit
14. Takahashi H, Masuya G, Hirota M (2009) Effects of injection and main flow conditions on supersonic turbulent mixing structure. In: 16th AIAA/DLR/DGLR international space planes and hypersonic systems and technologies conference
15. Watanabe J, Kouchi T, Takita K, Masuya G (2013) Characteristics of hydrogen jets in supersonic crossflow: large-eddy simulation study. J Propul Power 29:661–674
16. Santiago JG, Dutton JC (1997) Velocity measurements of a jet injected into a supersonic crossflow. J Propul Power 13:264–273
17. Vanlerberghe W, Dutton J, Lucht R, Yuen L (1994) Penetration and mixing studies of a sonic transverse jet injected into a Mach 1.6 crossflow. Fluid Dyn Conf
18. Everett DE, Woodmansee MA, Dutton JC, Morris MJ (1998) Wall pressure measurements for a sonic jet injected transversely into a supersonic crossflow. J Propul Power 14:861–868
19. Menter FR (1994) Two-equation eddy-viscosity turbulence models for engineering applications. AIAA J 32:1598–1605
20. Gao ZX, Lee CH (2011) Numerical Research on mixing characteristics of different injection schemes for supersonic transverse jet. Sci China Technol Sci 54:883–893
21. Huang W, Liu J, Jin L, Yan L (2014) Molecular weight and injector configuration effects on the transverse injection flow field properties in supersonic flows. Aerosp Sci Technol 32:94–102
22. Zhang Y, Liu W, Wang B (2015) Effects of oblique and transverse injection on the characteristics of jet in supersonic crossflow. Acta Astronaut 115:356–366
23. Ground CR, Drozda TG, Cabell KF, Axdahl EL (2018) Comparison of several global mixing performance metrics for high-speed fuel injectors. In: 22nd AIAA international space planes and hypersonics systems and technologies conference
24. Doster JC, King PI, Gruber MR, Carter CD, Ryan MD, Hsu K-Y (2009) In-stream hyper mixer fueling pylons in supersonic flow. J Propul Power 25:885–901
25. Fuller RP, Wu P-K, Nejad AS, Schetz JA (1998) Comparison of physical and aerodynamic ramps as fuel injectors in supersonic flow. J Propul Power 14:135–145

Effect of Turbulence Intensity on Soot Formation in Confined Turbulent Jet of Methane-Air Diffusion Flame

Shiv Kumar, Debi Prasad Mishra, and Vishnu Hariharan

Abstract The present numerical study investigates the air turbulence intensity effect on soot formation in a confined methane-air turbulent jet diffusion flame. The numerical simulations are conducted using RANS-based modified k-ε concept, whereas the combustion is modeled using a steady laminar flamelet with β-probability density function with GRI 2.11 detailed chemical kinetics mechanism. The radiative heat loss is calculated using P-1 model along with the non-gray weighted sum of gray gases model (WSGGM), while soot is computed using the Moss-Brookes method. The numerical results show excellent compatibility with the experimental measurements. The validated model is further extended to analyze the various free stream turbulence intensities' impact in the range of 5–40%. The study reveals that temperature distribution becomes more uniform due to increased oxidizer entrainment and enhanced fuel-oxidizer mixing. Besides this, the total soot volume fraction decreases with an increase in turbulent level. The study also provides a comparative analysis between WSGGM-based two radiative models.

Keywords Turbulence intensity · Flame characteristics · Soot volume fraction · WSGGM

1 Introduction

Most practical combustors such as jet engines, gas turbine engines, and furnaces operate under turbulence conditions for rapid fuel-oxidizer mixing. In such cases, the turbulence intensity, defined as the ratio of the root-mean-square velocity fluctuations to the mean flow velocity, characterizes the level of turbulence.

S. Kumar (✉) · V. Hariharan
Department of Aerospace Engineering, Indian Institute of Technology, Kanpur, India
e-mail: shivkumarskd3@gmail.com

D. P. Mishra
Department of Mechanical Engineering, NITTTR, Kolkata, India

D. P. Mishra et al. (eds.), *Recent Trends in Thermal and Fluid Sciences*, Lecture Notes in Mechanical Engineering, https://doi.org/10.1007/978-981-19-3498-8_3

Various past researches [1–5] have studied the effect of free-stream turbulence intensity and pointed out several advantages in terms of mixing and pollution minimization. In this regard, Lee et al. [1] and Hwang et al. [2] have reported different types of turbulence generators in their experimental studies. Lee et al. [1] investigated the influence of four kinds of swirler with varying intensities of turbulence ranging from 0 to 12% for methane-air combustion. They concluded that the mixing between fuel and oxidizer is significantly enhanced in the radial direction with increased turbulence intensity. It was also noted that temperature distribution becomes more uniform throughout the cross-section with the rise in turbulence level. But, the author did not report any quantitative estimation and did not discuss the pollutant emission behavior. Later, Hwang et al. [2] experimentally studied the effect of air turbulence intensity on flame structure and NOx formation in H_2-air diffusion flame. The study reported a significant reduction in NOx emission in the upstream region due to an increase in strain rate effect and a drop in the downstream area due to the rapid mixing effect. This strain rate and mixing effect are resulted due to an increase in turbulence level. In a different study, Saqr et al. [4] numerically studied the effect of free-stream turbulence on CH_4-air diffusion flame using standard k-ε turbulence model with eddy dissipation model. The author concluded that turbulence intensity adversely impacts the size of the reaction zone and flame propagation velocity. Both decreased with elevated values of intensity. In his successive numerical study [5], the same author studied the effect of turbulent intensity (20–60%) on NOx and soot products. A decreasing trend with turbulence intensity is observed in both the pollutants. In a similar study, Hashemi [3] analyzed the effect of varying turbulent intensity (15%–45%) using a realizable k-ε turbulence model and laminar flamelet model for turbulent jet diffusion flame of pure H_2 and different composition of CH_4–H_2 mixture in a confined combustor. The study revealed that pollutants' emission decreases with air turbulence intensity.

However, all these numerical studies were carried out using single-step chemistry and without considering the radiation effect. Such relaxations in radiation effect lead to a significant deviation in temperature distribution, species mass fraction, and mixture fraction [6, 7]. Also, most of the studies mentioned above are related to flame characteristics and NO_x formation; however, studies about soot formation considering detailed kinetics and participating media radiative heat transfer remain scarce as per the author's best knowledge. Soot formation is one of the major concerns associated with non-premixed combustion, directly affecting health, environment, and vegetation [8–10]. Hence, the present numerical study is aimed to study the turbulence intensity influence on flame characteristics and soot formation in a confined turbulent jet of CH_4-air diffusion flame with detailed kinetics and radiative heat loss.

Table 1 Experimental operating conditions

Absolute pressure (atm)	1
Fuel inlet temperature (K)	290
Air inlet temperature (K)	290
Mass flow rate at fuel inlet (g min^{-1})	10.3
Mass flow rate at air inlet (g min^{-1})	708

2 Numerical Details

2.1 *Flame Configuration*

For the numerical validation, the pre-existing experimental data of Moss and Brookes [11, 12] are chosen to carry out the analysis. The experimental setup consists of a fuel tube with a radius of 2.035 mm surrounded by an oxidizer tube with a 77.5 mm radius. The length of the burner is 980 mm. The operating conditions are described in Table 1.

2.2 *Computation Domain*

A 2-D axisymmetric model with the same geometrical configuration described above is adapted to simulate using commercial software ANSYS FLUENT, based on finite volume techniques (Fig. 1).

For the present analysis, a pressure-based solver is chosen in which pressure field is obtained using continuity and momentum equation. The following density-weighted Favre-averaged governing equations are solved using commercial software ANSYS FLUENT [13].

Conservation of mass

$$\frac{\partial \rho}{\partial t} + \nabla.\left(\rho \vec{V}\right) = 0 \tag{1}$$

Conservation of momentum

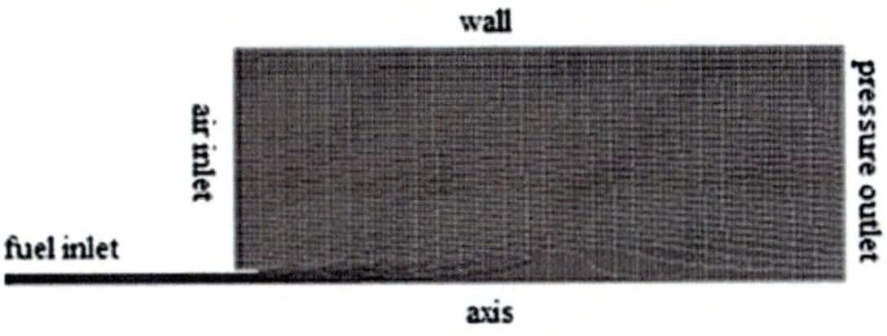

Fig. 1 Schematic diagram of the computation domain

$$\frac{\partial}{\partial t}(\rho\vec{V}) + \nabla\cdot(\rho\vec{V}\vec{V}) = -\nabla P + \nabla\cdot(\overline{\overline{\tau}}) \tag{2}$$

In the above formulation, gravitational body force and external body forces are neglected.

Conservation of energy

$$\frac{\partial}{\partial t}(\rho E) + \nabla.\left[\vec{V}(\rho E + P)\right] = \nabla.\left[k'\nabla T + \left(\overline{\overline{\tau}}.\vec{V}\right)\right] + S \tag{3}$$

In the equations mentioned above, ρ stands for density, $\vec{V}$ is the velocity vector, P is the static pressure, E is the specific total energy, k' is the thermal conductivity, $\overline{\overline{\tau}}$ is the viscous stress tensor, and S is the chemical reaction heat.

Again, the expressions for viscous stress tensor and total energy can be written as:

Viscous stress tensor

$$\overline{\overline{\tau}} = \mu[\left(\nabla\vec{V} + \nabla\vec{V}^{T}\right) - \frac{2}{3}\left(\nabla\vec{V}\right)I]$$

Where μ and I represent the molecular viscosity and the unit tensor, respectively.

Total energy

$$E = h - \frac{P}{\rho} + \frac{V^2}{2}$$

where h is the sensible enthalpy, moreover, for non-adiabatic non-premixed combustion case, the following equation for total enthalpy is solved [9]:

$$\frac{\partial}{\partial t}(\rho H) + \nabla.\left(\rho\vec{V}H\right) = \nabla.\left(\frac{k_t}{c_p}\nabla H\right) + S \tag{4}$$

where H refers to the total enthalpy and source term, S also includes radiation effect. it is also assumed that Lewis number is 1.

2.3 Turbulence Model

In the present study, the turbulence flow field is captured with the modified k-ε turbulence model. The constant C_1 in the dissipation equation is changed to 1.60 from default value 1.44 based on the conclusion made by Ziani et al. [14, 15] where three different numerical models, standard and modified k-ε, and RSM models are compared. The numerical results with modified k-ε [34] with C_1=1.60 are reported best with other unchanged constant C_2=1.92. Moreover, the expression for turbulence intensity is given by the following equation where k represents the turbulence kinetic energy [13]:

$$I = \frac{V}{V_{avg}} = \sqrt{\frac{\frac{2}{3}k}{u^2 + v^2}} \tag{5}$$

2.4 *Combustion Model*

The flamelet model is accurate and effective [14, 15] to model the combustion for the current configuration. The flamelet model computes two transport equations, species mass fraction and temperature distribution [16].

$$-\rho \frac{\chi}{2} \frac{\partial^2 Y_i}{\partial Z^2} = \dot{\omega}_i \tag{6}$$

$$\rho \frac{\chi}{2} \frac{\partial^2 T}{\partial Z^2} = \frac{1}{C_P} \sum\nolimits_{i=1}^{N} h_i \dot{\omega}_i - \rho \frac{\chi}{2C_P} \left[\frac{\partial C_p}{\partial Z} + \sum\nolimits_{i=1}^{N} C_{p,i} \frac{\partial Y_i}{\partial Z} \right] \frac{\partial T}{\partial Z} \tag{7}$$

2.5 *Radiation Model*

To precisely predict flame characteristics, it is essential to model participating media radiative heat transfer accurately. Two radiation models, P-1 and DO, are widely used to calculate radiative properties considering emission, absorption, and scattering [13]. The present study implements the P-1 radiation model, which solves the radiation transport diffusion equation with less computation effort.

2.6 *Radiative Properties Model in Combusting Flow*

Radiative property like absorption is highly dependent on the wavelength of incident radiation. Hence, it would be a poor approximation to model the complete spectrum with a constant absorption coefficient value. So, to model radiative properties in combustion flow, it is recommended to use a weighted sum of the gas gray model, also termed WSGGM [13].

2.7 Soot Modeling

The Moss-Brookes model computes the following two transport equations for soot mass fraction and normalized radical nuclei concentration.

$$\frac{\partial}{\partial t}(\rho Y_{soot}) + \nabla.\left(\rho \vec{V} Y_{soot}\right) = \nabla.\left(\frac{\mu_t}{\sigma_{soot}} \nabla Y_{soot}\right) + \frac{dM}{dt} \tag{8}$$

$$\frac{\partial}{\partial t}(\rho b_{nuc}) + \nabla \cdot \left(\rho \vec{V} b_{nuc}\right) = \nabla \cdot \left(\frac{\mu_t}{\sigma_{soot}} \nabla b_{nuc}\right) + \frac{1}{N}\frac{dN}{dt} \tag{9}$$

3 Results and Discussion

The flame structure and species mass fraction directly affect the soot formation phenomena. Therefore, a good agreement of numerical results with experimental data is essential for the better prediction of soot. Thus, in the first segment of this section, numerical results regarding temperature and mixture fraction are compared with experimental measurements published by Moss and Brookes [7]. Subsequently, results are presented for various free stream turbulence intensities (FSTI) in the range of 5–40% for temperature profile and SVF, and its effects are discussed.

3.1 Model Validation

For numerical model validation, predicted temperature and mixture fraction in the axial and radial direction are compared with the experiment data [12] using the P-1 radiation model with two different weighted-sum-of-gray-gases models (WSGGM); gray band and domain-based, as shown in Fig. 2. The peak temperature of 1666 K is noted at the axial location of 0.52 m for the gray band, whereas it is 1757 K at 0.59 m for the domain-based model. The deviations are 2.63% and 2.75%, respectively, with respect to the measured temperature. These negligible deviations indicate the excellent agreement of predicted values with measurements for both the models; however, the location of peak temperature shifts downstream in the case of domain-based, as shown in Fig. 2a. Moreover, the mean mixture fraction that characterizes the local fuel–air ratio at any location in non-premixed combustion is predicted more accurately in radial and axial directions using a gray band-based model than the domain-based model as shown in Fig. 2b and d. Although numerical predictions are consistent with both models, the gray band-based radiation model results are more accurate than the domain-based WSGGM model. Hence, further analysis is conducted using a gray band-based model.

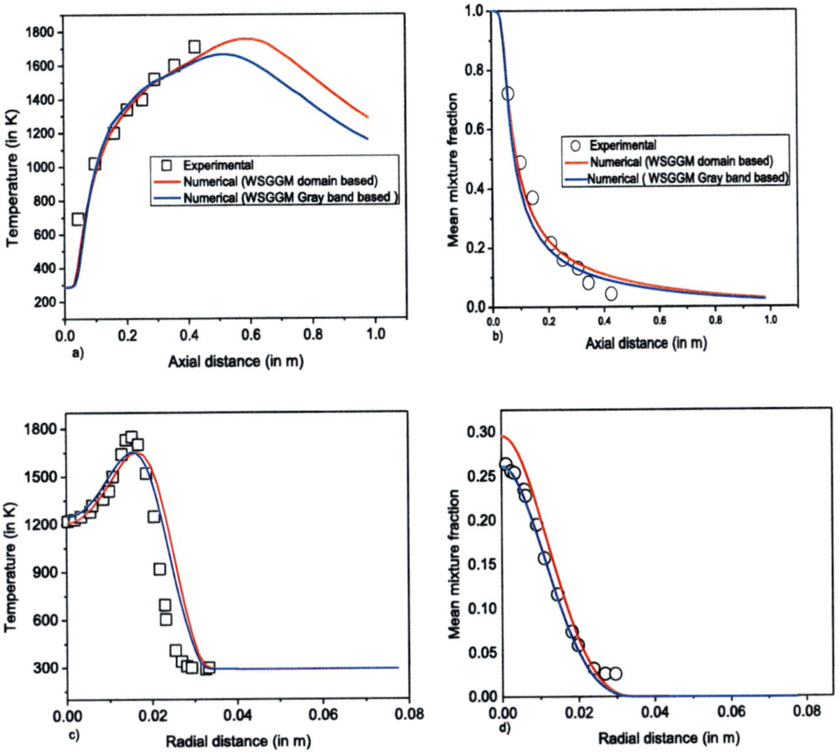

Fig. 2 **a** One temperature distribution, **b** axial mean mixture fraction distribution, **c** radial temperature distribution, and **d** radial mean mixture fraction distribution at 150 mm

3.2 *Effect of Turbulence Intensity*

Temperature contours for different free stream turbulent intensities varying from 5 to 40% are shown in Fig. 3. It can be observed that the high-temperature region (dark red labeled in contours) becomes wide as the turbulence level increases, which indicates that the high-temperature zone enlarges at elevated values of turbulence. However, the peak temperature remains almost the same for all intensities. It means turbulence intensity only affects the temperature distribution and uniformity but does not alter the maximum temperature in the region. This finding supports the argument made by Lee et al. [1], where the author made similar observations of enhancement in temperature distribution and negligible increment in peak temperature.

Figure 4 illustrates the O_2 mass fraction in axial direction spanning from 0.4 m to the outlet where significant temperature distribution changes occur. It shows that the oxidizer mass fraction consistently increases with the turbulence level and hence accelerates the oxidizer entrainment. A maximum of 25% increase in O_2 mass fraction is observed for 5–40% turbulence increment.

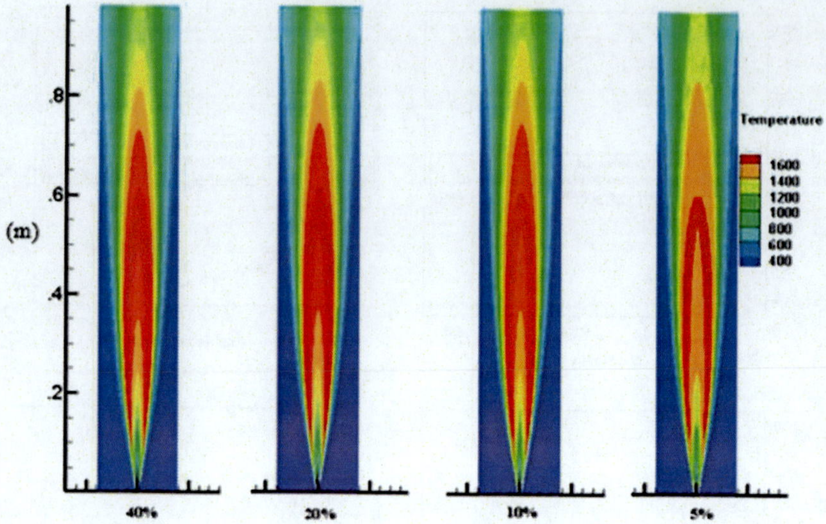

Fig. 3 Temperature contour for different free stream turbulent intensities

Fig. 4 Axial distribution of oxidizer mass fraction

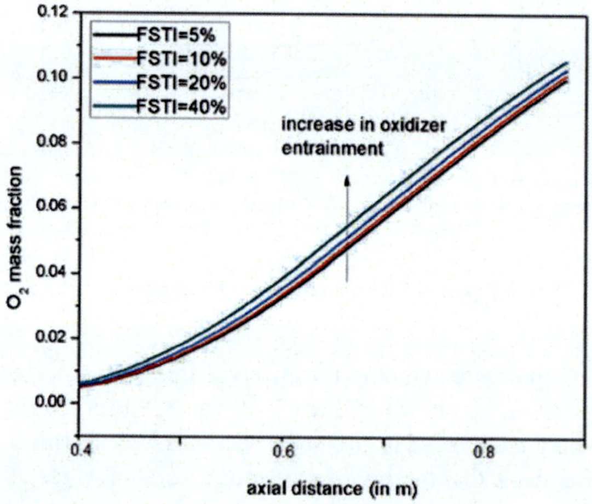

3.3 *Soot Volume Fraction (SVF)*

Figure 5 indicates the radial profiles of computed soot volume fraction (SVF) at the different axial locations with respective experimental data [12]. These are calculated using the soot mass fraction transport equation and normalized radical nuclei concentration associated with the Moss-Brookes model. The predicted values follow a similar trend as experimental measurements. At 300 mm, the predicted peak value of SVF is 6.3×10^{-8} as compared to the measured value of 5.4×10^{-8}, which leads

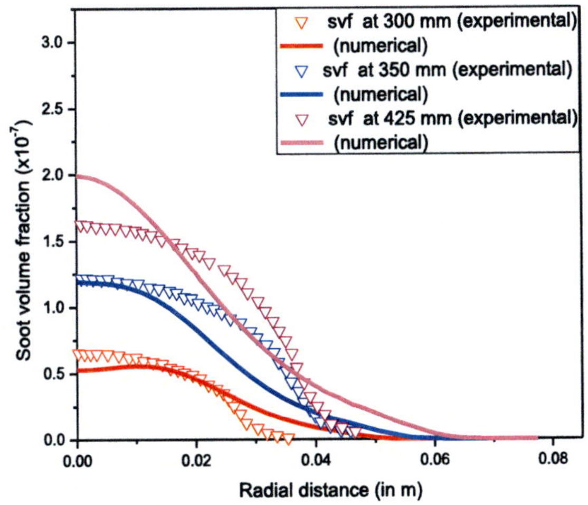

Fig. 5 Radial distribution of soot volume fraction at different axial locations

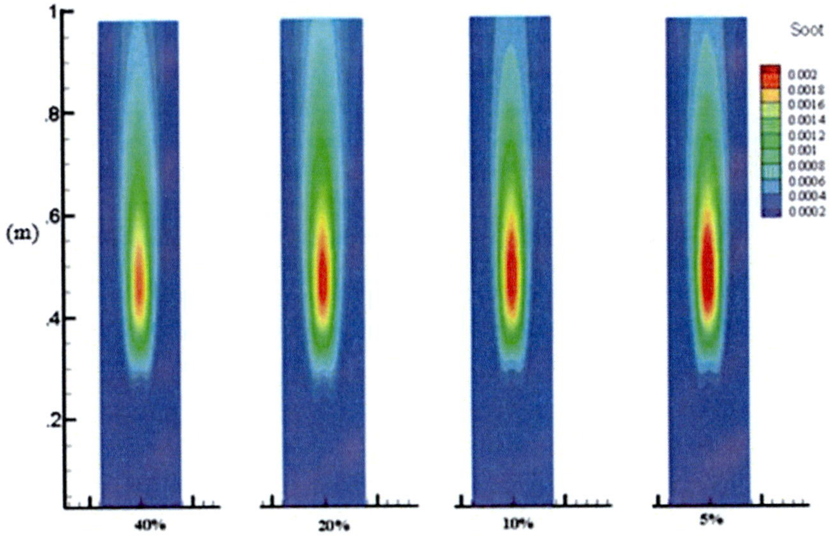

Fig. 6 Soot mass fraction contour for different free stream turbulent intensities

to a deviation of 14.3%. Similarly, at 350 mm and 425 mm, these deviations are 1.6% and 15.8%, respectively, as shown in Fig. 5. Also, the trend shows a high soot decay rate in downstream regions, which could be attributed to computational limitations. But, these deviations are in a reasonable range, and hence the model is benchmarked to carry out further analysis to study the effect of free-stream turbulence intensity.

Figure 6 illustrates the numerically obtained soot mass fraction contours for various turbulence intensities varying from 5 to 40% and displays its monotonic decay with the intensity increment.

The maximum soot mass fraction of 40% free stream turbulence intensity (FSTI) flame is noted to be 0.0016 compared to 0.002 for 5% FSTI, leading to a soot reduction of 20%. Additionally, the effective region of soot becomes thin and narrow, indicating that the soot formation weakens with the increase in turbulence. Moreover, Fig. 7 shows the radial and axial distribution of soot volume fraction (SVF). It can be observed that the SVF decays with an increase of FSTI irrespective of location, and also, the peak location of SVF shifts towards upstream regions, as shown in Fig. 7a and d. For example, in the radial distribution of SVF at $x = 300$ mm, the peak occurs at $r = 11.1$ mm, which moves to $r = 9.5$ mm and causes almost 14.4% upstream shifting. Apart from this, a maximum reduction of 15.5% in SVF is observed in the axial direction at $x = 0.475$ m, where the SVF is maximum. In contrast, the minimum reduction occurs in the region where SVF is the least. Hence, it can be said that the intensity has a significant impact in the areas where soot density is high.

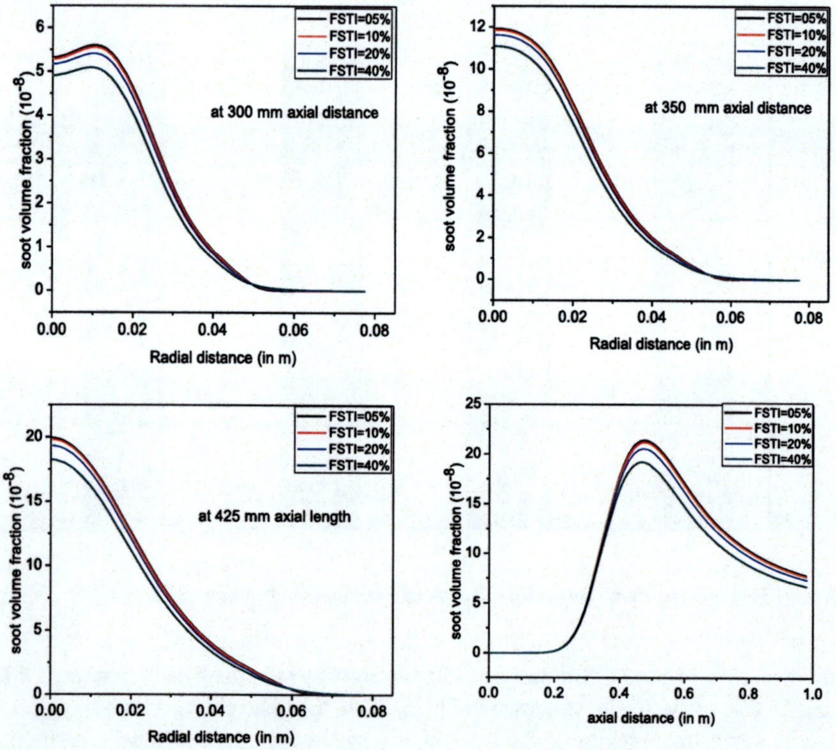

Fig. 7 Radial soot volume fraction at axial locations of **a** 300 mm, **b** 350 mm, **c** 425 mm, and **d** axial soot volume fraction

4 Conclusion

The present numerical study is conducted to study the air turbulence study influence on flame characteristics and soot formation in a confined turbulent methane-air jet diffusion flame. The study shows that the gray band-based WSGGM is more accurate than the domain-based WSGGM to model the radiative heat loss in the participating media. Moreover, the turbulence intensity has an insignificant effect on the maximum temperature as it remains almost unchanged. However, the temperature distribution becomes more uniform with turbulence intensity increment. Therefore, the extent of uniformity in temperature distribution improves due to the enhancement of the oxidizer entrainment, leading to the augmentation of the reaction rate. Additionally, the rise in O_2 mass fraction and reaction rate suppresses the soot formation. Therefore, the soot dominant region becomes narrow with an increase in turbulence intensity. The peak of soot volume fraction (SVF) moves towards the upstream regions for flame associated with higher turbulence intensity because of thinning of the reaction zone. Also, turbulence intensity affects the high-density soot regions.

The present study can contribute to developing and designing a combustion-based system integrated with a turbulence generator to reduce soot emission and other pollutants.

References

1. Lee CW, Moon SY (2002) An experimental investigation of the effects of turbulence generators attached to an axial swirler nozzle on mixing and combustion. Aerosp Sci Techno 6:517–520
2. Hwang CH, Lee S, Lee CE (2008) The effect of turbulence intensity of ambient air flow on NOx emissions in H2/air nonpremixed jet flames. Int Hydrogen Energy 33:832–841
3. Hashemi SA, Fattahi A, Sheikhzadeh GA, Mehrabian MA (2011) Investigation of the effect of air turbulence intensity on NOx emission in non-premixed hydrogen and hydrogen-hydrocarbon composite fuel combustion. Int J Hydrogen Energy 36:10159–10168
4. Saqr KM, Sies MM, Wahid MA (2009) Numerical investigation of the turbulence–combustion interaction in non-premixed CH_4-air flames. In J Appl Math Mech 5:69–79
5. Saqr KM, Aly HS, Sies MM, Wahid MA (2010) Effect of free stream turbulence on NOx and soot formation in turbulent diffusion CH_4-air flames. Int Commun Heat Mass Transf 37:611–617
6. Carvalho MG, Farias TL (1998) Modelling of heat transfer in radiating and combusting systems. Chem Eng Res Des 76:175–184
7. Mohamed S, Ahmed I (2013) Numerical simulation of methane combustion with and without radiation models. Port-Said Eng Res J 17:1–7
8. Zecca A, Chiari L (2010) Fossil-fuel constraints on global warming. Energy Policy 38:1–3
9. Perera F, Ashrafi A, Kinney P, Mills D (2019) Towards a fuller assessment of benefits to children's health of reducing air pollution and mitigating climate change due to fossil fuel combustion. Environ Res 172:55–72
10. Perera F (2018) Pollution from fossil-fuel combustion is the leading environmental threat to global pediatric health and equity: solutions exist. Int J Enviro Res Public Health 15
11. Brookes SJ, Moss JB (1999) Measurements of soot production and thermal radiation from confined turbulent jet diffusion flames of methane. Combust Flame 116:49–61

12. Brookes SJ, Moss JB (1999) Predictions of soot and thermal radiation properties in confined turbulent jet diffusion flames. Combust Flame 116:486–503
13. ANSYS FLUENT 13 (2013) Ansys fluent theory guide. ANSYS Inc., USA. 15317, 724–746
14. Ziani L, Chaker A, Chetehouna K, Malek A, Mahmah B (2013) Numerical simulations of non-premixed turbulent combustion of CH_4–H_2 mixtures using the PDF approach. Int J Hydrogen Energy 38:8597–8603
15. Ziani L, Chaker A (2016) Ambient pressure effect on non-premixed turbulent combustion of CH_4–H_2 mixture. Int J Hydrogen Energy 41:11842–11847
16. Veynante D, Vervisch L (2002) Turbulent combustion modeling. Prog Energy Combust Sci 28:193–266

Modeling and Analysis of Shell and Tube Heat Exchanger with Disc and Doughnut Baffles

Rupanshu Singh and Sanskar Sogani

Abstract The disc and doughnut baffles in the shell and tube heat exchangers can be superior alternatives over the conventional segmental baffles. In this paper, CFD modeling and analysis were carried out to determine the attributes of transferring heat in STHX with disc and doughnut baffles. The simulation results were obtained by varying the inlet mass flow rates (0.5, 1 and 2 kg/s). The results showed that shell side outlet temperature decreased and the heat transfer coefficient increased as the inlet mass flow rate was increased. The simulation results were validated by the analytical calculations. This investigation shows that the disc and doughnut baffles provided a better alternative with superior features.

Keywords Disc and doughnut baffles · Shell and tube heat exchangers · Heat transfer

1 Introduction

Heat exchangers are devices utilized for exchanging heat between fluids at dissimilar temperatures, without mixing. It has various utilizations in various industries; the most standard heat exchanger used is a shell and tube heat exchanger, in which two liquids at varied temperatures are contained on different sides (shell and tube side) which are segregated by a solid wall. The wall acts as a barrier, stopping the fluids to mix with each other. The rate of heat transfer depends on several parameters like arrangement and number of tubes, type and number of baffles, shell and tube diameter and flow characteristics of fluids [1, 2]. A decrease in heat transfer rate is observed as the mass flow rate is increased [3], and a considerable amount of heat is lost because of radiation, conduction and convection processes [4]. A vibration failure in tubes can occur due to the conventional zig-zag flow of fluid inside the heat exchanger, which can lead to corrosion followed by cracking of tubes [5].

R. Singh · S. Sogani (✉)
Department of Mechanical Engineering, Delhi Technological University, Delhi, India

R. Singh
e-mail: singhrupanhu@gmail.com

D. P. Mishra et al. (eds.), *Recent Trends in Thermal and Fluid Sciences*, Lecture Notes in Mechanical Engineering, https://doi.org/10.1007/978-981-19-3498-8_4

Baffles play a vital role in STHX, as it forces the shell side fluid to move over the tubes, for obtaining an enhanced heat transfer [6]. It was observed that on adding more baffles, recirculation zones were not developed, and enhanced heat transfer characteristics were obtained [7].

Several research have been carried out, which analyzed the effect on heat transfer characteristics using different baffle designs. Peng et al. [8] developed and proposed the helical baffle's use, and the experimental results show that the heat transfer coefficient on comparing with the usual segmental baffles gives an increase of about a decile.

The computational fluid dynamics method has provided ease in testing and experimenting the processes. Simulation provides identical results on the application of optimal governing equations and implementation of appropriate boundary conditions.

Biçer et al. [9] investigated the performance of three zonal baffles using CFD simulations and compared it with the results of conventional baffles. The results showed that no stagnant zones were formed, pressure loss was reduced and an improved rate of heat transfer was acquired. Lie et al. [10] investigated the thermal behavior of shell and tube heat exchanger with louver baffles by computational simulations. Results depict a uniform shell side temperature distribution, and a higher heat transfer coefficient of a unit drop in pressure was observed. Mohammadi et al. [11] conducted investigations and observed the effect of segmental porous baffles and the effect of permeability and porosity on the STHX model. The optimum results were obtained corresponding to $\varepsilon = 21.68\%$, $\varphi = 10\text{–}9\ \text{m}^2$ and $Bc = 27.58\%$.

A superior alternative to the segmental baffles is the disc and doughnut (D&D) baffles. It is the arrangement of alternate disc and doughnut-shaped baffles across the STHX. Due to its manufacturing complexity and lack of comparable results of thermal properties and pressure drop, it is not being utilized up to its potential [12]. Wusiman and Zhou [13] investigated the heat transfer characteristics of STHX with doughnut and disc-shaped baffles, and the results showed that optimum configuration was achieved with 7 baffles and 0.7 baffle cutting ratio.

The motive of this research is to develop an STHX model with disc and doughnut baffles and conduct a numerical simulation to determine the transferring of heat characteristics and drop in pressure across the whole heat exchanger. The simulation results obtained will be compared to the conventional segmental baffles and analytical results. This will provide the researchers an insightful information for a better understanding of the thermal characteristics of disc and doughnut baffles.

2 Methodology

2.1 Geometric Model

The experimental setup was designed on Solidworks software. To simulate the flow field and thermal variation inside the STHX on analysis software, a CAD model was developed as depicted in Fig. 1. The counter flow one-shell-pass and one-tube-passes type STHX with disc and doughnut baffles were designed. All the parameters

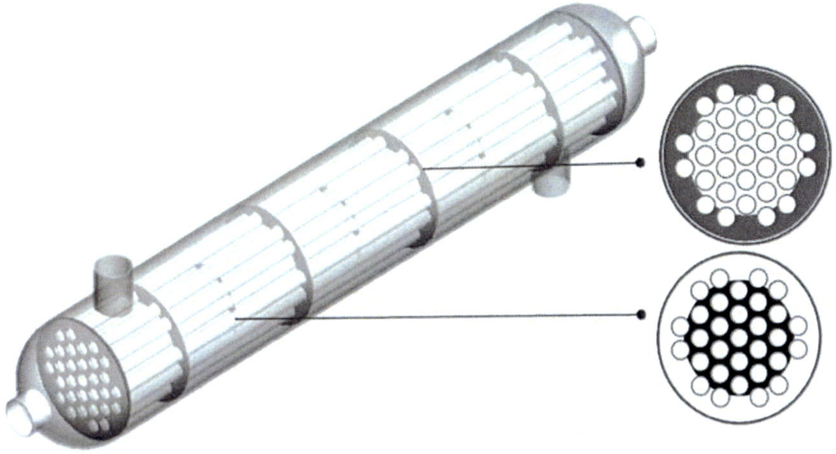

Fig. 1 CAD model of STHX with disc and doughnut baffles

Table 1 Geometric dimensions and configuration considered in designing STHX

Geometrical parameters	Dimension and description	Geometrical parameters	Dimension and description
Inside diameter of shell (D_i)	290 mm	Tube inner diameter (d)	30 mm
Shell length (L)	1600 mm	Baffle spacing (S)	200 mm
Doughnut outer diameter (D_1)	286 mm	No. of baffles	7
Doughnut hole diameter (D_c)	200 mm	Pitch ratio (a)	1.17
Disc outer diameter (D_d)	200 mm	Tube arrangement	Perforation
No. of tubes (N)	31	Baffle cutting ratio	0.7

considered for designing the STHX are shown in Table 1. The results will be simulated by applying the parametric conditions and governing equations to the CAD model.

2.2 Governing Equations

The simulation analysis was carried out using the ANSYS Fluent software. For obtaining accurate results and predicting the fluid's flow, appropriate boundary conditions should be implemented [2, 14]. The flow is incompressible, unsteady and three directional [15]. Different viscous models were incorporated and compared. The equations involved in simulating the 3D flow field are stated below [10, 16, 17]:

Continuity equation.

$$\frac{\partial u}{\partial x} + \frac{\partial u}{\partial y} + \frac{\partial u}{\partial z} = 0 \text{ or } \nabla \cdot \vec{V} = 0 \tag{1a}$$

Momentum conservation equation.

$$\frac{\partial}{\partial t}(\rho v) + \nabla \cdot (\rho v v) = -\nabla p + \rho g + F \tag{1b}$$

Energy conservation equation.

$$\rho c_v \left(u\frac{\partial T}{\partial x} + v\frac{\partial T}{\partial y} + w\frac{\partial T}{\partial z} \right) = \mathrm{K}\left(\frac{\partial^2 T}{\partial x^2} + \frac{\partial^2 T}{\partial y^2} + \frac{\partial^2 T}{\partial z^2} \right) + \phi \tag{1c}$$

2.3 Meshing and Model Setup

A refined mesh is generated over the model to simulate the accurate results. Meshing develops nodes, elements on the governing equations are implemented and results are obtained [14, 18]. For obtaining the optimum results, mesh independence was checked by conducting a grid independence test as shown in Table 2, and it depicts that on altering the elements and nodes, higher accuracy was obtained. Thus, the model with the highest elements and nodes was selected for computational analysis.

Table 2 Grid independence test results

Elements	Nodes	h (W/m^{2}K)	Nu
4,314,903	876,112	2812	128.55
3,693,667	797,871	2755	125.95
3,099,202	719,283	2686	122.79

For obtaining conclusive and comparative results, a realizable k-epsilon model was selected, and inlet thermal boundary conditions were assigned. Numerical results were obtained by varying the mass flow rate of the inlet cold liquid. The results obtained by the simulation were compared to the analytical calculations. Inlet temperature on the inlet end of the shell was 300 K, and on the side of the tube, temperature was 335 K.

2.4 Data Processing

The process of transferring heat in STHX occurs with two convective thermal resistance and one conduction resistance. The HT coefficient of the overall model is obtained as

$$\frac{1}{UA_s} = \frac{1}{U_iA_i} = \frac{1}{U_oA_o} = R = \frac{1}{h_iA_i} + R_{wall} + \frac{1}{h_oA_o} \quad (1)$$

where h_o and h_i are convective heat transfer coefficients of the outer and inner area of the tube.

The fluids in the STHX's inlet are at different temperatures. The outlet temperatures of the cold and the hot fluids can be calculated by implications of equations [19]

$$(\text{heat})\dot{Q} = C_c\left(T_{\text{cold,outside}} - T_{\text{cold,inside}}\right) \quad (2)$$

$$(\text{heat})\dot{Q} = C_h\left(T_{\text{hot,inside}} - T_{\text{hot,outside}}\right) \quad (3)$$

A variation in the temperature difference between two fluids is observed as the fluid moves inside a heat exchanger [20]. Thus, a log mean temperature difference ΔT_m is considered to develop the following relation [21]:

$$Q = UAs\Delta Tm \quad (4)$$

$$\Delta T_{lm} = \frac{\Delta T_1 - \Delta T_2}{\ln(\Delta T_1/\Delta T_2)} \quad (5)$$

where $\Delta T_1 = \text{Temp}_{\text{hot,inlet}} - \text{Temp}_{\text{cold,outlet}}$, $\Delta T_2 = \text{Temp}_{\text{hot,outlet}} - \text{Temp}_{\text{cold,inlet}}$

The relation between convective heat transfer coefficient and the Nusselt number inside a pipe is [1, 18, 22]

$$h = \frac{k}{D_h}Nu \quad (6)$$

Table 3 Thermal properties of the fluid

Property	Water	Air
Density (kg/m^3)	971.8	1.27
Specific heat capacity (J/kgK)	4195	1005
Thermal conductivity (W/mK)	0.6562	0.024

The effectiveness of heat exchangers is dependent on the number of transfer units (NTU) and the ratio of capacity. From the effectiveness of the heat exchanger and N T U relation, the heat transfer coefficient can be demonstrated as [23] (Table 3)

$$NTU = \frac{UA_s}{C_{min}} = \frac{UA_s}{(\dot{m}C_p)_{min}} \tag{7}$$

3 Results and Discussion

Computational code for the STHX model was developed on the ANSYS software. CFD fluent analysis mode was utilized to obtain satisfactory results. Figure 2 shows the streamlined flow inside a STHX with disc and doughnut baffles and double segmental baffles, and it is observed that shell side fluid uniformly flows over the tubes with disc and doughnut baffles. The resulting coefficient of heat transfer, the temperature on the outlet of the shell and drop in pressure on the side of the shell were obtained corresponding to different inlet mass flow rates as shown in Table 4.

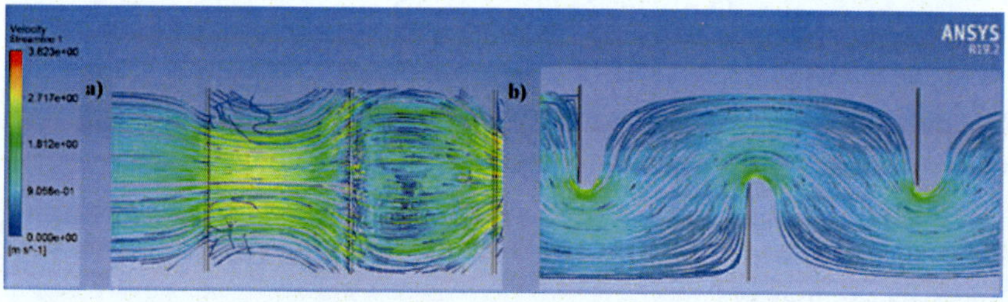

Fig. 2 Result streamlines with **a** disc and doughnut baffles and **b** segmental baffles

Table 4 CFD simulation results

Mass flow rate (kg/s)	Shell side outlet temperature (K)	Heat transfer coefficient (W/m^{2}K)	Shell side pressure drop (ΔP) (Pa)
0.5	333.46	2812	3288
1	328.25	4536	11,426
2	326.86	8652	42,565

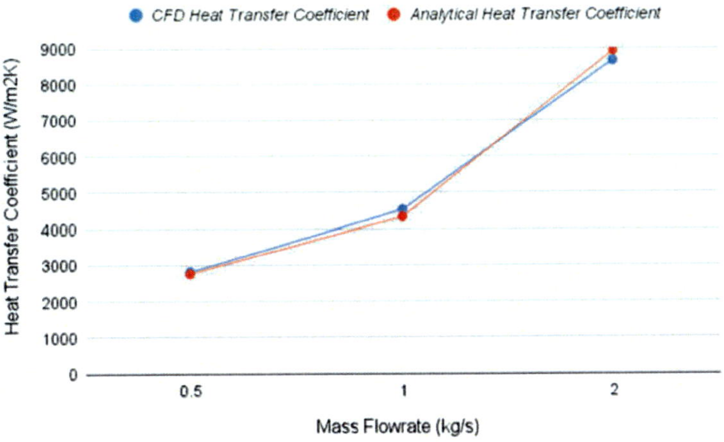

Fig. 3 Comparative graph between simulations and analytical (hand calculations) results corresponding to given mass flow rates

It is evident that as the inlet flow rate increased, the shell side outlet temperature decreased, while the heat transfer coefficient increased.

The resulting average temperature difference is the log mean temperature difference ΔTm which was found to be 4.87, 6.73 and 7.68 K for mass flow rates (0.5, 1 and 2 kg/s).

The graph (Fig. 3) demonstrates the difference between the outputs obtained from CFD and analytical results (obtained from hand calculations). For the given inlet mass flow rates, it was observed that identical results were obtained from both the methods. Thus, the simulation results are validated by the analytical calculations.

4 Conclusion

The CFD modeling and simulation of STHX with disc and doughnut-shaped baffles were carried out. The thermal behavior and flow characteristics of the designed heat exchanger were developed and investigated on ANSYS software. The resulting heat transfer coefficient, shell side's outlet temperature and pressure difference corresponding to different mass flow rates (0.5, 1 and 2 kg/s) were obtained by CFD simulation, which were further validated by the analytical calculations. It was observed that as the inlet flowing mass rate increased the shell side outlet temperature decreased, while the heat transfer coefficient increased. The conclusions can be made that the STHX performs better with D&D baffles than with the conventional segmental baffles. This research provides insightful information for analyzing the heat transfer characteristics of STHX with D&D-shaped round baffles.

References

1. Jamil M, Goraya T, Shahzad M, Zubair S (2020) Exergoeconomic optimization of a shell-and-tube heat exchanger. Energy Convers Manag 226:113462
2. Li N, Chen J, Cheng T, Klemeš J, Varbanov P, Wang Q et al (2020) Analysing thermal-hydraulic performance and energy efficiency of shell-and-tube heat exchangers with longitudinal flow based on experiment and numerical simulation. Energy 202:117757
3. Kallannavar S, Mashyal S, Rajangale M (2020) Effect of tube layout on the performance of shell and tube heat exchangers. Mater Today Proc 27:263–267
4. Kasmir J, Joshi S (2015) Experimental study of a shell and tube heat exchanger for performance enhancement. In: International conference on technologies for sustainable development (ICTSD)
5. Panahi H, Eslami A, Golozar M, Ashrafi Laleh A (2020) An investigation on corrosion failure of a shell-and-tube heat exchanger in a natural gas treating plant. Eng Fail Anal 118:104918
6. Wang S, Wen J, Li Y (2009) An experimental investigation of heat transfer enhancement for a shell-and-tube heat exchanger. Appl Therm Eng 29(11–12):2433–2438
7. Ozden E, Tari I (2010) Shell side CFD analysis of a small shell-and-tube heat exchanger. Energy Convers Manage 51(5):1004–1014
8. Peng B, Wang Q, Zhang C, Xie G, Luo L, Chen Q, Zeng M (2007) An experimental study of shell-and-tube heat exchangers with continuous helical baffles. J Heat Transf 129(10):1425–1431
9. Biçer N, Engin T, Yaşar H, Büyükkaya E, Aydın A, Topuz A (2020) Design optimization of a shell-and-tube heat exchanger with novel three-zonal baffle by using CFD and Taguchi method. Int J Therm Sci 155:106417
10. Lei Y, Li Y, Jing S, Song C, Lyu Y, Wang F (2017) Design and performance analysis of the novel shell-and-tube heat exchangers with louver baffles. Appl Therm Eng 125:870–879
11. Mohammadi M, Abbasi H, Yavarinasab A, Pourrahmani H (2020) Thermal optimization of shell and tube heat exchanger using porous baffles. Appl Therm Eng 170:115005
12. Li H, Kottke V (1999) Analysis of local shell-side heat and mass transfer in the shell-and-tube heat exchanger with disc-and-doughnut baffles. Int J Heat Mass Transf 42(18):3509–3521
13. Wusiman K, Zhou ZL (2020) Investigation of shell and tube heat exchanger with disc-and-doughnut baffles. Open Access Lib J 7:e6762
14. Labbadlia O, Laribi B, Chetti B, Hendrick P (2017) Numerical study of the influence of tube arrangement on the flow distribution in the header of shell and tube heat exchangers. Appl Therm Eng 126:315–321
15. Hosseini M, Ranjbar A, Sedighi K, Rahimi M (2012) A combined experimental and computational study on the melting behavior of a medium temperature phase change storage material inside shell and tube heat exchanger. Int Commun Heat Mass Transf 39(9):1416–1424
16. Yang D, Khan T, Al-Hajri E, Ayub Z, Ayub A (2019) Geometric optimization of shell and tube heat exchanger with interstitial twisted tapes outside the tubes applying CFD techniques. Appl Therm Eng 152:559–572
17. Yehia M, Attia A, Abdelatif O, Khalil E (2014) Computational investigations of thermal simulation of shell and tube heat exchanger. Vol 3 Eng Syst Heat Transf Therm Eng Mater Tribol Mechatron Robot
18. Amalia R, Safitra A, Ubudiyah H (2019) Numerical study on heat transfer and friction factor characteristics of transition flow in shell and tube heat exchanger. In: 2019 international electronics symposium (IES)
19. Miansari M, Jafarzadeh A, Arasteh H, Toghraie D (2020) Thermal performance of a helical shell and tube heat exchanger without fin, with circular fins, and with V-shaped circular fins applying on the coil. J Therm Anal Calorim 143(6):4273–4285
20. Zhou J, Li Y, Gu B, Shao C (2013) Temperature field prediction of rectangular shell-and-tube heat exchanger. J Press Vessel Technol 135(6)
21. Fares M, AL-Mayyahi M, AL-Saad M (2020) Heat transfer analysis of a shell and tube heat exchanger operated with graphene nanofluids. Case Stud Therm Eng 18:100584

22. Rad S, Afshin H, Farhanieh B (2014) Heat transfer enhancement in shell-and-tube heat exchangers using porous media. Heat Transf Eng 36(3):262–277
23. Ghaith F, Izhar A (2015) Thermal performance enhancement of an industrial shell and tube heat exchanger. ASME power conference

CFD Simulation of an Annular Diffuser with a Converging Hub and Diverging Casing

Pratima Vishwakarma, Anubhav Vishwakarma, and Devesh Kumar

Abstract The geometry and inlet conditions of diffusers determine their performance. Longitudinal velocity and without swirl velocity pro-files have been checked along the different diffuser passage heights × 0.1, × 0.2, × 0.4, × 0.6, × 0.8, and × 0.9 at different traverses along the length. In this present work, the CFD approach has been used to find the results by using the RNG k-ϵ turbulence model. Due to closeness with experimental results, the RNG k-ϵ turbulence model was adopted. The efficiency of an axial annular diffuser through a converging hub and diverging casing (CHDC) is investigated using Computational Fluid-Dynamics software with a cone angle of 10^{θ} and different area ratios 2.25, 3.25, and 4.25. An investigation is being conducted for different experimentally obtained inlet velocity profiles in the flow regime. The flow inside an annular diffuser was investigated, and the calculated pressure coefficient was found to be nearly equal to the experimental/available data. In this research, the pressure recovery coefficient and axial velocity are considered.

Keywords Annular diffuser · Converging hub and diverging casing (CHDC) · CFD software · Pressure recovery coefficient · Area ratio

1 Introduction

A diffuser is a mechanism that is used to reduce a fluid's kinetic energy (velocity) flowing through it. Diffusers are classified under stationary fluid devices, which increase fluid pressure, usually, diffusers do not have any heat interaction similar to nozzles, and diffuser is designed to be adiabatic. A diffuser is a device that reduces fluid flow while increasing static pressure. Diffusers are frequently used in turbomachine applications to recover kinetic energy at the compressor and turbine discharge,

P. Vishwakarma (✉) · D. Kumar
Department of Mechanical Engineering, Madan Mohan Malaviya University of Technology, Gorakhpur, India
e-mail: erpratima00@gmail.com

A. Vishwakarma
Department K-7, Jožef Stefan Institute, Jamova cesta 39, 1000 Ljubljana, Slovenia

D. P. Mishra et al. (eds.), *Recent Trends in Thermal and Fluid Sciences*, Lecture Notes in Mechanical Engineering, https://doi.org/10.1007/978-981-19-3498-8_5

thereby increasing total static isentropic efficiency [1]. Due to the presence of adverse pressure shields, designing an efficient diffuser is a difficult task. If the unfavorable pressure gradient is large enough, the boundary layer near the wall will separate and reverse flow to reduce pressure loss. Energy production is a major issue for the country. As new progress is made in the natural production of energy, more and more machine development is required.

Many simple and complex geometry machinery parts are being manufactured. Turbomachinery plays an important role in the energy transfer between rotor and fluid [2]. The compressor and turbine are two important components of turbomachinery. The overall performance of turbomachinery depends on the efficiency of the various parts. A diffuser is a major part of turbomachinery. It is a very common device used for fluid flow systems. The emergence of any fluid flow system determines its overall performance. There are so many diffusers designed, that they belong to the mounting area or divert at the exit. As an example, they are designed to reduce the exit velocity compared to the inlet, so the minimum pressure loss can be achieved with the maximum possible recovery of pressure. It is designed to achieve uniform flow at the exit [3]. The diffuser can be classified on the basis of geometry. The design of a diffuser can be done as per the requirement.

To achieve the best combustor performance, it is difficult to build and optimize the diffuser geometry. The exhaust annular diffuser is typically made up of an inner wall with an outer surface and an outer member with an inner surface, with the inner and outer crew, joined to form an annular channel. The surface of the inner member must be supported within the outer member and must direct the flow of the exhaust. According to the literature, in the late twentieth century, many theoretical studies were conducted to better understand the flow behavior in diffusers. Arora B.B. [4] used an experimental diffuser for various parameters, which were Rh1 = 38.00 mm, Rc1 = 77.50 mm, h = 0, c = 20, L 158.67 mm, and the pressure recovery coefficient evaluated with and without swirl [4]. According to the investigation, the pressure recovery coefficient (Cp) improved as the diffuser passage length increased independently without swirl. After further investigation, for flow without swirl, the pressure recovery coefficient (Cp) was found to increase as the diffuser passage length increased.

Prakash et al. [5] investigated Swirl flow on the annular diffuser preference, swirl quantities run through the diffuser of 0.0, 0.4, and 0.8, and compared different performance characteristics with and without struts. The pressure recovery coefficient of diffusers without strut is extremely high as compared to struts with almost the same swirl numbers. In recent research works related to annular diffusers, Keerthana and Jamuna Rani [6] determined the annular diffuser flow analysis and the performance of a series of annular diffusers with different divergence angles of (9, 15, 21 and 27). In this study, the pressure coefficient increases when the angle of divergence increases; while an increase beyond 21 degrees reduces the pressure recovery, an increase beyond 27 degrees is of no use. Kumar Manoj et al. [7] used CFD to study S-shaped diffuser output characteristics with rectangular combined horizontal, and vertical offsets.The researchers used two S-shaped Diffuser outlet cross-section geometries, one rectangular and the other semi-circular, with multiple one-dimensional

(z) vertical offset heights of 0.25B, 0.5B, 0.75B, and 1B in the rectangular case and the same in the semi-circular case. Singh S.N et al. [8] state that as vertical offset increases, outlet pressure recovery decreases. An intake swirl in the output to the annular diffuser with equal cone angles was investigated in this work. This study looks into the influence of an inlet swirl output of annular diffusers with the similar equal cone angle. This study is focused on determining (PDHC, UDHC, SHDC, and CHDC) achieve best with a distance of 278 mm and an equivalent cone angle of 15. This was discovered by the PDHC annular diffuser which performed best when swirl was introduced, and that the ideal swirl value was between 20 and 30 degrees [9].

The CHDC diffuser has the worst swirl flow performance, Kumar et al. [10]. An overall increase in static pressure recovery was found in experimental studies of annular diffuse in increasing the inlet swirl up to 25. Arun Gupta et al. [11] studied the Effect of Change in the Curvature of Flow and Performance Characteristics of Straight and S-shaped Diffusers. The centerline length is 4.0625 and the radius of curvature is 1.1938D. The rectangular inlet has a cross-section of a width of 0.3125D and a length of 0.625D, as well as a diameter of 160 mm. The offset cases of 0.3125D, 0.625D, 0.9375D, and 1.25D were investigated. In this analysis, the coefficients of static pressure gain (Cps) were investigated, and the efficacy decreased as the offset distances increased. The flow behavior of axial annular diffusers through a converging hub and diverging casing is investigated by Computational Fluid Dynamics (CFD) using ANSYS 15.0 with a casing cone angle of 10° and area ratio of 2.25, 3.25, and 4.25.

2 CFD Modeling

The performance of annular diffusers is influenced by the inlet geometrical parameters like area ratio (AR), axial length (L), casing and hub shape, as well as dynamic parameters like Reynolds number (Re), velocity profile, swirls condition, boundary layer blockage factor, and so on. We can set up an annular diffuser in two ways, as shown in Fig. 1, the standard parameter for evaluating the hub and casing inlet diameter improves the effectiveness of these setups for a given set of design parameters. The different parameters selected for the experimental diffuser are Rhi = 3.8 cm, Rci = 7.75 cm, h = -5.6222, c = 3.531, Li = 5 cm, and L = 38.60181828 cm. The diffuser's geometry was sketched in ANSYS 15.0.

The geometry of an annular diffuser was sketched using ANSYS software and a proper meshing scheme. The pre-study was implemented for k-ϵ turbulence equations including such conventional, renormalization group (RNG), and realizable on geometries with experimental/Simulation data, and the obtained results were compared. The velocity profile supplied at the inlet is identical to that produced experimentally using a turbulence specification of 5% and the strength of a hydraulic diameter diffuser inlet design has been specified.

The diffuser outlet design specifies the hydraulic diameter, and the outlet boundary is defined as the zero gauge pressure of the usual outlet boundary, through a turbulence

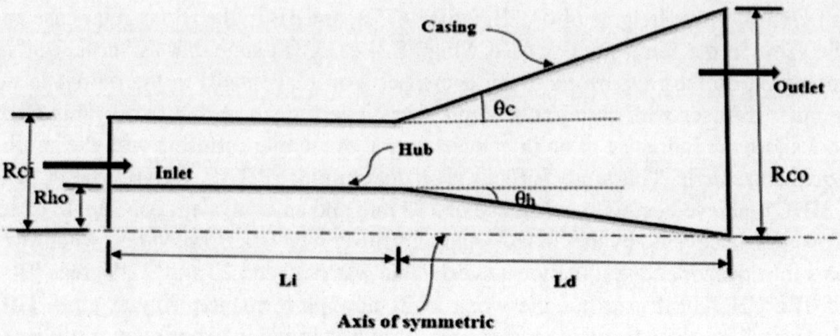

Fig. 1 Converging hub and diverging casing

Table 1 Geometry of diffusers for computational investigation

AR	Rci (cm)	Rhi (cm)	L (cm)	Rco (cm)	Rho (cm)	θc	θh
2.25	7.75	3.8	38.60181828	10.13166447	0	3.530573713	−5.622137489
3.25	7.75	3.8	56.51698449	12.17674525	0	3.530573713	−3.84657
4.25	7.75	3.8	77.20363656	13.9246409	0	3.530573713	−2.81785

intensity of 5%. Second-order up unwinding controls govern momentum, turbulent flow kinetic energy, and turbulence rate of dissipation. In terms of continuity, velocity components vx, vr, vz, k, and ϵ, the current study's findings were found to be stable k-ϵ. We achieve grid independence by repeating the analysis for several mesh sizes ranging from 57,786 to 129,987 mesh sizes. In a 2D axisymmetric RNG k-ϵ turbulence simulation having divisions along the centerline, the model was found to be accurate and came extremely close to the experimental data. Table 1 shows the geometries that were studied using CFD.

2.1 Governing Equation of Fluid

In order to solve a fluid-dynamics problem, only a few fundamental governing equations are used. There are three of them. These are fundamental physics equations given below from the reference [2]. (1) The conservation of mass equation. (2) Newton's second law of conservation of momentum. (3) The energy equation must be conserved.

(a) The Continuity Equation (FVM) Principle: Mass is conserved Differential Mass Conservation Equation

$$\text{dpdt} + \nabla(\text{pV}) = 0 \tag{1}$$

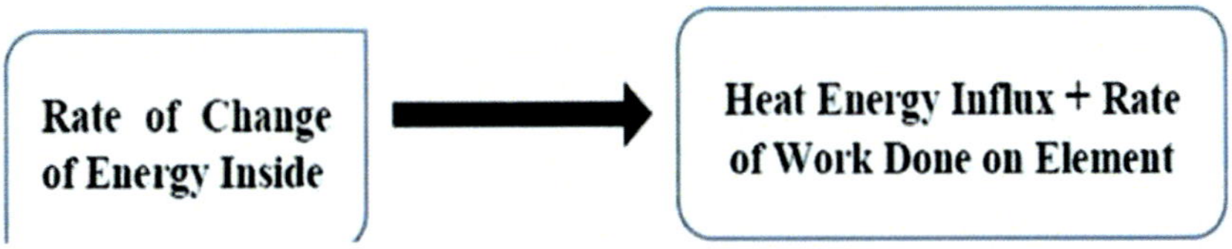

Fig. 2 Geometry of the diffuser with parallel diverging hub and casing

Integral Mass Conservation Equation

$$ddt \oiiint p.dV + ddt \oiiint p.(V.dS) = 0 \quad (2)$$

where dV = Elemental Volume, dS = Elemental Area.

(b) The Momentum Equation (FVM)
Principle: F = ma.
Navier stoke equation

$$pdVdt = pg - \nabla p + \mu \nabla 2V \quad (3)$$

(c) The Energy Equation

The amount of energy that enters the body/Mass is equal to the amount of energy that leaves the body/Mass. Differential energy equation (Fig. 2) [12]

$$pdudt + p(\nabla.v) = \nabla.(k\nabla T) + \psi \quad (4)$$

where ψ = viscous dissipation function.
Integral Energy Equation

$$dedt = ddt \oiiint p.e.dV + \oiint p.e.(V.dS) = O \quad (5)$$

where e = specific internal energy.

2.2 *Pressure Recovery Coefficient*

The first is a pressure recovery coefficient Cp, it is the dimensionless parameter and the difference between exit static pressure Pout, and inlet static pressure into the throat dynamic head and used mostly for subsonic flows

$$C_p = \frac{P_{out} - P_{in}}{\frac{1}{2} PU^2} \tag{6}$$

3 Grid Independence Test

The meshing generated in the specified geometry influences the result of every computer model. The mesh is made up of nodes or a grid. The grid's dimensions have a significant impact on the production of the physical model's performance [13]. A small number of mesh points implies that the calculated values may not be exact; however, fine mesh or a significant number of mesh points may provide improved results. The calculation time is considerably improved where the mesh is smoother. As a result, grid size must be optimized to produce better performance in less time without sacrificing precision. Grid independence was investigated in several divisions using various models in this study. Here's an example of a grid independence test, the experimentally obtained inlet velocity profile was adopted for the computer model. The RNG k-ϵ model is the turbulence model used in this experiment. The sensitivity was investigated using five different grid sizes [14]. In the current study, the grid size was used to establish symmetry seen between numerous annular diffusers, and utilize the symmetry parameter. The size of a quadrilateral cell's side is used to calculate the element size. In the current study, the grid size was used to establish symmetry between the numerous annular diffusers. The size of the quadrilateral cell side is used to calculate the element size. In comparison, the effect of element size on the reliability of the research observations, and the processing time necessary to get conclusions were investigated using mesh sizes ranging from 0.4 to 0.8 mm. Mesh independence was tested by changing the mesh size throughout the model and four solutions were obtained by four different mesh types as follows. Table 2 shows the Grid independence in terms of node count.

Figure 3 shows mesh size results of 0.4, 0.5, 0.6, 0.7, and 0.8 mm with the RNG k-ϵ model. Results of the mesh sizes of 0.4 and 0.5 are nearly comparable; consequently, in the current CFD analysis, the size of 0.5 mm has been considered to save computational time while maintaining accuracy. The quality of different geometries of axial annular diffusers was predicted using the same turbulence model.

Table 2 Grid independence in terms of node count

S.N	Mesh size (mm)	Number of nodes	Number of elements	Cp at outlet
1	0.8	58,376	57,786	0.618379
2	0.7	74,417	73,773	0.62427074
3	0.6	96,959	96,226	0.63190824
4	0.5	130,805	129,987	0.64151317
5	0.4	203,847	202,870	0.6434391

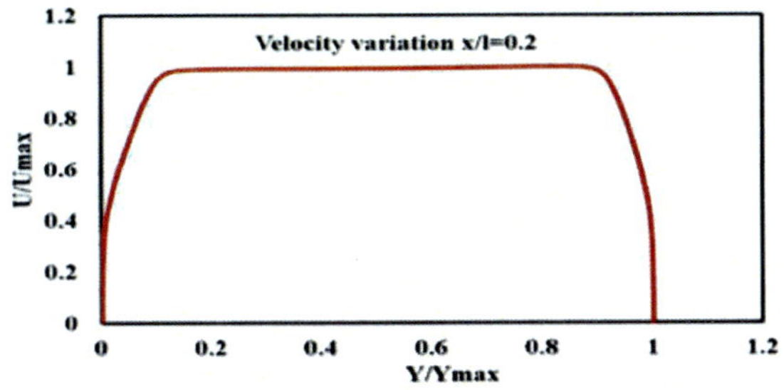

Fig. 3 Representation of Mesh independency on size of the element

3.1 Turbulence Model

In the various research papers, Singh [8], the RNG k-ϵ model produced better swirling performance flows, for non-swirling flows, the regular k-ϵ model per-formed better, according to exhaustive investigation on diffusers. These models have been evaluated by applying the inlet velocity profiles that have been fully experimentally obtained. Figures 4 and 5 show the translational velocity profile and pressure recovery coefficient, indicating that the RNG k-ϵ model is more closely coupled to the turbulence model, and thus was used for further investigation in this study.

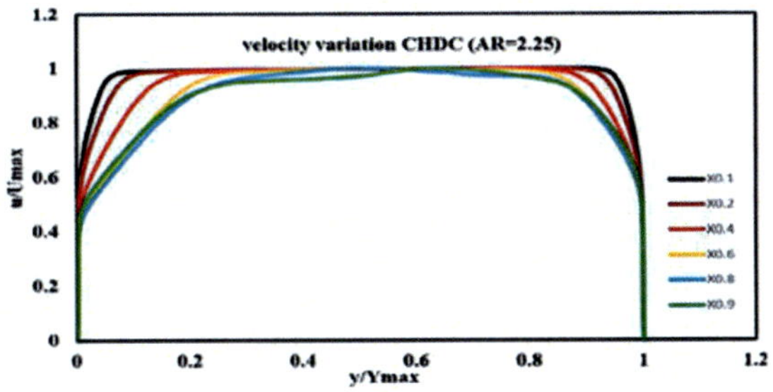

Fig. 4 NDL velocity variation profile for AR = 2.25 on the several transverses x/L = 0.1 to 0.9

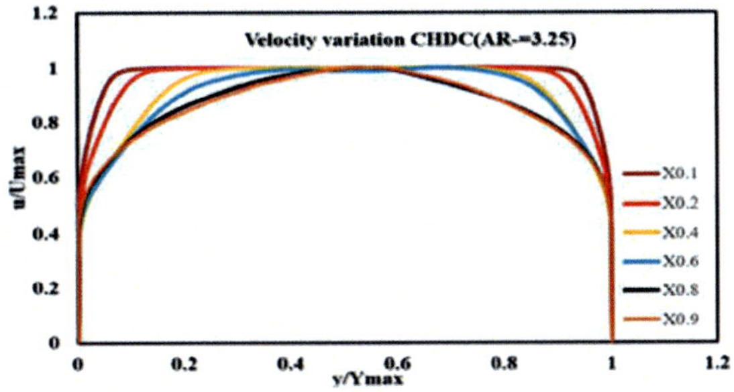

Fig. 5 NDL velocity variation profile for AR = 3.25 on the several transverses x/L = 0.1 to 0.9

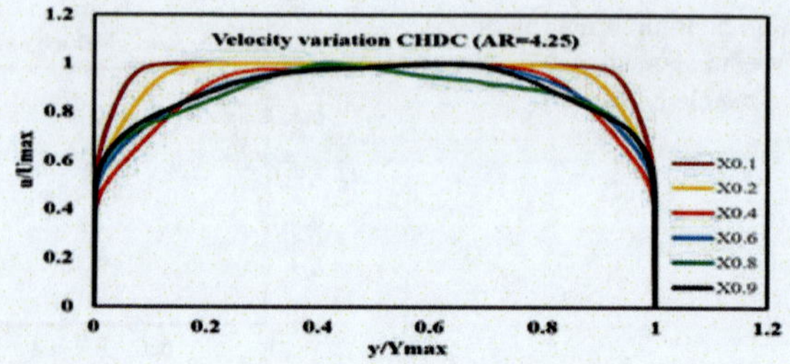

Fig. 6 NDL velocity variation profile for AR = 4.25 on the several transverses x/L = 0.1 to 0.9

4 Results and Discussion

4.1 Velocity Profile

Figures 4, 5, 6 show the transverse velocity profile and the pressure recovery coefficient, respectively, suggesting that k-ϵ model of RNG is more closely coupled to the turbulence model and used for further investigation in the present work. The pseudo velocity (y/Ym) was measured due to the traverse's passing height of a non-dimensional diffuser. The hub position is y/Ym = 0, and the direction of the traverse's casing is y/Ym = 1. For the entire area ratio, graphs are x/L = 0.1, 0.2, 0.4, 0.6, 0.8, and 0.9, various traverses of the diffuser passage are shown.

The flow is forced in the casing's direction. For Converging Hub and Diverging Casing (CHDC), AR = 2.25, and inlet velocity of 50 m/s, the reversal occurs at x/L = 0.8. The flow has been inverted to achieve an area ratio of 3.25. For x/L = 0.2, 0.4, and 0.6, y/Ym = 0.32, 0.66, and 0.67, and for x/L = 0.8 and 0.9, y/Ym = 0.54, 0.62 respectively. The flow is reversed in AR = 4.25 as well as y/Ym = 0.41 and 0.62 for x/L = 0.8 and 0.9, similarly. Figures show that because of the growth of the boundary layer since fluid flows down the diffuser passage, the velocity decreases faster near the hub and even near the casing surface. Along the hub wall, the pseudo velocity decreases, while near the casing wall, it increases.

The graph between the pressure coefficient and the length of the annular diffuser with the casing and hub at different area ratios of 2.25 to 4.25 is shown in Figs. 7 and 8. The pressure coefficient decreases first with the casing and then increases the pressure coefficient as shown in Fig. 7. Other broadcasters have followed the

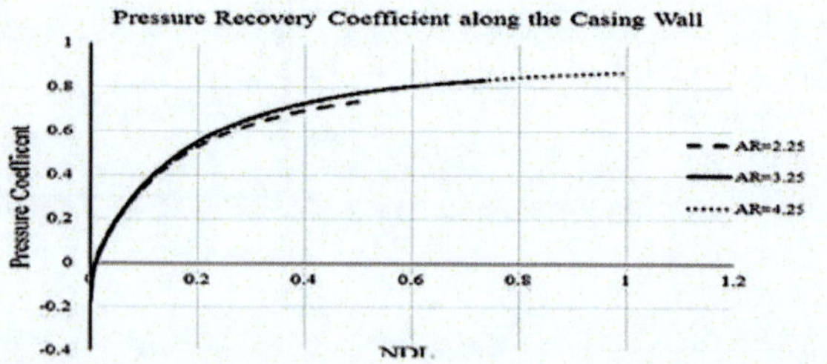

Fig. 7 Coefficient of Pressure Recovery distribution at casing wall (AR = 2.25,3.25, 4.25)

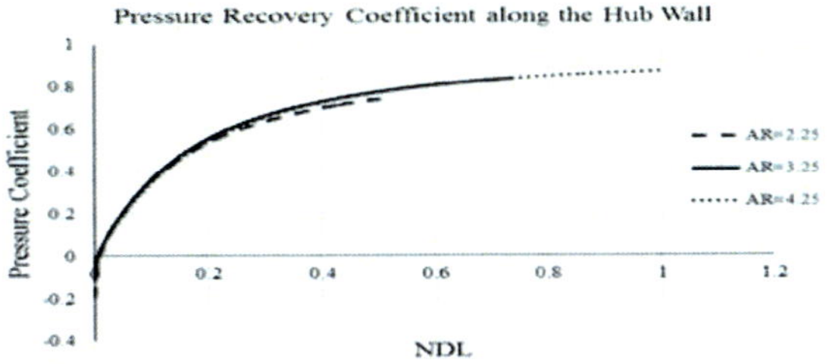

Fig. 8 Coefficient of Pressure Recovery distribution at hub wall (AR = 2.25, 3.25, 4.25)

Table 3 Variation of maximum Cp along casing wall and maximum Cp along the hub wall

S.N	Area ratio	Maximum Cp along casing wall)	Maximum Cp along hub wall
1	2.25	0.73688	0.737695
2	3.25	0.835102	0.835281
3	4.25	0.870399	0.870457

same pattern. Furthermore, it can be seen from Fig. 8 that the pressure coefficient decreases first with the hub wall and then increases. The annular diffuser is subjected to a CFD analysis in this research, by using different area ratios from 2.25 to 4.25 and the effect of pressure coefficient and axial velocity on the area ratio recovery was observed. The calculation was done using three different diffuser area ratios.

The position along the x-axis was kept similar for all the models, though the position on the y-axis was varied as per the value. The result shows that within the case of a Converging Hub and Diverging Casing, the result demonstrates that (CHDC) (Table 3), there is a recovery of a pressure coefficient maximum of area ratio = 4.25 and a pressure coefficient minimum of area ratio = 2.25, but the reverse flow of the fluid is also maximized in area ratio = 4.25 and reverse flow in area ratio = 2.25 is the minimum. According to the recovery of pressure coefficient and reverse flow of fluid in the same geometries with different area ratios, with area ratio = 2.25 are better because here the recovery of pressure coefficient is minimized but the reverse does not flow. We will be able to save even more energy as a result of this.

5 Conclusion

Here a confirmed Validation CFD RNG k-ϵ prototype model was used to show the performance analysis of the annular axial diffuser. The compute result conclusions are drawn from area ratios 2.25 to 4.25 at several inlet non swirl angles. The analysis was conducted on the computational tool named as ANSYS 15.0. The results were represented in the form of the contour of static pressure, contour of velocity magnitude, and contours of pressure recovery. The flow occurs toward downstream; the longitudinal velocity initially increases then decreases consistently in the case of non-swirling flows. Pressure recovery coefficient increases with the diffuser passage. As flow occurs toward downstream, the pressure recovery coefficient increases at a given flow. Swirl flow increases and the pressure recovery coefficient decreases along the casing and hub walls. A velocity profile has various shapes to different locations x/L to diffuser passage (0.1–0.9) due to the boundary layer formation. In the flow without swirl, there was no flow reversal and separation found along the hub wall. On the basis of the results, it is concluded that when the area ratio rises, the pressure recovery coefficient increases.

References

1. Shukri E, Abdullah B, Ali O (2017) Velocity distribution in an annular diffuser fitted with twisted hub for different area ratios. Energy Proc 141
2. Rasidi S, Seri S, Nordin N, Shariff M, Nasir N, Adzila S, Othman R (2020) Numerical investigation of 180 curved diffuser performance by varying geometrical and operating parameters. CFD Lett 12
3. Sullerey R, Chandra B, Muralidhar V (1983) Performance comparison of straight and curved diffusers. Department of Aeronautical Engineering, Indian Institute of Technology, Kanpur
4. Arora B (2014) Performance analysis of parallel hub diverging casing axial annular diffuser with 20 equivalent cone angle. Engineering 12(2)
5. Prakash R, Srinivas K, Anand H, Lakshmi Narayanan N (2015) Effect of swirl on the performance of an annular diffuser. Appl Mech Mater 787. ISSN: 1662-7482
6. Keerthana R, Rani G (2012) Flow analysis of annular diffusers. Int. J. Eng. Res. Appl. (IJERA) 2(3)
7. Gopaliya M, Goel P, Prashar S, Dutt A (2010) CFD analysis of performance characteristics of S-shaped diffusers with combined horizontal and vertical offsets Elsevier Ltd. All rights reserved, 2010
8. Singh S, Seshadri V, Saha K, Vempati K, Bharani S (2006) Effect of inlet swirl on the performance of annular diffusers having the same equivalent cone angle. Proc IMechE Part G: J Aerosp Eng 220
9. Mohan R, Singh SN, Singh DP (1997) Optimum inlet swirl for annular diffuser performance using CFD. Indian J Eng Mater Sci 5
10. Gopaliya MK, Kumar M, Kumar S, Gopaliya SM (2007) Analysis of performance characteristics of S-shaped diffuser with offset. Aerosp Sci Technol 11:130135
11. Gupta A, Singh H, Raghav AK (2020) Effect of Change in the curvature of flow and performance characteristics of straight and S-shaped diffuser
12. Stevens SJ, Williams QJ (1980) The influence of inlet conditions on the performance of annular diffusers. J Fluids Eng 102/357
13. Salim B (2013) Effect of geometrical parameters on the performance of wide-angle diffusers. Int J Innovat Res Sci Eng Technol 2
14. Alvarenga1 MA, Andrade1 CR, Zaparoli EL (2013) Compress ible subsonic flow in gas turbine annular diffusers. Int Rev Mech Eng (IREME) 10

An Approach to Enhance the Performance of Hybrid Solar Dryer

Rahul Sharma, Ashish kumar, Aviral Mathur, and Dadhish Kumar

Abstract Solar drying is an ancient technique and is present for ages. Due to the shortage of natural resources, it is necessary to adopt non-conventional energy resources to reduce the exploitation of natural resources and save the environment. Through this project, we have aimed to get the maximum efficiency from indirect solar drying and integrate hybrid properties to bring out the model as an optimal alternative. This report demonstrates the performance of the model under various weather conditions on various components and the improvement in efficiency figures if complementing it with hybrid properties. The results were found to be fairly good.

Keywords Solar dryer · Absorber · Drying chamber · Solar panels · Battery · Relay · LDR · Blower · LCD

1 Introduction

The purpose of sun drying is for drying various food items, i.e., tomatoes, bananas, and chilies. The process of sun drying has been a popular method for drying food for centuries and different types of solar dryers like natural convection, force convection, and hybrid solar dryers are used for the same. Here the performance of the indirect solar dryer is analyzed for drying up common food materials like bananas, tomatoes, and chilies under observation and the external power source has been used during night-time. The use of solar energy during the daytime, the use of fans for transmitting air into drying chamber, and the use of a heating coil during night-time showed a significant effect on the drying rate. The drying efficiency has been observed to be much higher than just the simple solar drying method. The maximum and the minimum (or ambient) temperatures observed were 36 °C and 84 °C, respectively. The change in weight and thickness of the food material has been analyzed with respect to temperature.

R. Sharma (✉) · A. kumar · A. Mathur · D. Kumar
Department of Mechanical Engineering, SRM Institute of Science and Technology, Modinagar, U.P, India
e-mail: rahul22sharma1998@gmail.com

D. P. Mishra et al. (eds.), *Recent Trends in Thermal and Fluid Sciences*, Lecture Notes in Mechanical Engineering, https://doi.org/10.1007/978-981-19-3498-8_6

2 Literature Review

2.1 Performance of Forced Convection Solar Dryer Integrated with Gravel as a Heat Storage Material for Chili Drying

This paper introduces the drying of 40 kg chili in Pollachi, India, utilizing a forced convection solar dryer. A flat plate solar air heater was used which consists of a drying chamber, heat storage, and a centrifugal blower. There is drying during the day, as well as at the night, so there is no time for re-hydration of the sample. Most of the moisture reduction occurs in the initial phase due to the evaporation in the outer surface of the sample and then it decreased due to removal of moisture from the inner layer of the sample, hence it leads to uniform dehydration. Moisture in the bottom tray is 4% to 6% greater than that of the top tray. The drier efficiency was estimated at about 21%. It has the following limitations such as: It is not reliable on rainy and cloudy days. As the blower needs electricity to run, therefore, it is expensive for the small farmers [1–3].

2.2 Design, Construction, and Performance Evaluation of Mixed-Mode Solar Dryer

This paper introduces the drying of yam chips by using a Mixed-mode solar dryer. In these dryers, direct solar radiation is used to dry samples. According to Ikejiofer, there are two types of water found in food items: one is chemically bound water and the other is physically held water; only physically held water is removed during drying. The time taken for the experiment is about 10 h. The percentage of moisture content removed is 85.4% with a drying capacity of 0.62 kg/hr. The system efficiency is found to be 57.5%. Limitations: The solar storage system is inadequate to store heat when insulation is insufficient in case of adverse weather conditions and at night when insulation is totally absent [4, 5].

2.3 Design and Performance Evaluation of New Hybrid Solar Dryer

This paper includes drying 30 kg banana slices on a sunny day for 8 h by utilizing a hybrid solar dryer. The necessities of storage for continuous operations of food drying are parameters such as variability and time dependency. The peak average air temperature was found to be 54.31 °C which was 30–40 °C above ambient temperature. The average collector efficiency was calculated at 58.23%. The initial moisture content was 82% and it was reduced by 18% at the end. It requires a large heat storage system and heat recovery equipment, which could result in excessive cost, hence making it unaffordable to small farmers [6, 7].

3 Experimental Set-Up

3.1 Design Procedure

The size of the dryer is determined as a function of the drying area needed per kilogram of fruit. Taking all the considerations in mind and collecting a few required data of the locality the dimensions of the dryer is determined using some formulas. These dimensions are used to build the base of the dryer and later the other essential equipment like solar panel, battery heating elements are installed to complete the design.

3.2 Design Calculations

The design calculations have been carried out to find out different parameter and to determine the required size of the dryer applicable to the location of Ghaziabad, Uttar Pradesh. Few of the conditions are made taking the ideal scenario and few assumptions (Table 1).

The moisture amount to be removed from the product, m_w [kg] has been calculated using the Eq. (1)

$$m_w = m_p(M_i - M_f)/(100 - M_f) \tag{1}$$

The drying rate, dr [kg/h] is calculated using the following equation:

$$d_r = m_w/t_d \tag{2}$$

Final relative humidity (RH) or Equilibrium Relative Humidity,4 ERH [%], is calculated using

$$M = M_f/(100 - Mf) \tag{3}$$

$$a_w = 1 - \exp\left[-\exp(0.914 + 0.5639 \ln M)\right] \tag{4}$$

$$ERH = 100\, a_w \tag{5}$$

Table 1 Design specifications and assumptions

S.N	Items	Condition and assumption
1	Location	Ghaziabad
2	Material to be Dried	Banana, Tomato and Chili
3	Drying Period	All year round
4	Loading Rate, m_p [kg/day]	5
5	Initial Moisture Content, M_i [%]	Banana—76% Tomato—94% Chili—84%
6	Final Moisture content, M_f [%]	10%
7	Ambient Air Temperature, Tam [°C]	40
8	Ambient Relative Humidity, RHam	0.31
9	Drying Time (Sunshine Hours) td [hours]	8
10	Incident Solar Radiation, Ih [KJ/m^2/day]	19,440
11	Average Wind Speed, Ws [m/s]	3
12	Density of air, ρ [kg/m^3]	1.2
13	Assumed Dryer Efficiency, η %%[%]	40%

where, a_w is water activity.

The mass of air needed for drying, ma [kg/h] is calculated using the following equation:

$$m_a = d_r/[(\omega_f - \omega_i) \times 100] \quad (6)$$

where ω_f and ω_i are the final and initial humidity ratios calculated from the psychometric chart, respectively.

The total heat energy required to evaporate water E [kJ] is calculated using the equation

$$E = m_a(h_f - h_i)t_d \quad (7)$$

where h_f [kJ/kg] and h_i [kJ/kg] are final and initial enthalpy of drying and ambient air calculated from the psychometric chart, respectively.

The area of the drying chamber can be calculated using the equation

$$A_d = E/I_h t_d \eta \tag{8}$$

Volumetric air flow rate V [m^3/s] can be calculated using the equation

$$V = m_a/\rho \tag{9}$$

The area of the solar chamber can be calculated using the equation

$$A_s = V/W_s \tag{10}$$

Using all the above equations, we can determine the required dimensions to design a solar dryer in any given set of conditions.

From the above set of equations, the dimensions of the dryer were calculated and tabulated (Table 2).

From the drying chamber area calculated above, the length, breadth, and height of the drying chamber could be determined. Assuming the length of the chamber as 46 cm and the height as 41 cm, the breadth of the drying chamber is calculated as 27 cm. Whereas from the solar chamber area calculated above, the length, bread, and height of the drying chamber are 46 cm, 21 cm, and 33 cm, respectively.

Therefore, the dimensions of the Solar Dryer [cm] are:

Drying Chamber—46 × 27 × 41

Solar Chamber—46 × 21 × 33

Table 2 Parameters and dimensions of the dryer

S.N	Parameters	Value
1	Initial Humidity Ratio, ω_i	0.014
2	Final Humidity Ratio, ω_f	0.0165
3	Initial Enthalpy, h_i [kJ/kg]	77.20
4	Final Enthalpy, h_f [kJ/kg]	96
5	Amount of moisture to be evaporated, m_w [kg]	4.667
6	Drying Rate, d_r [kg/h]	0.5884
7	Equivalent Relative humidity [%]	51.45
8	Heat Energy, E [kJ]	350.958
9	Drying Chamber Area, A_d [cm^2]	8462
10	Volumetric Air Flow, V [m^3/s]	1.944
11	Solar Chamber Area, A_s [cm^2]	6323

4 Procedure and Testing

The Arduino was coded using the embedded C language in order to run the LCD display and the relay which will provide the necessary connection to run the heating coil when the intensity of light will be insufficient to run the dryer (Fig. 1).

4.1 Working of Hybrid Solar Dryer

In the modern world, there is a growing popularity for a sustainable source of energy, especially solar thermal energy for farmers to increase productivity in the agricultural field. Solar air heaters are the devices for heating air with the help of solar radiation and utilizing this energy in crop drying and space heating. The drying process plays a major role in the preservation of foods.

Hybrid solar dryers are used for drying food items irrespective of any weather condition sunny, rainy, or cloudy. They are very efficient solar dryers. In this, the food items are kept in the drying trays. There are blowers, relay, solar panels, heating electrical coil, photodiode, resistor, temperature sensor, Arduino board, 12 V battery, display unit, transparent glass, and aluminum foil. In the sunshine hours, when solar radiation is available at its peak, the sun rays fall on the device, and at the same time, the solar heat falls on the solar panel whose work is to charge the battery with the help of the sun's heat. There are six solar panels used in the device, two solar panels are connected in series and three pairs of two solar panels each connected in series are connected in parallel. The battery is used to run the blowers, display screen, Arduino board, and other electrical gadgets in the device. The blowers fitted in the front of the apparatus force the flow of air stream inside the heating chamber, and due to the high temperature, the air also gets dried and goes to the drying chamber through the holes which connect the heating chamber to the drying chamber where three sliding trays are kept in which different food items are loaded for drying. The heating dry air dries the food items by removing moisture content from it and then the air goes out from the outlet hole located at the top of the drying chamber. Also, there is a temperature sensor that measures the temperature of the drying chamber. There is a resistor in the apparatus to prevent short-circuiting during the working of the dryer. There is also a photodiode whose work is to sense the intensity of the light and when it falls to a certain level that the solar intensity is insufficient to charge the solar panels, the relay attached with it switches on the heating coil which runs from the AC supply and this is the only gadget in the whole apparatus which works on the AC supply and it is present in the heating chamber. And this whole is guided and directed by Arduino board and it also controls the display screen which shows the temperature and light intensity. There is also coding given in the Arduino board to switch off the heating coil when the temperature rises up to a certain degree Celsius, i.e., above 85 °C.

4.2 *Testing of Hybrid Solar Dryer*

In this project, the testing of a hybrid solar dryer is done on three food items that are banana, chili, and tomato, respectively. Here we have compared the working of solar dryer under normal condition and hybrid condition for these three food items. Parameters for the testing are moisture content removed mass of dried product, temperature, and the drier efficiency.

4.2.1 Drying of Banana

The banana slices are chopped in a thickness ranging from 5–7 mm and the total weight of bananas taken for testing is 1.5 kg; 500 gm each in the three trays. The testing in normal condition is between 9 a.m.–5 p.m., and in the hybrid condition, it is kept for 4 h (12–4 p.m.) under normal condition and 4 h (4–8 pm) under heating coil condition. After keeping the banana slices in the trays, the temperature and mass variation is noted on an hourly basis.

4.2.2 Drying of Chili

The chilies were taken in a tray having 90% perforations, and the total weights of chilies taken were 300; 100 gm in each three trays. The testing in normal condition is between 9 a.m.–5 p.m., and in hybrid condition, it is kept for 4 h (12–4 p.m.) under normal condition and 4 h (4 p.m.–8 p.m.) under heating coil condition. After keeping the chilies in the trays, the temperature and mass variation is noted on an hourly basis.

4.2.3 Drying of Tomato

The tomato slices are chopped in a thickness ranging from 5–7 mm, and the total weight taken is 1.5 kg; 500 gm each in three trays. The testing in normal condition is between 9 a.m.–5 p.m., and in hybrid condition, it is kept for 4 h (12–4 p.m.) under normal condition and 4 h (4–8 p.m.) under heating coil condition. After keeping the tomatoes in the trays, the temperature and mass variation was noted on an hourly basis.

5 System Evaluation and Calculation

5.1 System Evaluation

The performance of the system has been thoroughly evaluated by taking the observations on different food items (banana, tomato, and chili) under different weather conditions. We have also made use of the heating coil as an integrated hybrid source to observe the system's performance. We have observed different food items having different moisture content and the results were quite different from each other.

Banana, chili, and tomatoes were chopped into thin slices and kept under the normal conditions for a day under 8 h of observation from 9:00 a.m. to 5:00 p.m. For finding the efficiency in hybrid conditions, it was kept 4 hours under normal conditions (i.e. 12:00 p.m.–04:00 p.m.) and after that 4 hours in hybrid conditions. The initial moisture content of bananas, tomatoes, and chilis was 76%, 94%, and 84%, respectively.

5.2 Observation

Formulas used to calculate system efficiency are as follows:

Dryer Efficiency (η_d): It is the ratio of the collector efficiency (ηc) and the system efficiency (η_s).

$$\eta_c = Q_u/A_d I_h \tag{11}$$

$$Q_u = m_p C_p \Delta T \tag{12}$$

$$\eta_s = WL/A_d I_h \tag{13}$$

where, m_p = Loading rate.

C_p = Specific heat at constant pressure.

ΔT = Difference between the minimum and maximum temperature.

A_d = Drying chamber area.

I_h = Incident solar radiation.

W = Moisture content left after drying.

L = Latent heat of evaporation at the maximum temperature.

Moisture Content:

$$M_{wb} = [(m_i - m_f)/m_i] \times 100 \tag{14}$$

M_{wb} = Moisture percentage removed.

Table 3 Moisture percentage removed and dryer efficiency of solar dryer under normal conditions

S.N	Material	Moisture percentage removed, M_{wb} [%]	Dryer efficiency, η_d [%]
1	Banana	63.66	13.5
2	Chili	46.89	10.03
3	Tomato	74.88	20.36
Average dryer efficiency			14.63

Table 4 Moisture percentage removed and dryer efficiency of solar dryer under hybrid conditions

S.N	Material	Moisture percentage removed, M_{wb} (%)	Dryer efficiency, η_d (%)
1	Banana	78.46	44.67
2	Chili	57.57	23.70
3	Tomato	85.10	58.71
Average dryer efficiency			42.36

m_i = Initial mass of material.
m_f = Final mass of material.

5.3 *Calculations*

Using the formula (11–14) to determine moisture removed and efficiency of the solar dryer in both normal and hybrid conditions is tabulated in Tables 3 and 4 respectively.

From Figs. 2 and 3, it is quite evident that the moisture removal rate was at its peak between 12:00–02:00 p.m. under normal conditions. It is because of the peak temperature which is observed during this time as the intensity of light is highest during this duration. After 02:00 p.m., the moisture removal rate starts decreasing as dawn begins and the temperature starts falling (Fig. 6).

From Figs. 4 and 5, it can be observed that the moisture removal rate increased between the time duration of 01:00–02:00 p.m. and then started falling during the dawn till 04:00 p.m. It was due to the decrease in temperature witnessed during this time. Also, after 04:00 p.m., the slope suddenly increased as the solar dryer started working under hybrid conditions and the temperature increased which intensified the moisture removal rate as well.

6 Results

From this study, it has been observed that the indirect solar dryer was far more efficient than the direct sun drying method. The food dried was of high quality and

Fig.1 Diagram of set-up

free from any kind of contamination and uneven drying. The method's efficiency and effectiveness were further improved after integrating hybrid properties and the results obtained were found to be quite good. Under continuous drying in both normal and hybrid conditions, the results were observed to find out effect of suggested modifications.

The testing was done on tomato, banana, and chili. Since all of them had different moisture conditions, the results differed from each other in both conditions. The moisture content removal for normal condition from banana, chili, and tomato are 63.66, 46.89, and 74.88, respectively and the dryer efficiency for banana, chili and tomato are observed to be 13.5, 10.03 and 20.36 respectively. The average dryer efficiency under normal conditions is 14.63. Under hybrid conditions, the moisture content observed for banana, chili, and tomato is 78.46, 57.57, and 85.10, respectively, and the dryer efficiency for banana, chili, and tomato is 44.67, 23.70, and 58.71, respectively. The average dryer efficiency under hybrid conditions is 42.36.

The results were quite different in both conditions and the implementation of hybrid properties improved the efficiency. This concept can be used to increase the speed of the complete drying process with better efficiency and moisture removal capacity.

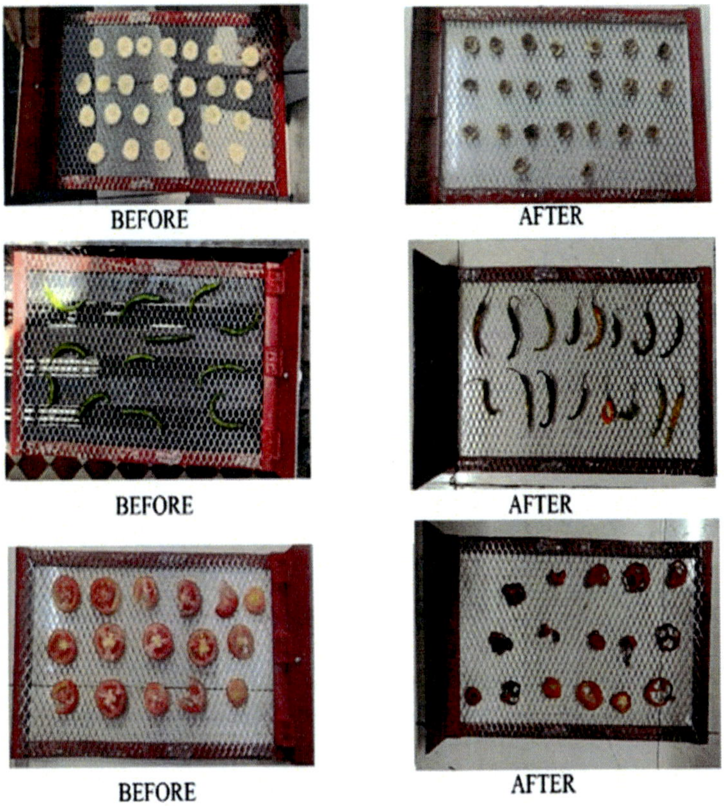

Fig. 2 Food materials before and after drying

7 Conclusion

The efficiency of hybrid solar dryer shows significant improvement as compared with conventional solar dryer. The tests were conducted during the months of April and May. The purpose of using the indirect method of sun drying was to protect the food from contamination, rain, etc. The maximum and minimum temperature inside the model was recorded at 600 C and 350 C, respectively, under normal conditions. Under the hybrid conditions, the maximum temperature was recorded as 85 °C. The efficiency observed during the normal conditions was found to be 14.63% and the efficiency observed during hybrid conditions was found to be 42.36%. The concept of solar drying is quite beneficial and several advancements can be made to improve its efficiency and effectiveness in the future.

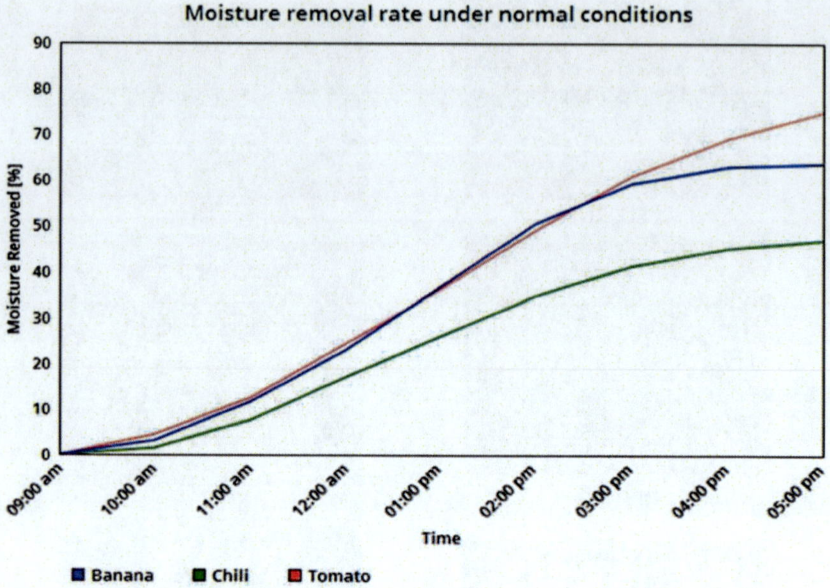

Fig. 3 Moisture removal rate under normal condition

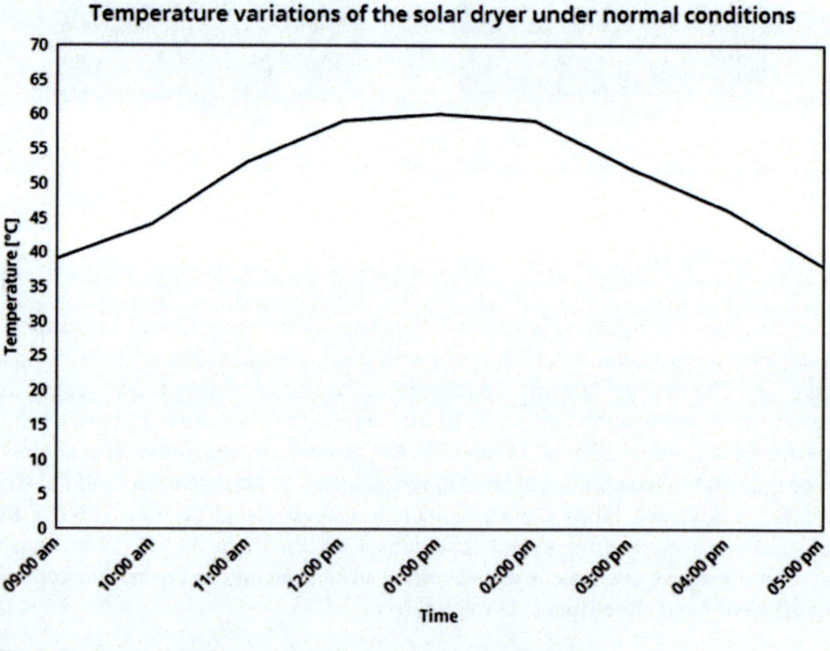

Fig. 4 Temperature variations of the solar dryer under normal condition

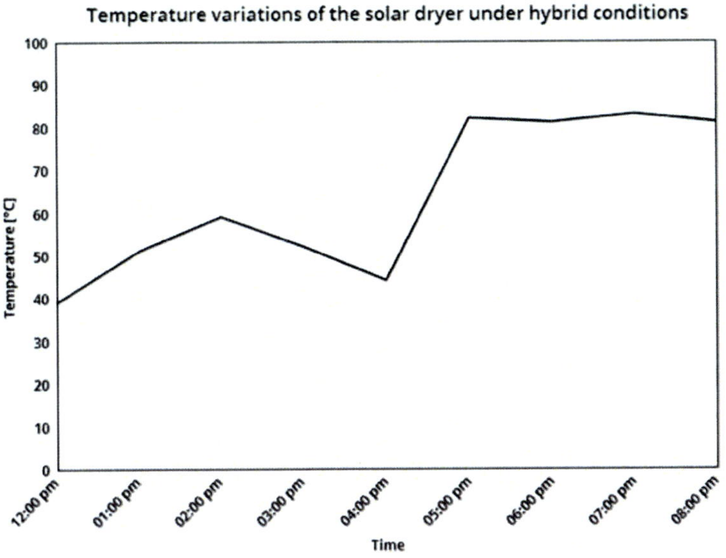

Fig. 5 Temperature variations of the solar dryer under hybrid condition

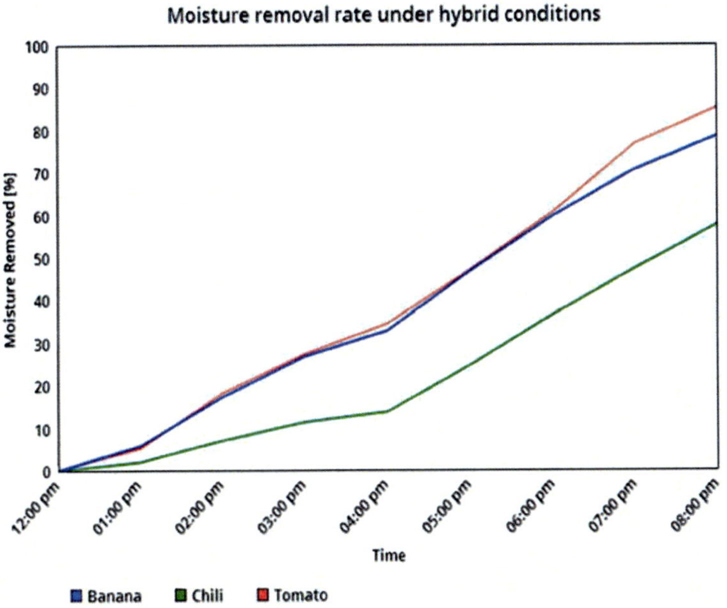

Fig. 6 Moisture removal rate under hybrid condition

References

1. Ogheneruona DE, Yusuf MOL (2011) Design and fabrication of a direct natural convection solar dryer for tapioca. University of Port Harcourt, P.M.B. 5323, Choba, Rivers State, Nigeria
2. Mohanraj M, Chandrasekar P (2009) Performance of a forced convection solar dryer integrated with gravel as heat storage material for chili drying. School of Mechanical Sciences, Karunya University, Coimnatore, India, School of Engineering and Sciences, Swinburne University of Technology (Sarawak Campus), Kuching Sarawak, Malaysia
3. Kumar A, Rai AK (2018) Comparative study of open sun drying and solar cabinet drying techniques for drying of green chilies. Department of Mechanical Engineering, SRMSCET, Bareilly, Department of Mechanical Engineering, SSET, SHIATS DU, Allahabad, India
4. Bolaji BO, Olalusi AP (2008) Performance evaluation of a mixed mode solar dryer. University of Agriculture Abeokuta, Ogun State, Nigeria, Department of Mechanical Engineering
5. Pardhi CB, Bhagoria JL (2013) Development and performance evaluation of mixed-mode solar dryer with forced convection. Int J Energy Environ Eng
6. Amer BMA, Hossain MA, Gottschalk K (2009) Design and performance evaluation of a new hybrid solar dryer for banana. Cairo University, Germany, Agriculture Engineering Department
7. Hedge VN, Hosur VS, Narayana KB (2015) Design, fabrication and performance evaluation of Solar Dryer for Banana. Sustain Soc, Energy

An Experimental Investigation of the Effect of Diethyl Ether as an Additive on the Performance of a Single Cylinder Diesel Engine

Aqueel Ahmad, Achhaibar Singh, D. K. Singh, and Ashok Kumar Yadav

Abstract The use of diethyl ether (DEE) as an oxygenated additive in a diesel engine is studied in this experiment. The effect of blending diethyl ether (DEE) with diesel at 6, 10, and 14% on the performance analysis of a single cylinder diesel engine running at a constant speed of 1500 rpm is proposed in this research. The results showed that adding DEE enhances brake thermal efficiency while significantly reducing brake specific fuel consumption and fuel flow rate. With a 14% blending of DEE with diesel, brake thermal efficiency provides optimal efficiency and low brake specific fuel consumption.

Keywords Diesel engine · Diethyl ether · Blends · Engine performance

1 Introduction

India is a fast-developing country with steady economic development, and its energy demand is growing by leaps and bounds every day. Fuel consumption is directly proportional to demand [1]. Over the last few years, the use of petroleum-based fuels has increased. Leading to a shortage of fossil fuel sources, India is mostly reliant on imported fuels. Importing gasoline has a significant economic impact, and petroleum-based fuels contribute to the sharp decline of traditional energy sources [2]. As a result, it is high time to develop alternative engine fuels. Recent experiments and studies have made it possible to boost engine efficiency while lowering emissions.

A. Ahmad (✉) · A. Singh · D. K. Singh
Mechanical Engineering Department, Netaji Subhas University of Technology, New Delhi, India
e-mail: Aqueelahmad191@gmail.com

A. Singh
e-mail: asingh@nsut.ac.in

D. K. Singh
e-mail: dksingh@nsut.ac.in

A. K. Yadav
Department of Mechanical Engineering, Raj Kumar Goel Institute of Technology, Ghaziabad, India

D. P. Mishra et al. (eds.), *Recent Trends in Thermal and Fluid Sciences*, Lecture Notes in Mechanical Engineering, https://doi.org/10.1007/978-981-19-3498-8_7

Many researchers believe that adding additives to a diesel engine can improve its performance. Diethyl ether (DEE) is a substitute fuel that improves the efficiency of diesel engines [3]. DEE has no C–C bonds, but it does contain oxygen, which aids in smokeless combustion. DEE is a renewable fuel that can be produced by dehydrating bioethanol. Due to its capabilities, DEE can be utilized as a standalone fuel or in a blending with diesel in compression ignition engines [4]. DEE has a lower autoignition temperature, a higher cetane number (125), and a higher latent heat of vaporization than diesel. DEE is a liquid at room temperature, which is easy to store and is also non-corrosive [5]. Fossil diesel inhibits DEE ignition, while adding DEE to fossil diesel lowers the cetane number [6]. The goal of this study is to see how adding Diethyl ether to a diesel engine affects its performance. Different DEE-added diesel blends (D94DEE6, D90DEE10, and D86DEE14) are constructed and compared to pure diesel performance.

2 Material and Methodology

2.1 Experimental Setup

Engine A single-cylinder, four-stroke, automated Kirloskar CI engine with water cooling was used to carry out this research. The compression ratio can be adjusted without interrupting the engine or adjusting the combustion chamber geometry. This system is made up of two pressure sensors, one for measuring combustion pressure and the other for measuring injection pressure. The gas calorimeter is used to determine how much heat is lost by specific gases. Temperature, airflow, fuel flow, and load measurement sensors are all measured by the test rig. The software connects all of the sensors to the CPU. Table 1 lists the specifications of the test rig.

Table 1 Nomenclature

T1 & T3	Water temperatures (°C) at inlet	DEE	Diethyl ether
T2	Engine jacket water temperature (°C) at outlet	BP	Brake power
T4	Calorimeter temperature(°C) at outlet	BSFC	Brake specific fuel consumption
T5	Temperature of the exhaust gas before the calorimeter (°C)	EGT	Exhaust Gas Temperature
T6	Temperature of exhaust gas after calorimeter (°C)	EGA	Exhaust gas analyzer
F1	Fuel flow differential pressure unit	D	Diesel
F2	Air intake differential pressure unit	PT	Pressure transducer
Wt	Load	N	rpm

Dynamometer The engine's performance is measured using a DC electrical dynamometer. The dynamometer is reversible and can be used as both a controlling and absorbing unit. The field current is used to control the load (Figs. 1 and 2).

In this study, various diesel blends with DEE additives were developed. The parameters of diesel fuel, as well as DEE blends with diesel fuel, are shown in Table 2. Because of the lower density differential between the diesel and the DEE, mixing is straightforward and no separation occurs. The additives have a lower autoignition temperature, a greater cetane value, and a lower density than the base gasoline. These diesel and DEE blends were created in the lab. In a single cylinder engine with a

Fig. 1 Experimental set up of computerized C.I. engine

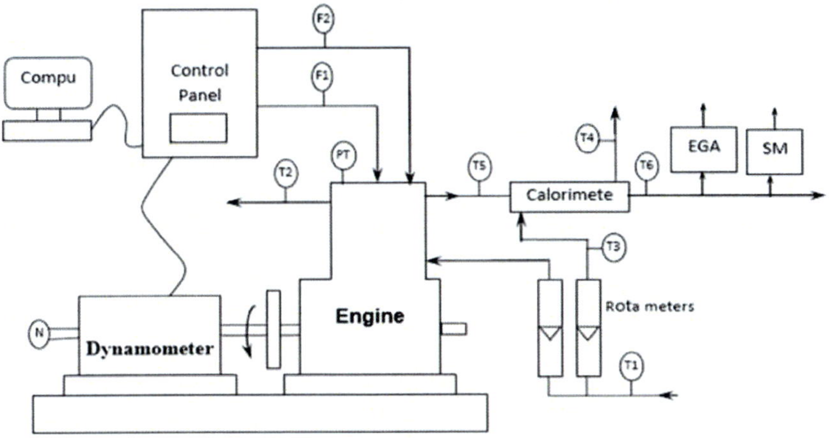

Fig. 2 Experimental arrangement as a line diagram [7]

Table 2 The test rig's specifications

No of cylinders	01
No of strokes	04
Rated capacity	5.2 KW/7 hp at 1500 rpm
Fuel and type of cooling	Diesel and water cooled
Bore and stroke of a cylinder	87.5 and 110 mm
Compression ratio	17.5:1
Type of dynamometer	Eddy current
Arm length of dynamometer	185 mm

Table 3 Properties of diesel with DEE blends

Properties	Diesel	DEE	6% DEE	10% DEE	14% DEE
Viscosity (cst)	2.7	0.23	1.63	1.01	0.735
Density (Kg/m^3)	830	713	807	774.4	760.5
Cetane number	49	125	55.1	53	51.5
Fire point (°C)	52	44	41	37	34
Flash point (°C)	46	40	38	35	32
Calorific value (KJ/Kg)	42,800	33,900	43,100	42,994	42,874

compression ratio of 17.5 and constant rpm, experiments were conducted with pure diesel and various mixtures of diesel and diethyl ether additives. Running the engine with diesel at a CR of 17.5 and adjusting the load from 10 to 100% while recording engine performance data yielded the results. The engine was also tested on a variety of diesel and DEE mixtures (Table 3).

3 Result and Discussion

The experiments were conducted using 100% diesel, as well as mixes of diesel and DEE in proportions of 6, 10, and 14%, all of which were thoroughly assessed across performance criteria. Brake thermal efficiency, BSFC, EGT, and fuel flow are all measured output parameters that are tested and compared to pure diesel.

3.1 Brake Thermal Efficiency

For various DEE/diesel blends, Fig. 3 demonstrates the link between brake thermal efficiency and braking power. The braking thermal efficiency of all blends appears to rise as the load is raised. The presence of oxygen in the DEE helps the fuel burn

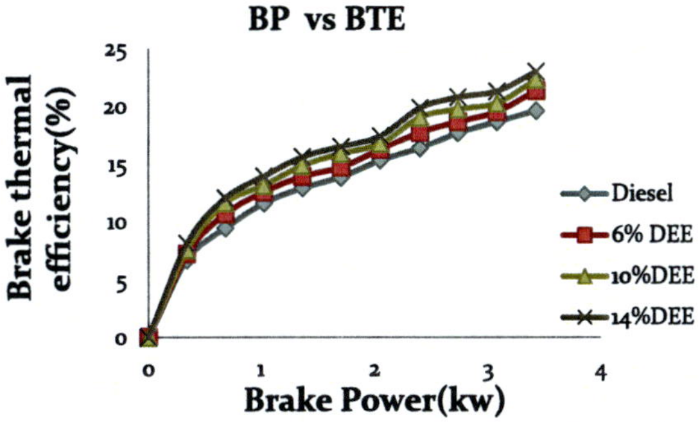

Fig. 3 Brake thermal efficiency varies with brake power for various blends

completely, boosting the break thermal efficiency. BTE is approximately identical for all combinations of diesel and DEE at low loads, and then increases gradually as the DEE/diesel ratio grows. On a 14% DEE/Diesel mix, BTE is at its best. DEE has the potential to lower the surface tension between two immiscible liquids, which improves diesel combustion. So BTE of 14% blend is more as compared to other blends.

3.2 *Brake Specific Fuel Consumption*

The difference in Brake Specific Fuel Consumption (BSFC) for various blends at various brake power is depicted in Fig. 4. When the load is low, the BSFC is higher, but it gradually drops as the load increases. The BSFC for diesel with a 14% DEE blend is lower. This is owing to the fact that diesel fuel burns more efficiently. As DEE concentrations rise, the BSFC drops.

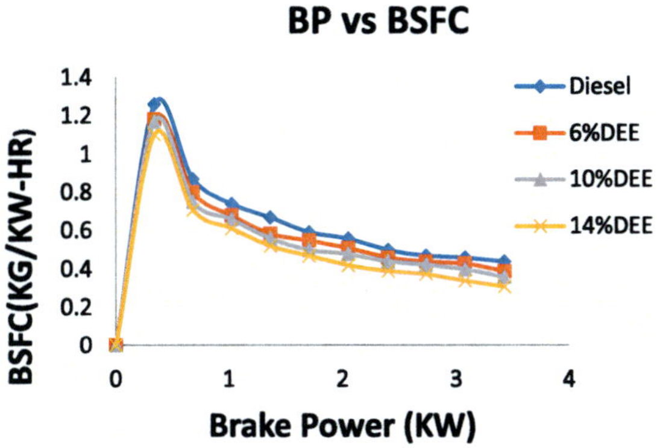

Fig. 4 BSFC varies with BP for various blends

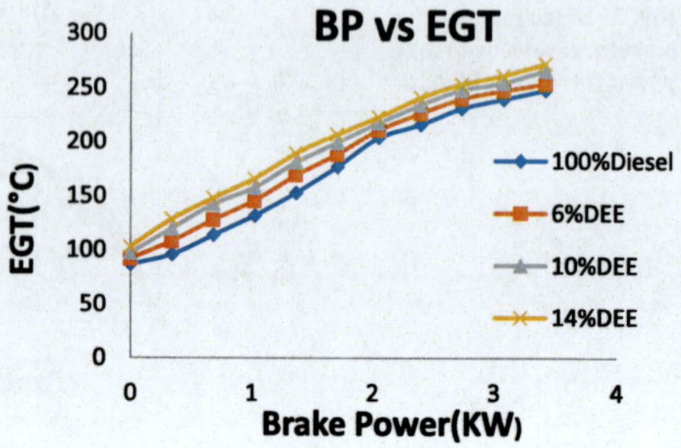

Fig. 5 EGT varies with break power for different blends

3.3 *Exhaust Gas Temperature*

Figure 5 depicts the differences in Exhaust Gas Temperature for Different Blends at Various Break Power. As the amount of DEE in diesel rises, the exhaust gas temperature virtually stays the same. At full load, the exhaust gas temperature for 14% DEE is slightly higher than for 10% DEE. This is due to the calorific content of the fuel in different blends varying.

3.4 *Fuel Flow*

Figure 6 depicts the difference in fuel flow at various loads for various blends. For all blends, the amount of fuel burnt per unit time increases as the load is increased. However, when DEE is added to diesel, the overall fuel flow decreases. On 14% DEE blends with Diesel, fuel flow is at a bare minimum.

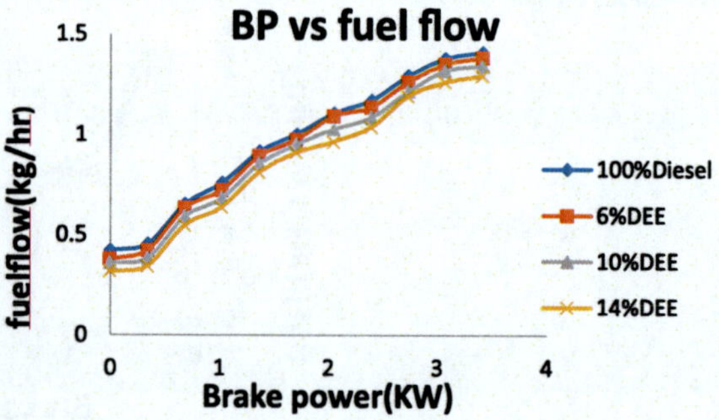

Fig. 6 Fuel flow varies with brake power for different blends

4 Conclusions

The output parameters of DEE (6, 10, and 14%) blends with diesel were studied in the experiments. Based on the results of the experiment, the following conclusions can be drawn:

The brake thermal efficiency improves as the proportion of DEE in diesel increases, peaking at 14% of the blend's brake thermal efficiency.

As the load is increased, the BSFC decreases linearly. On a 14% DEE mix, the BSFC is lower.

As DEE is added to diesel, the total fuel flow decreases. On a 14% DEE mix, fuel flow is at a bare minimum.

The viscosity of diesel reduces as the concentration of DEE in gasoline rises. Raising the concentration of DEE in diesel by more than 15%, the engine made a lot of high-pitched knocking noises.

DEE can be used as a diesel engine additive due to its low cost.

References

1. Chauhan B, Kumar N, Cho H, Lim H (2013) A study on the performance and emission of a diesel engine fueled with Karanja biodiesel and its blends. Energy 56:1–7
2. Ashok B, Denis Ashok D, Kumar C (2015) LPG diesel dual fuel engine—a critical review. Alex Eng J 54(2):105–126
3. Ashok M, Saravanan C (2007) Performance and emission of the emulsified fuel in a DI diesel engine using oxygenated additive diethyl ether with surfactant of span−80. Energy Fuels 21(4):1878–1882
4. Varpe B, Kharde Y, Rahman K, Khidmatgar A (2020) Effect of diethyl ether additive in Jatropha biodiesel-diesel fuel blends on the variable compression ratio diesel engine performance and emissions characteristics at different loads and compression ratios. Heat Transf 49(8):4427–4447
5. Venu H, Madhavan V (2017) Influence of diethyl ether (DEE) addition in ethanol-biodiesel-diesel (EBD) and methanol-biodiesel-diesel (MBD) blends in a diesel engine. Fuel 189:377–390
6. Clothier P, Pritchard H, Poirier M (1993) Synergy between additives in stimulating diesel-fuel ignition. Combust Flame 95(4):427–429
7. Clothier P, Moise A, Pritchard H (1990) Effect of free-radical release on diesel ignition delay under simulated cold-starting conditions. Combust Flame 81(3–4):242–250

Performance Analysis of a Jet Impinged Solar Air Heater Having the Corrugated Absorber Plate: A CFD Approach

Amitesh Sharma, Sushant Thakur, Devender Kumar, Ajay Kumar, and Prashant Dhiman

Abstract The present work has been presented to improve efficiency by using jet impingement through the corrugated absorber plate. Due to this, jet impingement turbulence enhancement takes place because of which the rate of thermal heat gain increases. The thermohydraulic efficiency of SAH was calculated at four different mass flow rates and 500 W/m^2 solar radiations. The fan power requirement to propel the air inside the section is small for a lower mass flow rate. Thus, the consequence of jet size, no. of jets (N), has been discussed to predict an optimum configuration of SAH. Current research shows that 3 mm jet size, 70 jets, and 0.015 kg/s offer the highest thermohydraulic efficiency. The maximum thermal efficiency and outlet temperature (T_O) were found at h/D = 0.40. From the analysis, it can be concluded that the thermohydraulic efficiency enhances with an increase in mass flow rate up to a specific value, but beyond it, η_{THE} starts declining.

Keywords Solar air heater · Artificial roughness · Computational fluid dynamics · Thermal efficiency · Thermohydraulic efficiency

Abbreviations

A	Area of the glass plate (mm^2)
H	SAH height (mm)

A. Sharma · S. Thakur (✉) · D. Kumar · A. Kumar · P. Dhiman
Department of Mechanical Engineering, National Institute of Technology, Hamirpur, Himachal Pradesh 177005, India
e-mail: sushantthakur@nith.ac.in

D. Kumar
e-mail: devenderkumar@nith.ac.in

A. Kumar
e-mail: ajaykumar@nith.ac.in

P. Dhiman
e-mail: prashantdhiman@nith.ac.in

D. P. Mishra et al. (eds.), *Recent Trends in Thermal and Fluid Sciences*, Lecture Notes in Mechanical Engineering, https://doi.org/10.1007/978-981-19-3498-8_8

I	Solar Radiations (W/m^2)
k	Thermal conductivity (W/m·K)
L	SAH length (mm)
N	Number of jets
P	Pitch (mm)
P_f	Fan Power (W)
T_i	Inlet Temperature (K)
T_o	Outlet Temperature (K)
W	SAH width (mm)
MFR	Mass Flow Rate (kg/s)
H	Height (mm) (in Fig. 10)
D	Width (mm) (in Fig. 10)
A	Absorptivity of glass plate
α_a	Absorptivity of absorber plate
ε	Emissivity of glass plate
T	Transmissivity of glass plate
P	Fluid density
η_{THE}	Thermohydraulic efficiency
η_{th}	Thermal efficiency

1 Introduction

Jet impingement methodology is a very effective technique for enhancement of the rate of heat transfer from absorber plate to flowing fluid. By this method, there is more mixing of flowing fluid inside SAH, due to which flowing fluid gains more heat from the plate and outlet temperature increases. The literature study originated that the η_{THE} and ηth of SAH are enhanced by significant amounts as there is more fluid mixing inside the SAH. When we compare a jet impingement SAH with conventional SAH, there is a rise of 13–14% in thermal efficiency at a gain of 3 K temperature (ΔT).

SAH is widely used in building space heating and cooling [14, 15]. It is a cheaper and more economical way of heating and cooling a building. Solar collectors [6] are used to collect the solar radiation emitted by the sun and use it for various purposes. A comprehensive literature survey has been carried out on solar air heaters. Chauhan and Thakur [3] investigated the efficiency of jet impingement SAH. The experimentation found that the maximum efficiency was achieved at 70%. Aboghrara et al. [1] analyzed the performance of the Solar Air Heater. The experimental investigation found that the thermal efficiency increase is 14% and the temperature increase is about 3 K compared to the traditional smooth duct at ambient temperature 308 K, solar radiation 500 W/m^2, and Mass Flow Rate 0.01–0.03 kg/s. Soni and Singh [13] performed an experimental investigation of geometrical consideration on the efficiency of an inline jet plate SAH. Rajaseenivasan et al. [11] performed experimental works on the performance of an impinging jet SAH. The SAH with jet impingement increases the outlet temperature of SAH. Higher Nusselt no. has been achieved at high MFR, which cause in additional heat transfer. Choudhury and Garg [5] performed work on the evaluation of a jet plate. The Author aims to enlarge the efficiency of SAH by the jet impingement method. They calculate the effect of geometrical parameters, for example, jet diameter or nozzle diameter, the distance between plates, and nozzle height. Nayak and Singh [10] performed experimental work to find out the consequence of geometrical features on the efficiency of jet plate SAH. The efficiency was examined for different mass flow rates. Shukla et al. [12] analyzed performance of solar air heater using computational fluid dynamics simulation. They found that the maximum temperature in case A is 395 K, while for case B, it is 402 K; this is because of the increment in outlet temperature (T_0) by the presence of hollow bodies over the absorber plate. Huber and Viskanta [7] implemented experiment work to ascertain consequences of jet pitch on convective heat transfer, impinging arrays of axisymmetric jets. Nadda et al. [9] performed experimentation on a jet impinged SAH with several arc protrusion obstacles.

The experimental investigation was executed to certify friction factor (f) and Nusselt number (Nu) characteristics of impinging jet SAH. Chauhan and Thakur [4] performed an experimental investigation of heat transfer and friction factor correlations for an impinging jet SAH. The experimental investigation was set up to know about the consequence of jet impingement on the SAH absorber plate. Belusko et al. [2] experimented to find out its efficiency and concluded the effectiveness of air impingement. Jet impingement is a method to raise the heat transfer from the absorber plate to the caring fluid (air). Studies have found that the efficiency of glazed air collectors can marginally increase by jet impingement. Matheswaran et al. [8] performed analytical investigation of SAH with jet using exergy along with energy study. They perform work on single pass double duct jet impingement SAH energy efficiency.

2 Description of the CFD Model

The computational domain of SAH is shown below in Figs. 1 and 2. In the present research work, the jet diameter of 3 mm is taken to achieve a high-velocity jet through a corrugated absorber plate to achieve more turbulence inside the computational domain. CFD investigation of the test section has been accomplished for different jet sizes and different numbers of jets. Therefore, outlet temperature and efficiency have been investigated. The proposed operating parameters for the current study are furnished in Table 1.

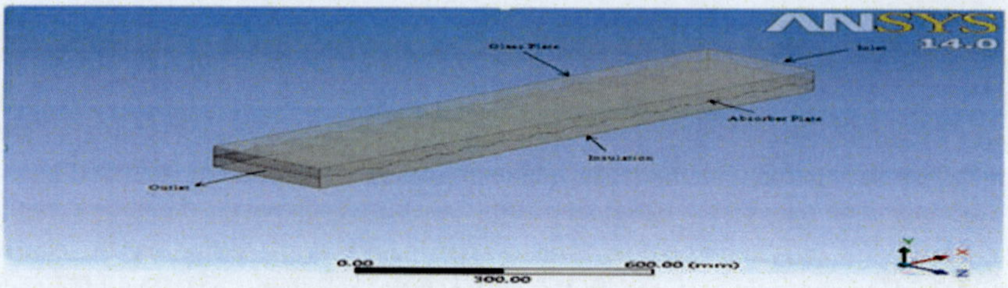

Fig. 1 Sketch of the computational domain

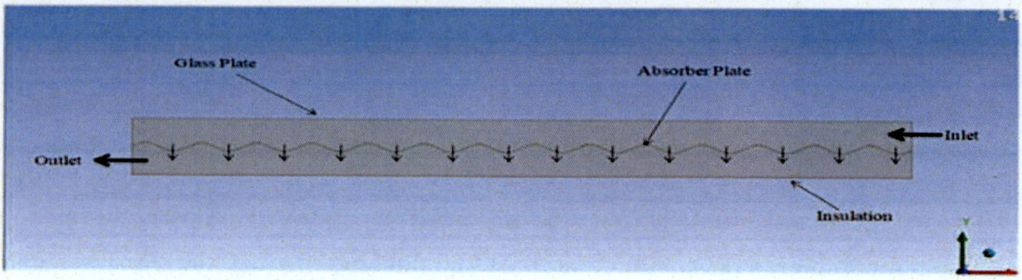

Fig. 2 Front view of the test section

Table 1 Operating parameter of the proposed SAH

Parameter	Value
Duct length (L)	1400 mm
Duct width (W)	300 mm
Duct height (H)	140 mm
Diameter of jet (D)	3 mm,5 mm,7 mm
Total no. of jets (N)	70,84,91,98
Solar radiations (W/m^2)	500
Mass flow rate (kg/s)	0.01, 0.015, 0.023, 0.028

2.1 Governing Equations

Governing Eqs. (1–3) of the fluid flow and heat transfer in the SAH are as follows:

(1) Continuity equation:

$$\frac{\partial u}{\partial x} + \frac{\partial u}{\partial y} + \frac{\partial u}{\partial z} = 0 \tag{1}$$

(2) Momentum equations

(a) X- Momentum equation

$$\left(u + \frac{\partial u}{\partial x} + v\frac{\partial u}{\partial y} + w\frac{\partial u}{\partial z}\right) = -\frac{1}{\rho}\left(\frac{\partial p}{\partial x}\right) + v\left(\frac{\partial^2 u}{\partial x^2} + \frac{\partial^2 u}{\partial y^2} + \frac{\partial^2 u}{\partial z^2}\right) \tag{2.a}$$

(b) Y-momentum equation

$$\left(u + \frac{\partial v}{\partial x} + v\frac{\partial u}{\partial y} + w\frac{\partial v}{\partial z}\right) = -\frac{1}{\rho}\left(\frac{\partial p}{\partial y}\right) + v\left(\frac{\partial^2 v}{\partial x^2} + \frac{\partial^2 v}{\partial y^2} + \frac{\partial^2 v}{\partial z^2}\right) \tag{2.b}$$

(c) Z- Momentum equation

$$\left(u\frac{\partial w}{\partial x} v\frac{\partial w}{\partial y} + w\frac{\partial w}{\partial z}\right) = -\frac{1}{\rho}\left(\frac{\partial p}{\partial z}\right) + V\left(\frac{\partial^2 w}{\partial x^2} + \frac{\partial^2 w}{\partial y^2} + \frac{\partial^2 w}{\partial z^2}\right) \tag{2.c}$$

(3) Energy equation

$$\left(u\frac{\partial T}{\partial x} + v\frac{\partial T}{\partial y} + w\frac{\partial T}{\partial z}\right) = \alpha\left(\frac{\partial^2 T}{\partial x^2} + \frac{\partial^2 T}{\partial y^2} + \frac{\partial^2 T}{\partial z^2}\right) \tag{3}$$

2.2 *Meshing of SAH Model*

The non-uniform unstructured grids are created for the test section of SAH using ANSYS 14.0 software (Figs. 3 and 4). The minimum sizes of the grids are at the edge, such as the jet of the absorber plate, and it means grids are fine near the edges. ANSYS provides a comprehensive tool for meshing.

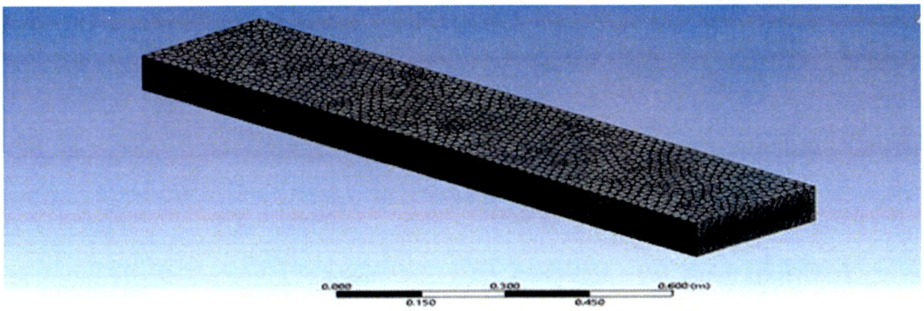

Fig. 3 Meshed computational domain with tetrahedral elements

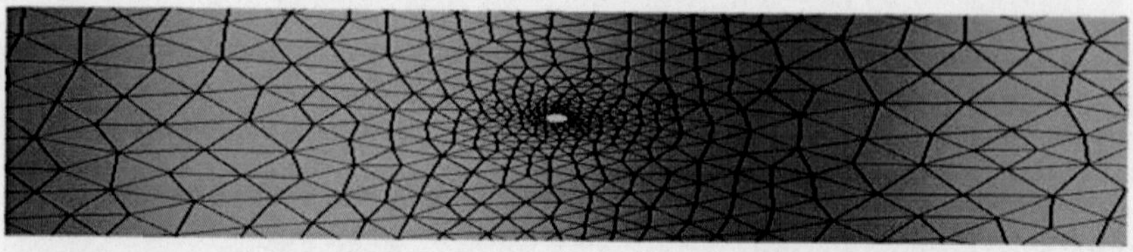

Fig. 4 Enlarged view of meshing

Table 2 Grid independence test

No. of elements	Temperature difference (K)
1,412,886	6.77
1,583,048	6.02
1,656,589	5.75
1,784,874	4.81
1,800,291	4.81

2.3 Boundary Conditions

1. No-slip boundary condition: the no-slip state is applied at all walls of SAH as the flow is viscid. Therefore, there will be the formation of a boundary layer.
2. Solar radiation on the Glass Plate: Solar radiations of 500 W/m^2 are applied on the glass plate.
3. Outlet and inlet boundary condition: Mass flow condition is applied at the inlet section. MFR at inlet section of SAH are varied as 0.01, 0.015, 0.023, and 0.028 kg/s. The air temperature at the inlet section varies from 306.1 to 309.8 K.
4. Insulated boundary conditions: The bottom and sides of the SAH are made of wood and provide insulated boundary conditions.

2.4 Mesh Validation

The present mesh of the computational domain contains non-uniform tetrahedral. For fine mesh, 1,800,291 elements, and for coarse mesh, 141,286 elements are obtained. Table 2 highlights the grid independence test.

2.5 Data Reduction

For data reduction, the following Eqs. (4–8) were used:

(1) Heat gain by SAH is given by

$$Q_1 = \dot{m}c_p(T_o - T_i) \quad (4)$$

(2) Solar heat input is given by

$$Q_2 = I * A \quad (5)$$

(3) The fan power is given by

$$P_f = \frac{(\Delta p)}{\eta} \quad (6)$$

Therefore

$$\eta THE = \frac{(Q1 - Pf)}{Q2} \quad (7)$$

$$\eta_{th} = m \cdot cp \cdot \Delta t/I * A \quad (8)$$

2.6 Validation of the Solar Air Heater Model

The results from the CFD analysis have been compared with the results obtained by the jet impingement on a corrugated plate [1] operating under similar boundary conditions. It is concluded that our design is in line with the earlier established results, as both the curves are in synchronization with each other as shown in Fig. 5.

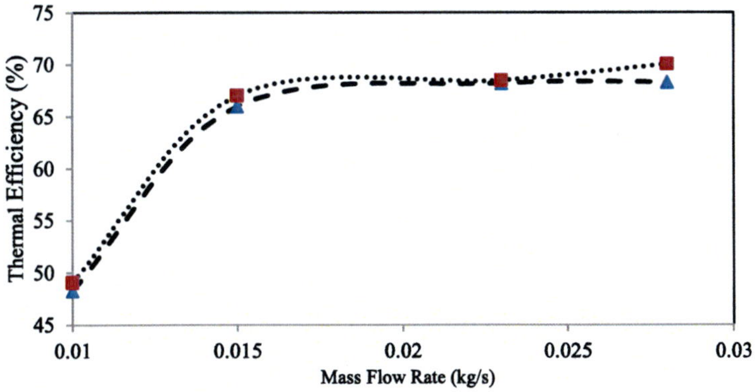

Fig. 5 Validation of solar air heater model

3 Results and Discussion

The thermal efficiency, thermohydraulic efficiency, and pressure difference character of the solar air heater with jet impingement through the corrugated plate are analyzed using the CFD approach. The jet impingement SAH thermal efficiency is firstly calculated for 84 jets at 0.01, 0.015, 0.023, 0.028 kg/s and incident solar radiation 500 W/m^2. Secondly, thermal efficiency is calculated at 70 jets, 91 jets, and 98 jets for the same conditions (see Figs. 6 and 7).

The jet diameter of jets is varied to find the optimum value at which outlet temperature is highest, and hence there will be maximum thermal efficiency. From Fig. 8, it has been obtained that the maximum value of outlet temperature is 316.65 K for 0.01 kg/s and 3 mm diameter jet because when jet size decreases, there is in jet velocity, turbulence, and heat transfer rate, due to which thermal efficiency increases.

The deviation of thermohydraulic efficiency with mass flow rate at different jets has been investigated. From Fig. 9, it is found that the thermal efficiency varies from 32.83 to 64.37%.

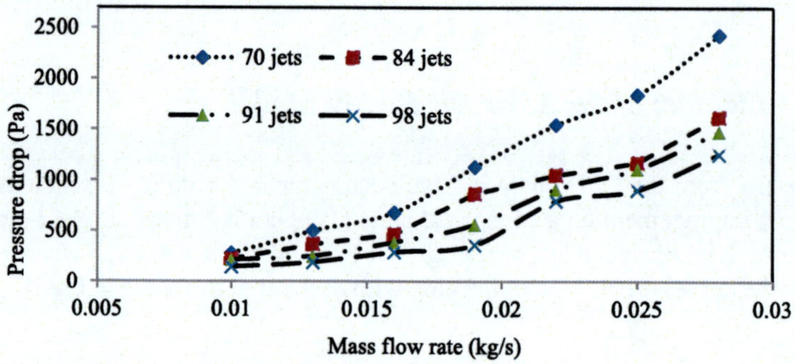

Fig. 6 Pressure drop versus mass flow rate at 3 mm jet size

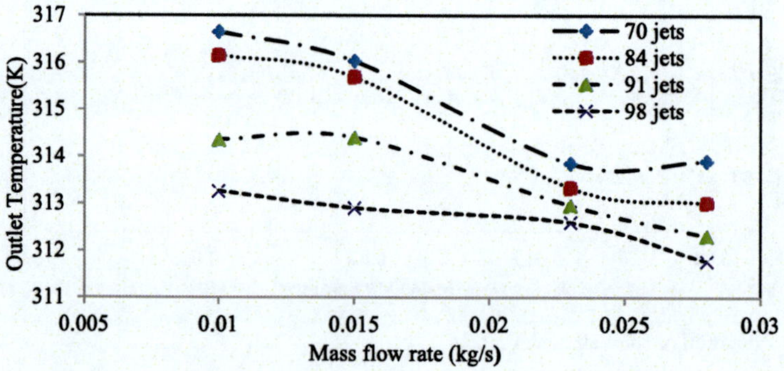

Fig. 7 Outlet temperature versus mass flow rate at 3 mm jet size

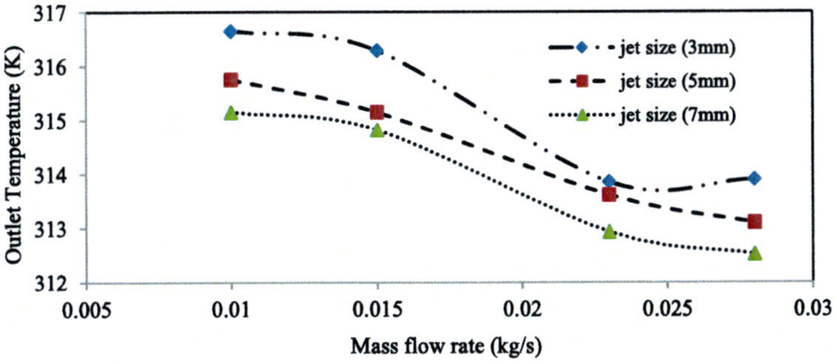

Fig. 8 Mass flow rate effect on outlet temperature at 70 jets

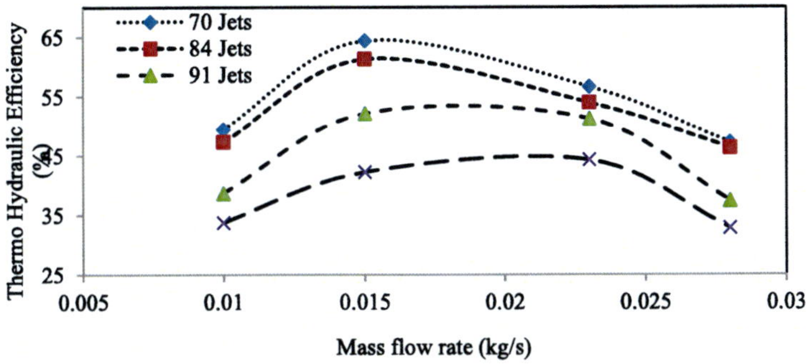

Fig. 9 Thermohydraulic efficiency versus mass flow rate at jet size 3 mm

The efficiency initially starts to increase, but beyond a specific value, it decreases. This occurs because with a rise in flow rate, the heat gain increases, but there is also an increase in fan power that has to apply at the inlet of the test section. Therefore, the optimum value of the flow rate is to be calculated to have maximum useful heat gain. Hence, there is an optimum value of jets, jet size, and MFR where the highest thermohydraulic efficiency is obtained. The test section with 70 jets and a 3 mm diameter gives a maximum jet efficiency at 0.015 kg/s.

The schematic showing the h/D ratio is shown in Fig. 10. The maximum outlet temperature (T_o) and thermal efficiency were found at h/D = 0.40, as evident in Fig. 11 and Fig. 12, respectively.

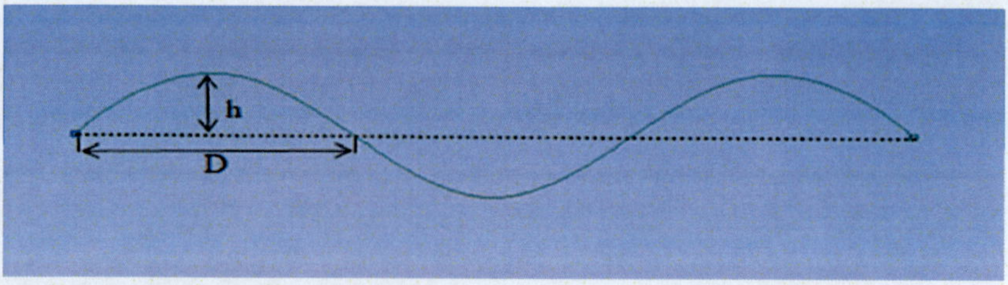

Fig. 10 Schematic showing h/D ratio

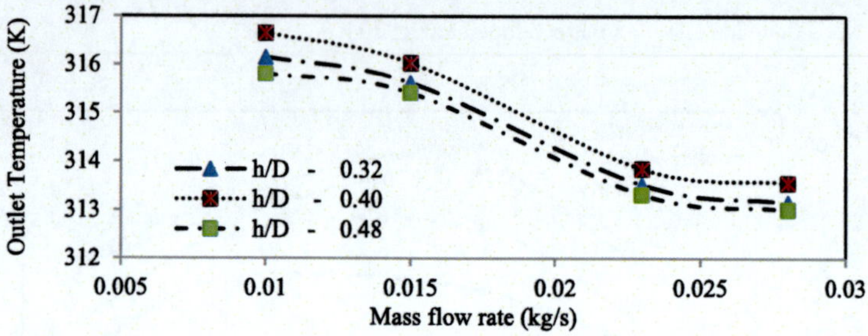

Fig. 11 Outlet temperature (T_o) versus mass flow rate (kg/s) at different h/D ratio

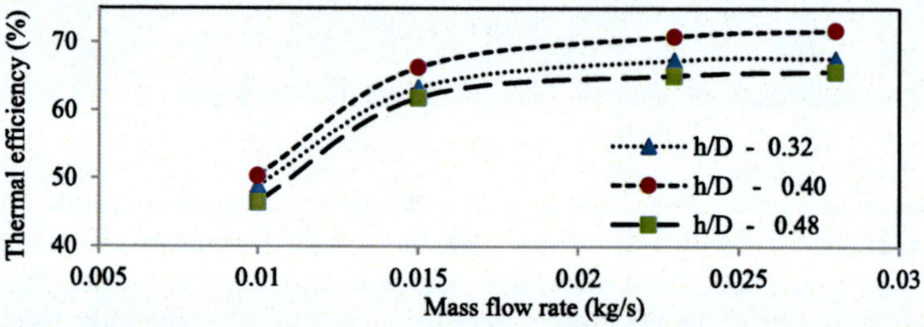

Fig. 12 Thermal efficiency versus mass flow rate (kg/s) at different h/D ratios

4 Conclusion

The main conclusion from the analysis of jet impingement through the corrugated plate is that it increases heat transfer augmentation by creating strong intermixing of the incoming jets after that impinges on the corrugated absorber plate. The following results are concluded:

I. The maximum thermo hydraulic efficiency is 64.03% for Solar radiation (I) = 500 W/m^2, 0.015 kg/s, 3 mm jet size and No. of jets (N) = 70.
II. The minimum thermo hydraulic efficiency is 32.83% for Solar radiation (I) = 500 W/m^2, 0.028 kg/s, jet size 3 mm, No. of jets (N) = 98.

III. An increase in the number of jets results in a decrease in thermohydraulic efficiency. Also, η_{THE} increases with an increase in mass flow rate up to a maximum value but beyond this value, η_{THE} decreases.
IV. The jet velocity, turbulence, and heat transfer rate increase as the diameter of the jet decreases therefore, η_{THE} also increases.
V. The maximum thermal efficiency and outlet temperature (T_O) were found at h/D = 0.40.

References

1. Aboghrara AM, Baharudin BTHT, Alghoul MA (2017) Performance analysis of solar air heater with jet impingement on corrugated absorber plate. Case Stud Therm Eng 10:111–120. https://doi.org/10.1016/j.csite.2017.04.002
2. Belusko M, Saman W, Bruno F (2008) Performance of jet impingement in unglazed air collectors. Sol Energy 82:389–398. https://doi.org/10.1016/J.SOLENER.2007.10.005
3. Chauhan R, Thakur NS (2014) Investigation of the thermohydraulic performance of impinging jet solar air heater. Energy 68:255–261. https://doi.org/10.1016/J.ENERGY.2014.02.059
4. Chauhan R, Thakur NS (2012) Heat transfer and friction characteristics of impinging jet solar air heater. J Renew Sustain Energy 4:1–12. https://doi.org/10.1063/1.4747821
5. Choudhury C, Garg HP (1991) Evaluation of a jet plate solar air heater. Sol Energy 46:199–209. https://doi.org/10.1016/0038-092X(91)90064-4
6. Duffie JAB, W.A. (1991) Solar engineering of thermal processes. Wiley, Hoboken
7. Huber AM, Viskanta R (1994) Effect of jet-jet spacing on convective heat transfer to confined, impinging arrays of axisymmetric air jets. Int J Heat Mass Transf 37:2859–2869. https://doi.org/10.1016/0017-9310(94)90340-9
8. Matheswaran MM, Arjunan TV, Somasundaram D (2018) Analytical investigation of solar air heater with jet impingement using energy and exergy analysis. Sol Energy 161:25–37. https://doi.org/10.1016/j.solener.2017.12.036
9. Nadda R, Kumar A, Maithani R (2017) Developing heat transfer and friction loss in an impingement jets solar air heater with multiple arc protrusion obstacles. Sol Energy 158:117–131. https://doi.org/10.1016/j.solener.2017.09.042
10. Nayak RK, Singh SN (2016) Effect of geometrical aspects on the performance of jet plate solar air heater. Sol Energy 137:434–440. https://doi.org/10.1016/J.SOLENER.2016.08.024
11. Rajaseenivasan T, Ravi PS, Salamon AM, Srithar K (2017) Experimental investigation on the performance of an impinging jet solar air heater. Alexandria Eng J 56:63–69. https://doi.org/10.1016/j.aej.2016.09.004
12. Shukla AP, Kushwaha R, Gupta B, Bisen A (2018) Performance analysis of solar air heater using CFD simulation. Int J Therm Technol 8. https://doi.org/10.14741/ijtt/v.8.1.2
13. Soni A, Singh SN (2017) Experimental analysis of geometrical parameters on the performance of an inline jet plate solar air heater. Sol Energy 148:149–156. https://doi.org/10.1016/J.SOLENER.2017.03.081
14. Thakur S, Thakur NS (2019) Investigational analysis of roughened solar air heater channel having W-shaped ribs with symmetrical gaps along with staggered ribs. Energy Sources, Part A Recover Util Environ Effects https://doi.org/10.1080/15567036.2019.1675815
15. Thakur S, Thakur NSL (2020) Impact of multi-staggered rib parameters of the 'W' shaped roughness on the performance of a solar air heater channel. Energy Sources Part A Recover Util Environ Eff. https://doi.org/10.1080/15567036.2020.1764672

Effective Cooling of Electronic Components Using Porous Filled Heat Exchangers

S. Venkata Sai Satyanarayana Sastry and S. Bhanu Prakash

Abstract The usage of porous media along with jet impinging technique is one of the efficient ways of cooling electronic equipments such as insulated gate bi-polar transistors. Analysis of porous filled heat exchanger is carried out in both stream and span wise directions by varying porous material (copper, aluminium and boron arsenide), porosity of the media, inlet jet cross section (square, circle and ellipse) at constant discharge of the heat exchanger and fluids of different heat capacity (water, liquid-ammonia and HFE-7100) for better heat removal at constant Reynolds number (laminar). Usage of elliptical inlet jet cross section for the heat exchanger, Boron Arsenide as porous material, porosity of 0.45 with water as cooling fluid dominates over square inlet jet cross section by 6% and almost nearer results around 1% higher for circular inlet jet cross section in effective cooling.

Keywords Electronic cooling · Jet impinging cooling technique · Porous media · Heat exchanger

1 Introduction

The electronic devices dissipate high amount of heat under the compact area by challenging the traditional systems of electronic cooling such as Heat pipes [1], Heat Sinks which laid a path to design innovative heat exchangers. There are several ways of addressing cooling of electronic systems like using heat exchangers [2, 3] when the heat removal is large, and for lesser heat removal one can go for thermoelectric modules [4], identifying specific areas/hot spots and applying novel methods [5, 6] one of which is introducing porous media in a heat exchanger which will also enhance the heat transfer with less weight, and large surface area for a given volume [7]. Stated after experimentation by investigating the heat exchanger parameters that the metal foamed heat exchanger enhanced the heat transfer by 2.5 times. Comparison was done in between the usage of bare-tube bundle, the open-cell metal foams

S. Venkata Sai Satyanarayana Sastry · S. Bhanu Prakash (✉)
Department of Mechanical Engineering, Amrita School of Engineering, Amrita Vishwa Vidyapeetham, Bengaluru, India
e-mail: b_prakash@blr.amrita.edu

D. P. Mishra et al. (eds.), *Recent Trends in Thermal and Fluid Sciences*, Lecture Notes in Mechanical Engineering, https://doi.org/10.1007/978-981-19-3498-8_9

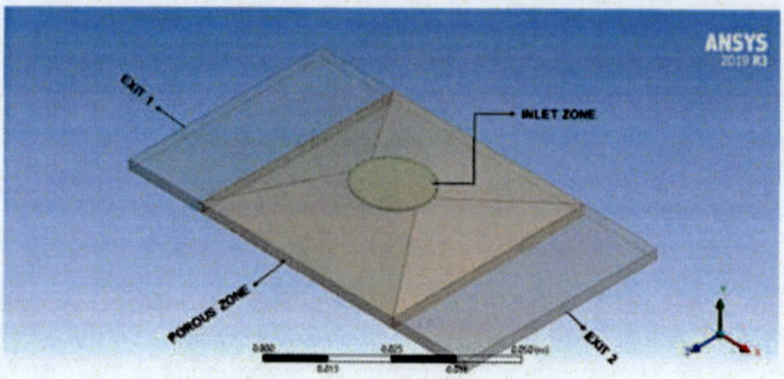

Fig. 1 Representation of Ellipse inlet model

and the louvred-fin in the heat exchanger for HVAC application [8] and stated that for a same fan power, the metal foamed heat exchanger dominated over the other two heat exchangers. Simulation of the temperature distribution and the cooling effectiveness for a novel designed compact jet impinging heat exchanger containing metal foam (Copper and APG)/porous zone is been studied [9]. In order to reduce the pressure loss that arises in between the inlet and exit of a system, the optimal height of the jet that to be employed should be in between 0.1 and 0.5 mm [10, 11] Experimentally studied by comparing the thermal conductivities in between Boron Nitride, Boron Arsenide(BAs) and Diamond and stated that due to high thermal conductivity (1300 W/(m k)) BAs can be utilized for thermal management [12]. Experimentally studied and stated that liquid-ammonia can be used as the coolant for high heat flux application [13]. Stated that di-electric fluid HFE-7100 can be utilized for micro-channel heat sinks and flow through porous media respectively. In the present work porous filled heat exchanger (shown in Fig. 1) is subjected to different porosity values, various inlet geometry shapes and various fluids of better heat transfer characteristics at constant Reynolds number (Re).

2 Governing Equations

The flow is considered as laminar and hydrodynamically fully developed in the porous region. Other walls, except the base of the heat exchanger is considered to be adiabatic and the flow enters from the, comes out in lateral direction.

2.1 Modified Darcy's Law (Inertial Effects)

$$\frac{dp}{dx} = -\frac{\mu_f}{k}u_D - \frac{F(K, R_e k, Geometry)}{\sqrt{K}}\rho_f \quad (1)$$

2.2 Inlet and Exit Channels

For the inlet and exit channels, the steady state, incompressible and conservation of mass, momentum, energy equations are as follows

$$\nabla.\overrightarrow{v} = 0 \tag{2}$$

$$\rho_f \vec{v}.\nabla\vec{v} = -\nabla p + \mu\nabla^2\vec{v} \tag{3}$$

$$\rho_f C_f(\vec{v}.\nabla T) = k_f \nabla^2 T \tag{4}$$

2.3 Porous Substrate

The volume averaged steady state governing equations such as Conservation of Mass, Momentum, Energy are as follows

$$\nabla.\langle\overrightarrow{v}\rangle = 0 \tag{5}$$

$$\frac{\rho f}{\varepsilon}\left\langle\left(\overrightarrow{v.\nabla}\right)\vec{v}\right\rangle = -\nabla\langle\rho\rangle^f + \frac{\mu}{\varepsilon}\nabla^2\langle\vec{v}\rangle - \frac{\mu}{k}\langle\vec{v}\rangle - \frac{\rho f F_e}{\sqrt{k}}[\langle\vec{v}\rangle.\langle\vec{v}\rangle]\mathrm{J} \tag{6}$$

$$\rho_f C_f\langle\vec{v}\rangle \cdot \nabla\langle T\rangle = k_f \nabla^2 \mathrm{T} \tag{7}$$

$$\mathrm{k_{eff}} = \varepsilon k_f + (1-\varepsilon)k_s \tag{8}$$

$$\mathrm{F} = \frac{1.75}{\sqrt{150\varepsilon^3}} \tag{9}$$

$$\mathrm{K} = \frac{\varepsilon^3 d^2 f}{150(1-\varepsilon)^2} \tag{10}$$

where ρ = the fluid density, ε = porosity, C = specific heat, k = thermal conductivity, K = permeability, u = fluid velocity, p = fluid pressure, T = temperature, F = geometric function, $\overrightarrow{v}$ = velocity vector and μ = coefficient of friction.

3 Modelling of Heat Exchanger Setup

The heat exchanger has an inlet jet. The different cross sections compatibility based on the hydraulic diameter (D = 17.3 mm) are square (17.3 mm × 17.3 mm), circle (17.3 mm diameter), Ellipse 1 (major axis = 18.15 mm and minor axis = 16.5 mm) and Ellipse 2 (major axis = 21.84 mm and minor axis = 14.56 mm) modelled. The fluids enter at a temperature of 300 K while the velocity depends on the inlet cross section of jet. The base of the heat exchanger is subjected to constant heat flux of 10^6 W/m^2, the inclination angle is considered as 13.74°. The three different porous materials are considered (copper, aluminium and boron arsenide with densities 8978, 2707, 5220 kg/m^3 and Thermal Conductivities 381, 204.2, 1300 W/mK, respectively) with different values of porosity of 0.45 and 0.90 having constant pore filament thickness 4.3154×10^{-4} m. The results are studied along the stream wise (X, non-dimensional position in x-direction = 2x/L) and span/transverse direction (Z, non-dimensional position in z-direction = 2z/L). The inlet fluids, such as water (ρ = 1000 kg/m^3, k = 0.6 W/mK, μ = 0.00089 kg/ms and C_p = 4182 J/Kg K), liquid-ammonia (ρ = 610 kg/m^3, k = 0.493 W/mK, μ = 0.000152 kg/ms and C_p = 4758 J/Kg-k), and HFE-7100 (ρ = 1528 kg/m^3, K = 0.066 W/mK, μ = 0.00074 kg/ms and C_p = 1180 J/Kg k), are considered by varying the inlet velocities based on the concept of jet Reynold's number ($Re = 4\frac{\dot{m}}{\pi D\mu}$) and the constant mass flow rate, $\dot{m}$.The quadrilateral grid with 7,65,800 elements for square, 8,58,646 elements for circle and around 8,00,000 elements were generated as meshing for different cross sections of heat exchanger.

Tables 1 and 2 show various jet inlet cross sections and the properties of the fluids used for the study. ANSYS Fluent is used for the analysis of design while the validation is done by comparing the results with the square inlet heat exchanger that was designed by [9] with water as inlet fluid at a velocity of 12.12 cm/s by

Table 1 Models with different geometries at constant inlet flow rate of 26 cm^3/s

Inlet-jet cross section	Inlet velocity (m/s)
Square	0.086
Circle	0.11
Ellipse 1	0.11
Ellipse 2	0.104

Table 2 Properties of fluids at constant Reynolds number = 1682 with ellipse 1 as inlet jet cross section

Inlet fluid	$\dot{m}$(kg/s)	Inlet velocity (m/s)	Heat Capacity rate ($\dot{m} \times C_p$) J/kg^{-s}
Water	0.0203	0.086	84.894
HFE-7100	0.0169	0.047	19.942
Liquid-NH_3	0.0034	0.024	16.1772

Fig. 2 Non-dimensional temperature distribution along the stream-wise direction

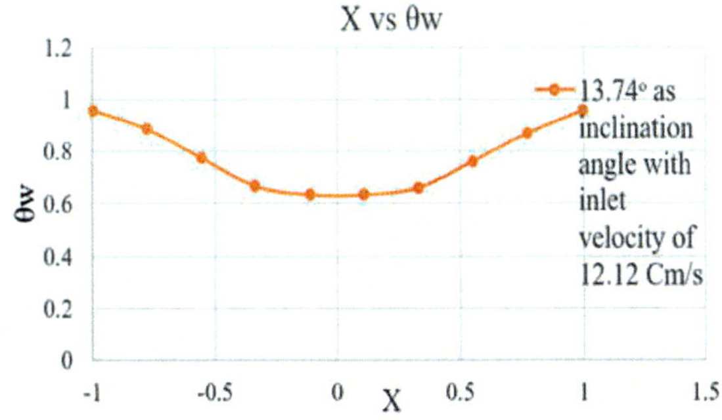

applying 10^6 W/m^2 as heat flux at the base with copper as porous material with porosity of 0.45, shown in Fig. 2, where $\theta_w = k_{cu}(T_w - T_{in})/q_{w.}t$, non-dimensional base temperature.

4 Results and Discussion

A compact heat exchanger (HX) by utilizing the concept of jet impinging cooling technique applying a constant high heat flux at the base is studied by incorporating high thermal conductivity semiconductor porous material (BAs) and is compared with existing conductors like copper and aluminium which are used as porous materials. Further comparison is done with square, circle, and elliptical jet inlet cross sections, and different single-phase fluids which were compared on the basis of the heat capacity rate. The comparison of percentage in cooling effectiveness is done by using = (higher base centre value-lower base centre value) × 100. When the fluid velocity increases, though the cooling effectiveness (on base centre line) increases which is due to increase in mass flow rate, but it also increases the power consumption of pump. As the thermal conductivity of the porous substrate increases, the cooling effectiveness (base centre line temperature) also increases due to enhancement of heat transfer by conduction. In comparison with different porous materials from Fig. 3, Boron Arsenide shows 46% effective cooling (reduction in base temperature) while Copper shows only 21% when compared to Aluminium. Clearly, the aluminium is dominated by the BAs and the copper, therefore the following sections are concentrated only by comparing in between the copper and BAs as porous materials. From Fig. 4, for a given velocity, as the porosity of the porous material (BAs) increases, lesser will be the amount of the porous material present which will reduce the heat transfer by the conduction, but the convective heat transfer increases which depends on the fluid that is flowing [14].

The inlet cross-sectional area and the inlet velocity play an important role, as the inlet area reduces, the velocity increase which is proportional to effective cooling. At constant inlet flow rate, the circular inlet jet cross section shows 5% more than

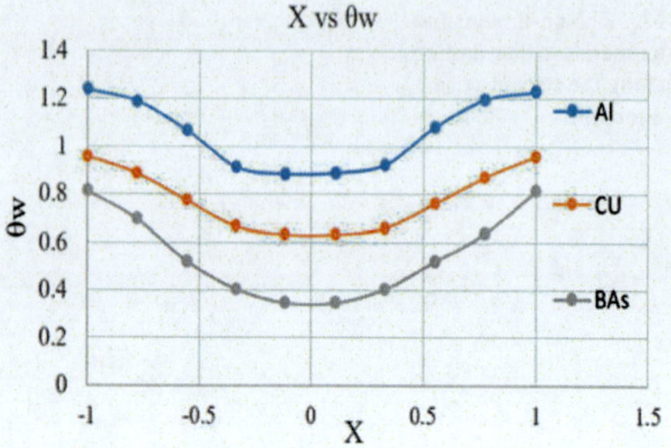

Fig. 3 Comparison of non-dimensional temperature distribution for al vs cu vs bas (porosity = 0.45, square inlet cross section, water inlet velocity of 0.1212 m/s)

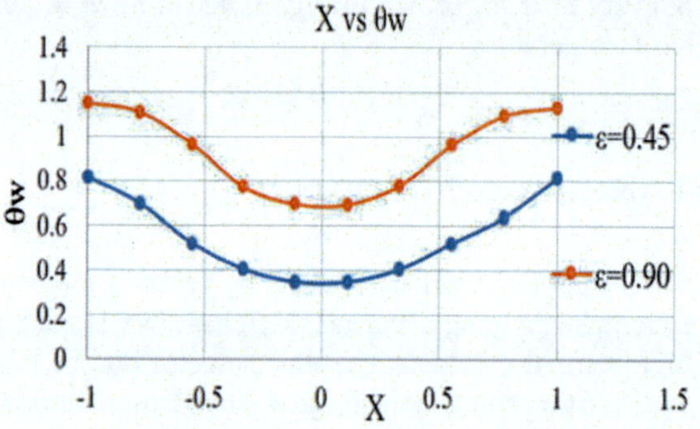

Fig. 4 Comparison of non-dimensional temperature distribution at porosity = 0.90 and 0.45, water inlet velocity of 12.12 Cm/s, square inlet cross section

square, hence can be treated as effective cooling (base centre line temperature) in both stream-wise and span (transverse)-wise directions. The jet impinging system should not only focus on the effective cooling at the centre of the base plate but also its exit, such that the cooling is uniform throughout. In comparison with circle and ellipse1, the cooling is observed slightly more (1%) in ellipse 1 cross section, in both directions in more uniform way. When compared with circle and ellipse 2, the cooling effective is slightly more in circular cross section (2%), which can be observed from Fig. 5 and Fig. 6, respectively.

From Figs. 7 and 8, at constant Reynolds number, for effective heat transfer application, the inlet velocity and the relation between density, dynamic viscosity of fluid play an important role. The heat capacity rate which depends on the C_p and mass flow rate values of the fluid creates an impact on the cooling due to the proportional relation that exists with the heat load (Q). For constant value of porosity, the water exhibits optimal mass flow rate as well as C_p value, by processing high heat capacity rate when compared with other fluids. At a porosity value of 0.45, in case of water, BAs shows 21% (stream-wise) better cooling when compared to copper. Usually, at a constant inlet velocity, as the thermal conductivity of the porous material increases, the cooling effect will also increase, but in case of the lower velocities, the reverse

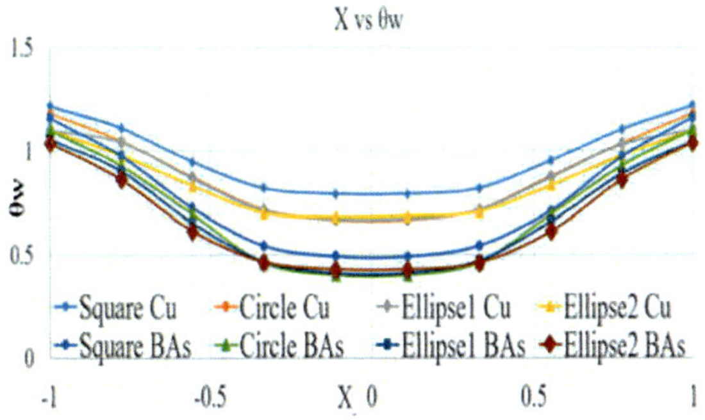

Fig. 5 Non-dimensional temperature distribution on centre line of different jet inlet cross section with water as coolant at porosity of 0.45

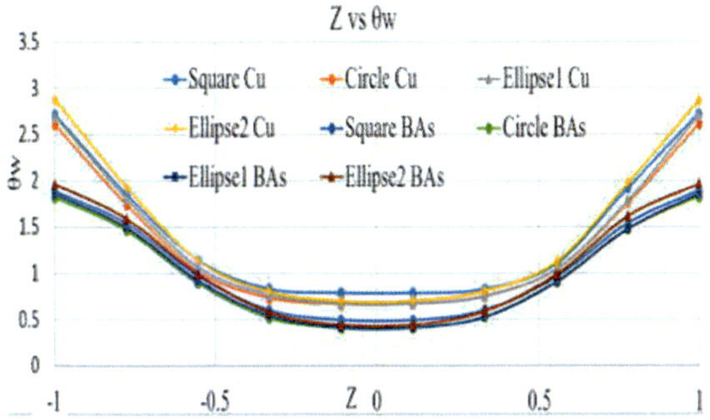

Fig. 6 Non-dimensional temperature distribution on centre line of different inlet jet cross section in the transverse direction with water as coolant at a porosity value of 0.45

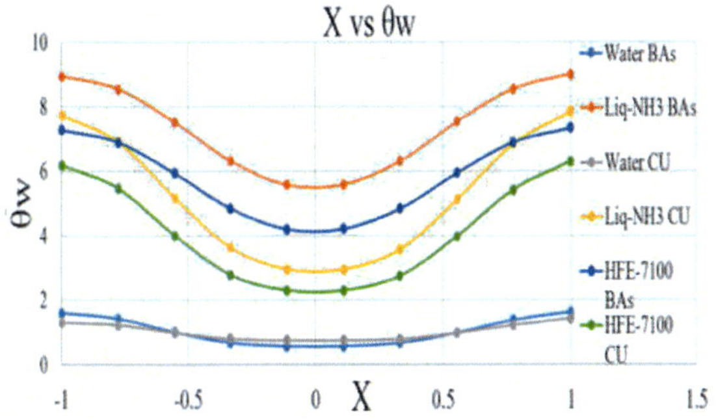

Fig. 7 Comparison of non-dimensional temperature distribution for ellipse1 inlet cross section, copper and BAs as porous material, porosity = 0.45

trend is observed (Figs. 7 and 8 for HFE-7100 and liquid-ammonia) in the behaviour of the two porous materials.

Fig. 8 Comparison of non-dimensional temperature distribution for ellipse 1 inlet cross section, copper and BAs as porous material, porosity = 0.45

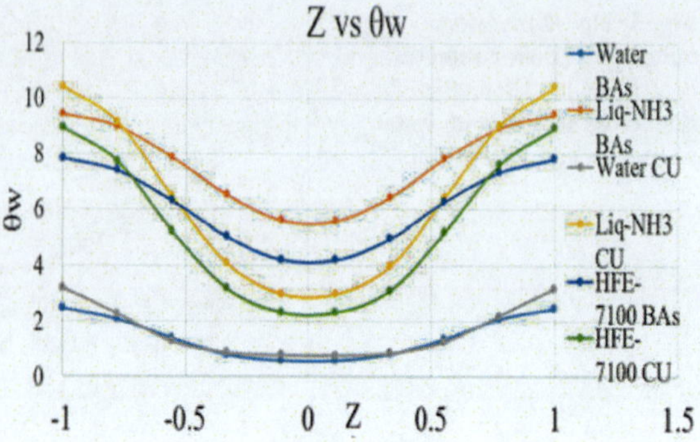

5 Conclusions

The inlet velocity plays an important role for effective cooling. In comparison with different porous materials at a specified velocity, the square inlet cross section, Boron Arsenide shows 46% effective cooling (reduction in base temperature) while copper shows only 21% when compared to aluminium. The circular inlet cross section with porosity of 0.45 shows cooling is effective (base centre line temperature) in both stream-wise direction and along span (transverse)-wise directions (5% more than square inlet cross section) due to decrease in cross-sectional area of jet. In comparison with circular inlet, BAs as porous material, porosity of 0.45, the effective cooling is slightly less (1%) when compared with ellipse 1 cross section, while it increased by 2% for ellipse 2 in both directions. When the porosity increases, permeability for the fluid is increasing at high inlet velocities of fluid and as the porosity of porous substrate increases, the effective cooling reduces. The cooling effectiveness is observed by employing high Thermal conductivity BAs at constant porosity value of 0.45. The fluid with high heat capacity rate shows good heat transfer rates due to the proportionality relation that exists in between the heat capacity rate and the heat load. Finally, by incorporating BAs as porous material having porosity value of 0.45 with the ellpse1 as inlet cross section of the jet to the heat exchanger with water as inlet fluid provides an effective cooling.

References

1. Rao RS, Prakash SB (2018) Thermal characteristics of a cylindrical heat pipe using multi-layer screen mesh wick. Int J Veh Struct Syst 10(5):347–350
2. Chaturvedi S, Bhanu PS (2019) Heat transfer enhancement using delta and circular winglets on a double pipe heat exchanger. AIP Conf Proc 2200:020064
3. Harikrishnan G, Ebin PS (2019) Numerical study of heat transfer enhancement in fin and tube heat exchangers using curved semi wavy vortex generators. IOP Conf Ser: Mater Sci Eng 577:012001

4. Rohit G, Manaswini D, Kotebavi V, Nagaraja SR (2017) Performance study of thermo-electric generator. AIP Conf Proc 1859:020094
5. Rupendra SPA, Venkata Krishna M, Jakkareddy PS (2021) Natural convection heat transfer on the strip heaters flushed on the vertical flat plate: a numerical study. Proceed Int Conf Thermofluids
6. Kumar S, Jakkareddy PS, Balaji C (2020) A novel method to detect hot spots and estimate strengths of discrete heat sources using liquid crystal thermography. Int J Therm Sci 154:106377
7. Boomsma K, Poulikakos D, Zwickv F (2003) Metal foams as compact high performance heat exchangers. J Mech Mater 35:1161–1176 (2003)
8. Huisseune H, De Schampheleire S, Ameel B, De Paepe M (2015) Comparison of metal foam heat exchangers to a finned heat exchanger for low Reynolds number applications. Int J Heat Mass Transf 89:1–9
9. Zing C, Mahjoob S, Vafai K (2019) Analysis of porous filled heat exchangers for electronic cooling. Int J Heat Mass Transf 133:268–276
10. Wei TW, Oprins H, Fang L, Cherman V, De Wolf I, Beyne E, Baelmans M (2020) Nozzle scaling effects for the thermohydraulic performance of microjet impingement cooling with distributed returns. Appl Therm Eng 180
11. Kang JS, Li M, Wu H, Nguyen H, Hu Y (2018) Experimental observation of high thermal conductivity in boron arsenide. Science 361:575
12. Wang H, Wu JJ, Yang Q, Zhu X, Liao Q (2016) Heat transfer enhancement of ammonia spray cooling by surface modification. Int J Heat Mass Transf 101:60–68
13. Yang M, Li M-T, Hua Y-C, Wang W, Cao B-Y (2020) Experimental study on single-phase hybrid microchannel cooling using HFE-7100 for liquid-cooled chips Int. J Heat Mass Transf 160:120230
14. Baloyo JM, Zhao Y (2020) Permeability, form drag coefficient and heat transfer coefficient of porous copper. Heat Transf Eng

Performance Evaluation of Thermoelectric Cooler

Palakala Sai Kiran and S. Bhanu Prakash

Abstract Thermoelectric cooler (TEC) which uses Peltier effect to create heat flux across its sides has been of interest to researchers because of its compact size, precise control of temperature stability, and of quiet operation. TEC with 127 couples of volume $4 \times 4 \times 3.5$ cu.cm is taken and its performance affecting parameters are analyzed. A comparative study is performed on various TEC models to obtain higher cooling capacity. Simulation is carried out by varying the geometry and the material used for the designed thermoelectric module. Results show cooling capacity has increased thrice with the decrease in leg height of TEC (linear square cross section, Bismuth Telluride) model from 0.3 to 0.1 cm. TEC model with circular cross section with the limiting taper angle 20° showed 50% decrease in its work input when compared with linear circular cross section legs. The Coefficient of Performance of the former model with Bismuth antimony as thermoelectric material showed 3.6 times higher value in comparison with Bismuth Telluride due to its figure of merit.

Keywords Peltier module · Thermoelectric material · COP · Figure of merit

1 Introduction

Owing to the advancement in semiconductor materials, thermoelectric industry has drawn more interest to research in the future years. Thermoelectric modules (TEM) are referred as solid-state heating/cooling modules and are able to convert Thermal energy to Electric Energy and vice versa with no presence of moving mechanical parts and therefore, recognized as durable, compact, and reliable. Thermoelectric generators use waste heat energy (automobiles, Solar PV cells, and other applications) to generate power [1–3]. Thermoelectric cooler works on Peltier effect and the two

P. S. Kiran · S. B. Prakash (✉)
Department of Mechanical Engineering, Amrita School of Engineering, Bengaluru, Amrita Vishwa Vidyapeetham, Bengaluru, India
e-mail: b_prakash@blr.amrita.edu

P. S. Kiran
e-mail: palakalasaikiran@gmail.com

D. P. Mishra et al. (eds.), *Recent Trends in Thermal and Fluid Sciences*, Lecture Notes in Mechanical Engineering, https://doi.org/10.1007/978-981-19-3498-8_10

sides of the module can be maintained at desired temperatures with the help of electrical work input. One p-type and n-type semiconductor that constitutes one couple is sandwiched between copper conductors and electrical insulation is provided on both sides of device to protect it from the condensation effect. Several couples can be incorporated into a device according to working requirements. Geometry optimization of thermoelectric leg helps in finding out the optimal values at which the performance of thermoelectric cooler could be the best [4, 5]. Unlike in conventional geometry thermoelectric module, tapered legs of TEC would help in improving its performance by reducing overall thermal conductance of the device, increasing the temperature gradient across the legs, (harnessing the Thomson effect). Performance factors also depend on the figure of merit of thermoelectric material which needs to be high to show better performance in service. Various geometrical parameters and thermoelectric materials that affect the performance of device are studied [6–8]. Thermoelectric devices can be used in applications where the temperatures are cryogenic [9]. Also the literature shows multi-stage TEC air cooling module gives the higher coefficient of performance (COP) than single stage TEC [10]. A novel air–water-based TEC is analyzed experimentally to suit specific applications [11]. In few of the applications like electronic cooling/PV Solar energy systems, a combination of heat pipes along with the TEM are installed [12, 13].

In the present work a Thermoelectric module is designed by varying thermoelectric material (based on Figure of merit, which gives how efficient the material is), leg height/shape of the p-type and n-type semiconductors used for a specified temperature difference and analysis is made. Figure of merit and the temperature range at which TEC has to be installed are the important considerations that have to be made. At times there can be trade-off between these two parameters.

2 Analysis of Thermoelectric Cooler

Semiconductor materials are chosen to manufacture the legs of any Peltier/Thermoelectric module as they have better thermoelectric properties. Both sides of module are provided with electrical insulation material like aluminum oxide (Al_2O_3), silica gel to protect device from the condensation effect. Dimension of the TEC module is taken as 4-cm × 4-cm. Analysis is made for a TEC of 127 couple for a fixed temperature difference, $\Delta T = 300–227$ K with an input voltage of 12 V. Performance parameter Coefficient of Performance (COP) has been evaluated with the help of cooling power and the electrical power input supplied using the Ref. [14].

Results obtained (Figs. 1, 2 and 3) are for a TEC, Bismuth Telluride as the thermoelectric material for a leg height of 0.3-cm and 0.15 cm thick copper plate. Figure 1 helps in explaining TEC with circular shaped legs can have better temperature distribution over the square c/s legs due to its lower thermal conductance.

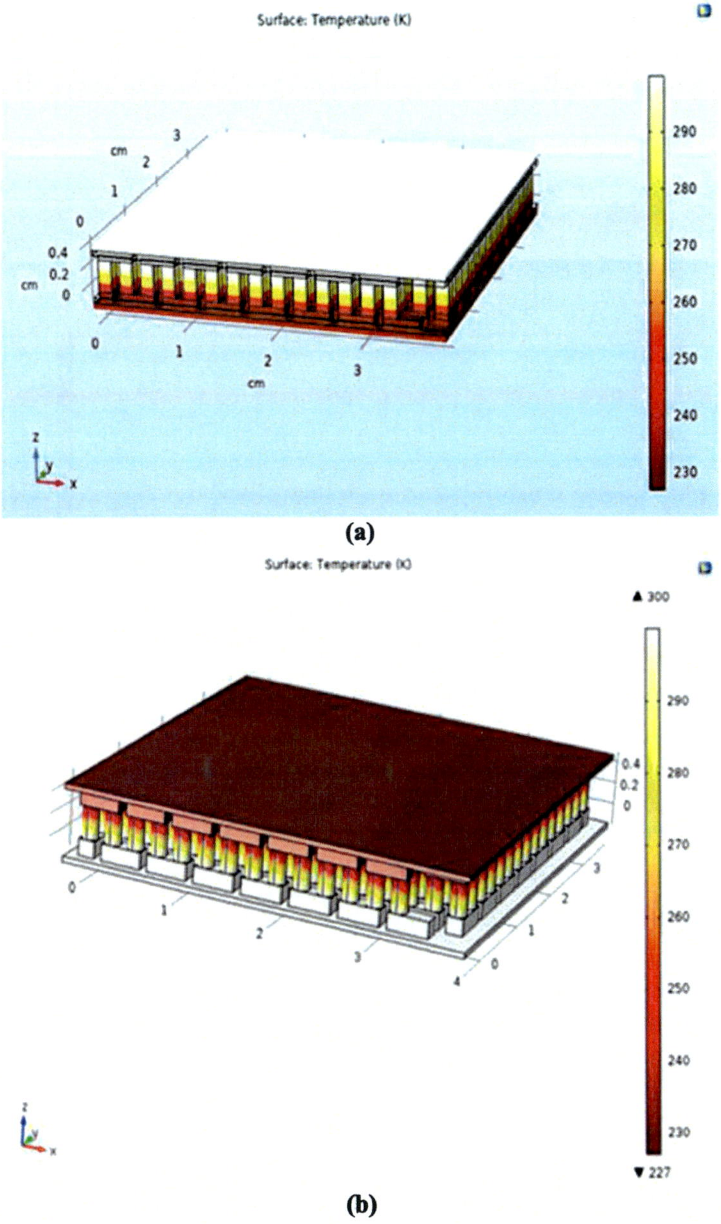

Fig. 1 Temperature variation across the TEC. **a** Square legs. **b** Circular legs

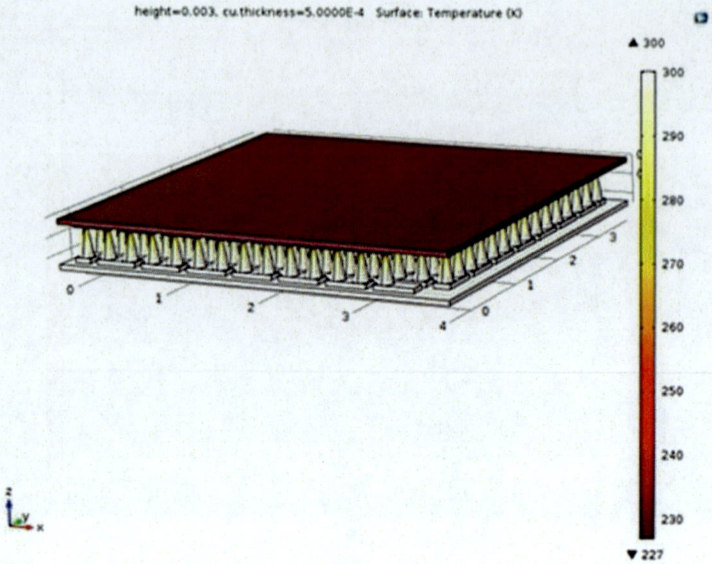

(a)

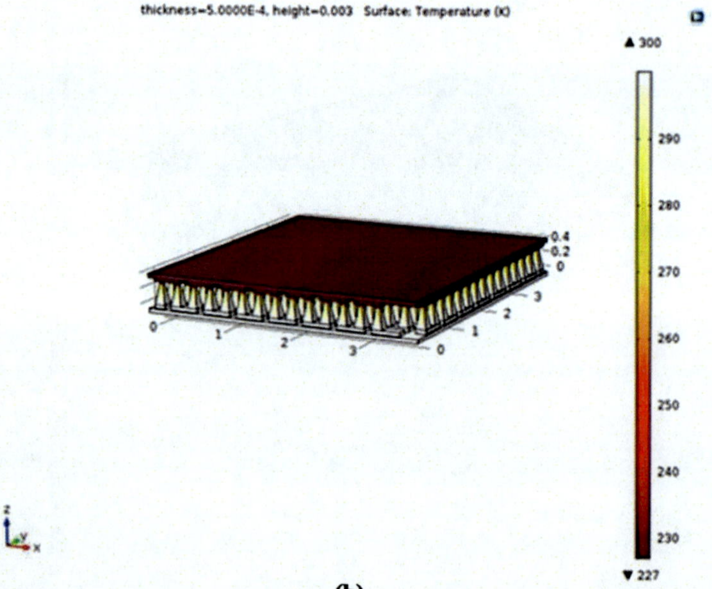

(b)

Fig. 2 Temperature variation across the TEC. **a** Tapered square legs. **b** Tapered circular legs

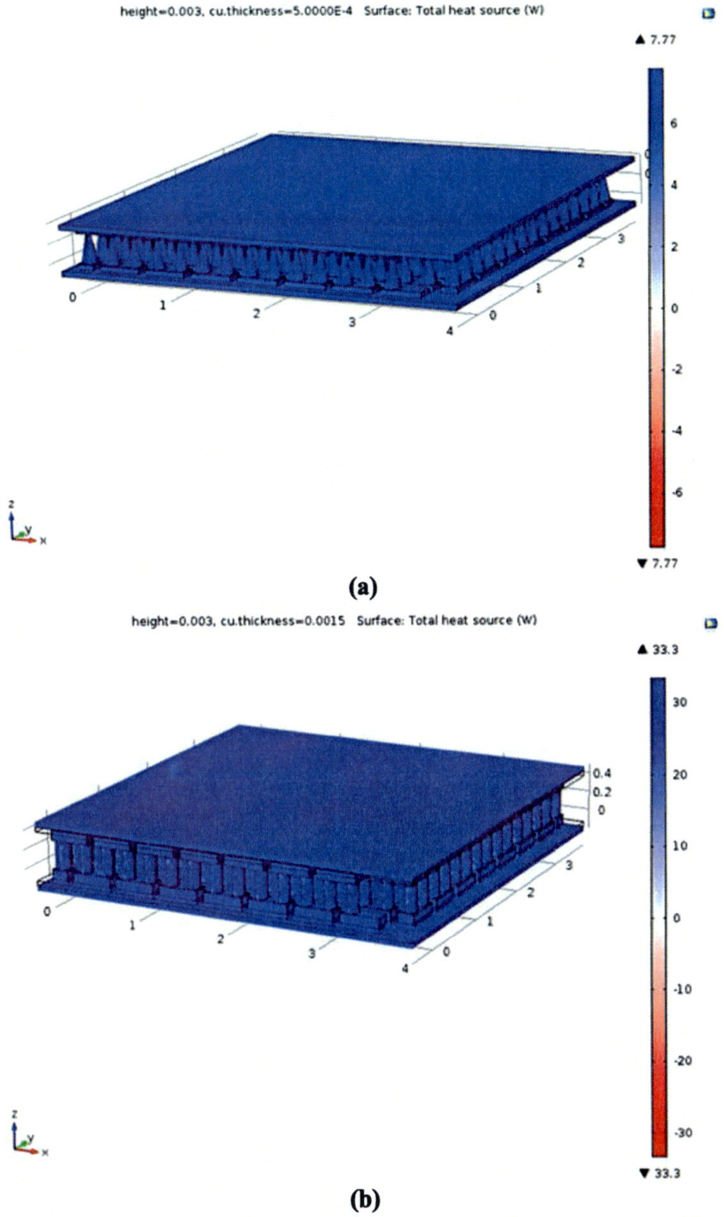

Fig. 3 Work Input of TEC. **a** Uniform circular legs. **b** Tapered circular legs

TEC with tapered square legs, in Fig. 2 shows better variation of temperature than uniform square legs, and tapered (non-linear) circular legs can do better over tapered square legs by harnessing Thomson effect.

Tapered circular legs show lesser consumption of work input relative to uniform circular legs, as shown in Fig. 3. Therefore the module with tapered circular legs would be a better option over the other two options, i.e., uniform and tapered square legs.

Lowering the leg height of module will increase its cooling capacity along with increased work input as shown in Fig. 4a. Changing of material to Bismuth Antimony shows increase of cooling capacity because of high Seebeck coefficient and smaller thermal conductance and electrical resistivity. Cooling capacity of 166.5 W for leg height of 0.1 cm square linear c/s and cooling capacity of 53.62 W for leg height of

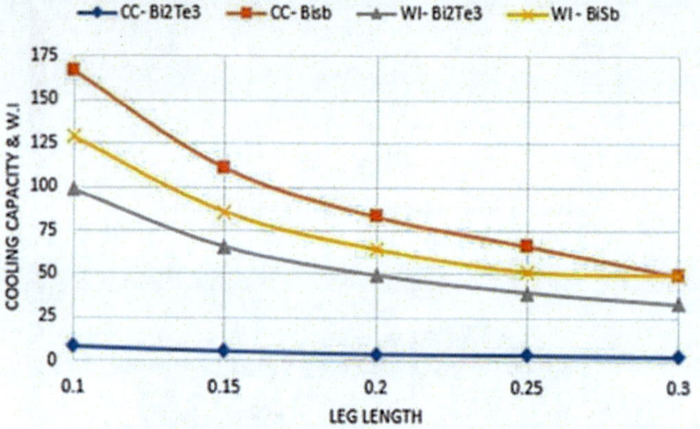

(a)

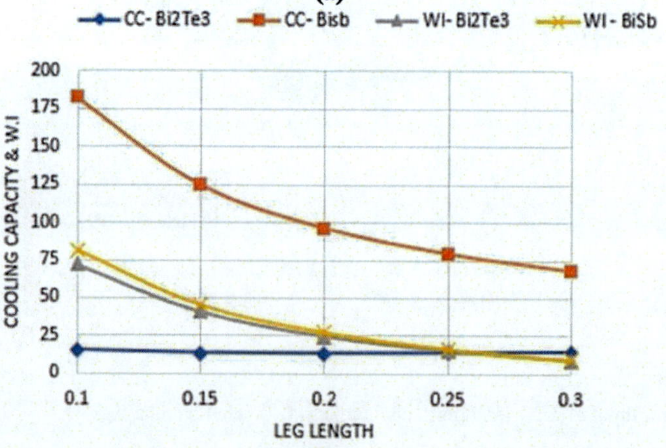

(b)

Fig. 4 Variation of cooling capacity and work input along the leg height of TEC. **a** Uniform square legs. **b** Tapered square legs with 20° taper angle

0.1 cm comparatively higher than module with Bismuth Telluride of cooling capacity 8.86 W and work input of 98.2 W for leg height of 0.1 cm and cooling capacity of 3.01 W and work input 33 W for leg height of 0.3 cm. Figure 4b shows varying c/s of module to non-linear can decrease the work input needed compared with linear c/s. changing of material to Bismuth Antimony shows larger cooling capacity value of 182.28 W for leg height of 0.1 cm square non-linear c/s and cooling capacity of 67.83 W for leg height of 0.1 cm comparatively higher than module with Bismuth Telluride of cooling capacity 15.52 W for leg height of 0.1 cm and cooling capacity of 14.4 W for leg height of 0.3 cm.

Figure 5a shows varying c/s of module from square to circular shape can increase in cooling capacity and slight decrease in work input. Simulation of module with Bismuth Antimony shows larger cooling capacity value of 169.46 W and work input of 119 W for leg height of 0.1 cm circular linear c/s comparatively higher than module

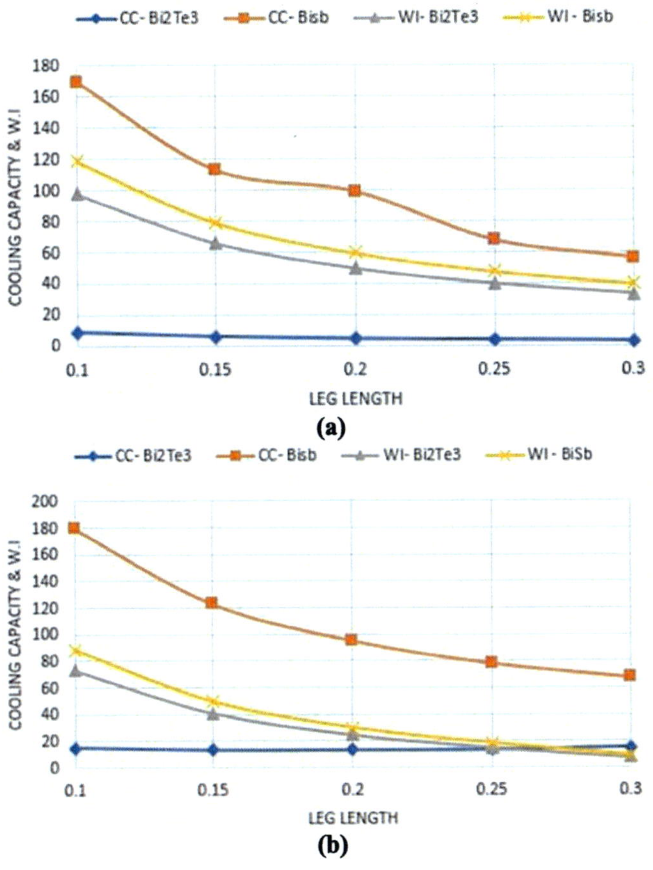

Fig. 5 Variation of cooling capacity and work input along the leg height of TEC. **a** Uniform circular legs. **b** Tapered circular legs with 20° taper angle

with Bismuth Telluride of cooling capacity 8.87 W and work input of 98.1 for leg height of 0.1 cm.

Figure 5b shows varying c/s of module to non-linear can decrease the work input needed compared with linear c/s. Changing of material to Bismuth Antimony shows larger cooling capacity value of 179.66 W and work input of 88.8 W for leg height of 0.1 cm cylinder non-linear c/s and comparatively higher than module with Bismuth Telluride of cooling capacity 15.14 W and work input of 73.5 W for leg height of 0.1 cm.

Figure 6a shows varying c/s of module to non-linear from linear c/s shows increase in COP value. Larger cop of 1.65 is obtained for square non-linear c/s module of

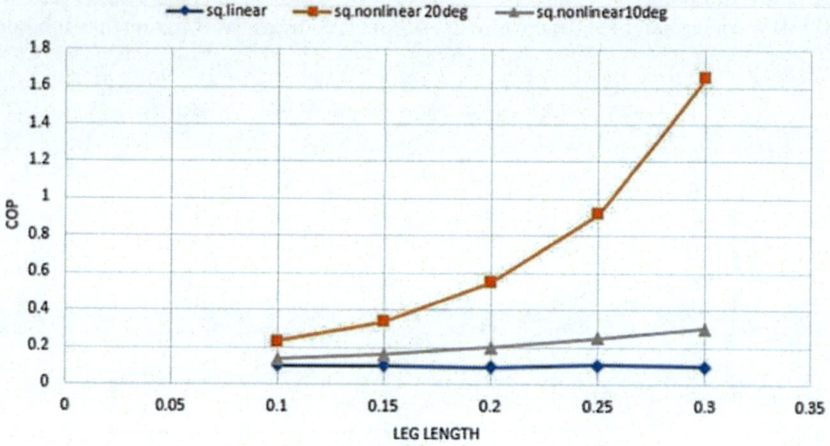

(a)

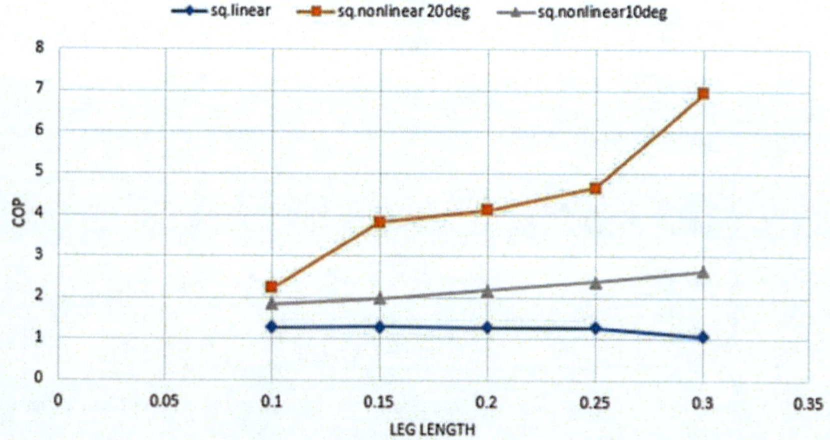

(b)

Fig. 6 Leg length versus COP of TEC for uniform and tapered square legs. **a** Bismuth telluride. **b** Bismuth antimony

0.3 cm with tapered angle 20° followed by COP of 0.3 for square non-linear c/s module with tapered angle 10° and COP value of 0.096 for square linear c/s module of leg height 0.3 cm. Lowering the leg height shows lesser COP values because of increase in work input to module.

Figure 6b shows changing material from Bismuth Telluride to Bismuth Antimony can show larger COP values because of higher see beck coefficient and lower electrical resistivity and thermal conductance. Higher cop of 6.942 is obtained for square non-linear c/s module of 0.3 cm with tapered angle 20° followed by COP of 2.65 for square non-linear c/s module with tapered angle 10° and COP value of 1.08 for square linear c/s module of leg height 0.3 cm.

Figure 7a shows varying c/s from square to circular and shows slight increase in COP. Higher cop of 1.95 is obtained for cylinder non-linear c/s module of 0.3 cm with tapered angle 20^0 followed by COP of 0.29 for cylinder non-linear c/s module with tapered angle 10° and COP value of 0.085 for cylinder linear c/s module of leg height 0.3 cm. Lowering the leg height shows lesser COP values because of an increase in work input to module.

Figure 7b shows higher cop of 7.187 is obtained for cylinder non-linear c/s module of 0.3 cm with tapered angle 20° followed by COP of 2.386 for cylinder non-linear c/s module with tapered angle 10° and COP value of 1.41 for cylinder linear c/s module of leg height 0.3 cm.

3 Conclusions and Future Works

Simulation of TEC has proved that non-linear cross section of thermoelectric leg, smaller leg length, were able to minimize the size of the thermoelectric cooler and improving its cooling capacity. Varying the thermoelectric material shows higher COP values and larger cooling capacity range because of higher Seebeck coefficient value and lower in thermal conductivity and electrical resistivity values. Multistage can still improve the performance of TEC keeping ΔT constant. Design of TEC along with heat pipe and heat sink can be applied to improve the COP of the module. Finding different material combination for various applications such as electronic satellite component which works at −233.15 °C can be one of the options. Arrangement of legs at varying distances is still unanswered.

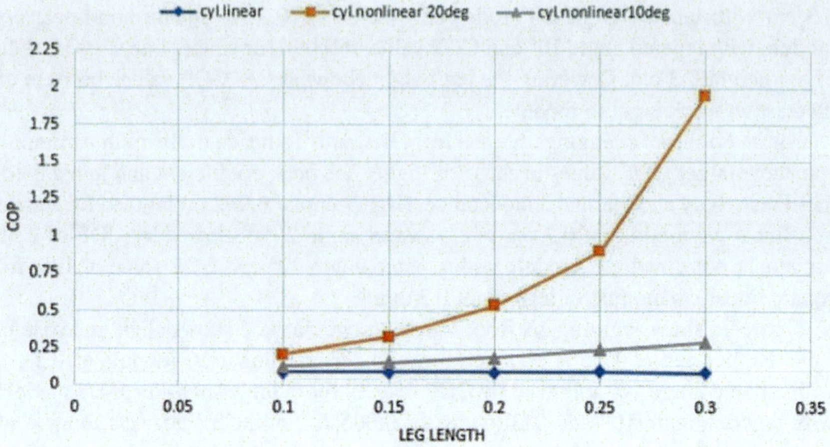

(a)

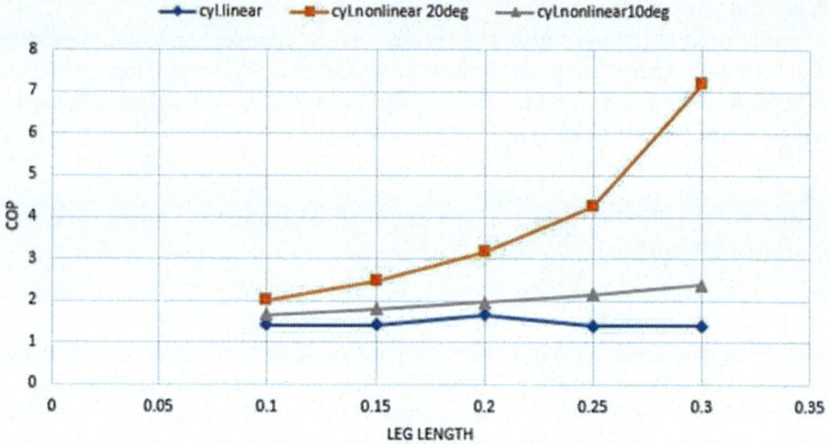

(b)

Fig. 7 Leg length versus COP of TEC with circular linear and non-linear c/s legs with. **a** Bismuth telluride and **b** Bismuth antimony as semiconductor material

References

1. Rohit G, Manaswini D, Kotebavi V, Nagaraja SR (2017) Performance study of thermo-electric generator. AIP Conference Proc 1859:020094
2. Hiranandani K, Aravind B, Ratna Kishore V, Kumar S (2020) Development of a numerical model for performance prediction of an integrated microcombustor-thermoelectric power generator. Energy 192:116624
3. Aravind B, Raghuram GKS, Ratna Kishore V, Kumar S (2018) Compact design of planar stepped micro combustor for portable thermoelectric power generation. Energy Convers Manage 156:224–234
4. Lin S, Yu J (2016) Optimization of a trapezoid-type two-stage Peltier couples for thermoelectric cooling applications. Appl Thermal Eng Int J Refrig 65:103–110

5. Wang X, Wang H, Su W, Zhai J, Wang T, Chen T (2016) Optimization of the performance of the SnTe uni-leg thermoelectric module via metallized layers. Appl Thermal Eng
6. Mijangos F, Gao MB, Alvarez-Quintana CJ (2017) Enhanced performance thermoelectric module having asymmetrical legs. Appl Therm Eng 148:1372–1381
7. Luo D, Wang R, Yu W, Zhou W (2019) Performance evaluation of novel TEM with BiSbTeSe-based material. Appl Therm Eng 238:1299–2131
8. Grabov VM, Uryupin ON, Komarov VA (1998) Thermoelectric properties of polycrystalline Bismuth and Bismuth antimony alloys. In: 17th international conference on thermoelectric
9. Aliabadi P, Mahmoud S, AL-Dadah RK (2014) Simulation of cascaded thermoelectric devices for cryogenic medical treatment. Int J Refrig 38:352–357
10. Patel J, Patel M, Patel J, Modi H (2016) Improvement in the cop of thermoelectric cooler. Int J Sci Technol Res 5(05)
11. Dizaji HS, Jafarmadar V, Khalilarya S, Moosavi A (2016) An exhaustive experimental study of a novel air-water based thermoelectric cooling unit. Appl Thermal Eng 181:357–366
12. Rao RS, Prakash SB (2018) Thermal characteristics of a cylindrical heat pipe using multi-layer screen Mesh Wick. Int J Vehicle Struct Syst 10(5):347–350
13. Koundinya S, Vigneshkumar N, Krishnan AS (2017) Experimental study and comparison with the computational study on cooling of PV solar panel using finned heat pipe technology. In: Mater Today Proc 4:2693–2700 (2017)
14. Chavan V, Prakash SB (2019) Compact design of thermoelectric cooler and its performance analysis. AIP Conf Proc

Plate and Frame Heat Exchangers for Offshore Oil and Gas Industry: Challenges in Design and Engineering

Jainendra Kumar and Saqib Zaidi

Abstract The content of the paper enables mechanical design engineers to address basic design and engineering issues in the early stages which will help to optimize the selection of PHE and to minimize the design changes during detail engineering. This paper will provide information on the classification of PHEs, selection criteria, different construction parts, working principles, flow patterns, and arrangement of passes, and comprehensive information on thermal plate effects of different chevron angles are explained explicitly. It also provides information for material selection. The objective of this paper is to provide detailed design engineering aspects and challenges in design engineering and its mitigation.

Keywords Plate and frame heat exchanger (PHE) · Thermal plates · Chevron angle · Port · Channel

1 Introduction

A plate heat exchanger is considered a revolutionized method of indirect heating and cooling of fluids, invented in 1923 by Dr. Richard Seligman and categorized under compact heat exchanger which uses metal plates for heat transfer between two fluids. Plate and frame heat exchanger (PHE) is one of the categories of the plate heat exchanger, widely used in offshore oil and gas industries due to its compactness, a wide range of thermal/hydraulic performance, ease of cleaning, and maintenance.

PHE demonstrates outstanding heat transfer features allowing compact designs in comparison with conventional S&T HE. PHE has a special feature that can be used for different heat duties by simply modifying the number of plates based on its frame size.

J. Kumar (✉) · S. Zaidi
Department of Mechanical Engineering, Samsung Heavy Industries (I) Pvt. Ltd, Noida, India
e-mail: jainendrasinha084@gmail.com

S. Zaidi
e-mail: Saqib.zaidi@samsung.com

D. P. Mishra et al. (eds.), *Recent Trends in Thermal and Fluid Sciences*, Lecture Notes in Mechanical Engineering, https://doi.org/10.1007/978-981-19-3498-8_11

As there are limited international standard practices for PHE, it is to be noted that PHE is a customized and proprietary product of the manufacturer and there are no such detailed codes or manufacturer guidelines/standards like TEMA for S&T HE. It is to be noted that the engineering consultant/EPC contractor is responsible to select and procure suitable PHE for the offshore facilities, which poses challenges due to the above fact that these exchangers are proprietary in nature.

There are several challenges in the design engineering and system integration of PHE in offshore oil and gas projects. Reliability of equipment in offshore industries is of paramount concern due to safety, environmental and economic implications. Failure of these equipments may lead to unplanned shutdowns and maintenance, thus increasing production costs.

Major challenges are limited space availability in offshore facilities, corrosion/erosion in plates due to fluid and marine environment, gaskets selection to avoid leakages, available pressure drop in system, routing of interconnecting piping, and material handling of exchangers. Also, it is seen that over the period of use of these exchangers can lead to a reduction in heat transfer efficiency, loss of flow and performance, process fluid leakage, and increased pressure drop from inlet to outlet.

Plates and gaskets are major parts of the PHE which affects its performance. These are mass-produced by a vendor with good research work. Selection of PHE requires past technical and performance data, and also requires strong technical and commercial evaluation, to select the suitable PHE.

These are typically used in liquid–liquid applications which are known for uniform and fast heat transfer. But there is also development in design for different services like evaporation and condensation. With the advancement of materials, PHEs are now widely used in the process and other industries.

PHE limitations are its operation range, as indicated in the table [2], pressure drop due to plate pattern, and limited use in phase change services. However, its several advantages like high heat transfer coefficients, compact constructions, flexibility to add/remove plates based on required thermal duties, suitability for easy maintenance, and in place cleaning (help to reduce fouling) makes it widely used in offshore oil and gas industries.

The content of this paper enables mechanical design engineers to address basic design and engineering issues in the early stages which will help to optimize the selection of PHE and to minimize the design changes during detail engineering.

1.1 Classification

PHE is classified on the basis of joints, flow arrangements, and plate corrugations. Based on joints, PHE is classified into four types: (i) Gasketed, which is the most commonly used PHE, used for almost all the purposes. Here, the flow is separated with the help of a gasket (glued or clipped) in the plate. (ii) Brazed, where the two endplates are brazed. There is no requirement for gaskets, bolts, carrying, and guide bars. (iii) Semi-welded PHE, which can take a higher design pressure compared to fully gasketed PHE exchangers. It resolves the problem where the medium on any one side is corrosive or limitation of fluid temperature and therefore gasket cannot meet the requirements. It does not require frequent replacement of gaskets which lowers the operating cost. Laser welding is used for sealing and replacing conventional gaskets on the process material side, thus largely reducing the procurement of spare parts. (iv) Fully welded PHE, which has high heat transfer efficiency and is smaller in size. It is convenient to clean and requires a small amount of maintenance. This design helps to save space and reduce your investment cost. A fully welded plate heat exchanger is designed to operate under further harsh conditions and high temperatures.

Based on flow arrangement, PHE is further classified as single pass (either U or Z arrangements) and multipass.

Based on plate corrugation, PHE is classified into various types. Some of them are provided in Fig. 4. Recent PHEs are designed based on chevron plate types majorly (Fig. 5).

1.2 Construction Parts

1. Mounting feet 2. Plate pack (thermal plates) 3. Fixed cover (frame plate) 4. Movable cover (pressure plate) 5. Support column 6. Tie bolt 7. Guide bar (Bottom) 8. Carrying bar (Top) 9. Connection, studded or flanged 10. Tie Nuts.

Plates are the most imperative and costly part of a PHE. The plates are either flat/corrugated (mostly), refer to the figure in [4]. The size of a plate can range from (0.1 m × 0.3 m side) to (1 m × 2.5 m side). Fixed cover (frame plate), a thick plate, that supports the carrying bar and guide bars. In most designs, the fluid nozzles are fitted on a fixed cover.

A movable cover (pressure plate), a thick plate, called the pressure plate is also hung on the carrying bar and it is movable as per frame size. In Z-type arrangement, multipass configuration, the piping connections are located on the pressure plate. Refer to Figs. 2 and 3. Support column and rear column to which carrying bar and guide bar are bolted (not on all models). Tie bolts are pressing the channel plates so that these come into metallic contact; also it is compressing the gaskets for sealing. The guide bar (bottom) and the thermal plates are aligned by this.

Carrying bar (top) and the thermal plates are aligned and, in most models, hang from. Gaskets are placed in a groove, along the plate and around holes for sealing. Refer to Fig. 6.

The gaskets have vents to prevent the mixing of fluids in the case of a failure, which also enables to locate leaks.

1.3 Working Principle

PHE comprises parallel plates placed in such a way that hot and cold fluids flow alternate between each of the plates. Inlet and outlet nozzles are connected at the junctions of the plates to permit hot and cold fluids through alternating channels. These special pattern plates (thickness range 0.5–3 mm thickness, with gapping of 1.5–5 mm) provide an extremely large heat transfer area to each fluid. The passages in between the subsequent plates cause spinning motion to the fluids, which leads to high turbulence, hence more heat transfer. The hot side heat energy to the cold side through the thin plates. According to the current flow direction, there are two types of flow: concurrent, this type of flow is generally not preferable because of a decrease in thermal efficiency and requires a large area to meet the heat duty, and counter-current, this type of flow is mostly chosen due to higher thermal efficiency. Based on the types of pass, single pass is the simplest form of arrangement in which both the fluids make just one pass. U-type, inlet, and outlet connections are fitted on a fixed plate, which allows access to equipment/piping for cleaning/maintenance, without disconnecting the piping network. This is extensively used in a single pass (refer to Fig. 1). Z-type, inlet, and outlet connections are fitted on both endplates (refer to Fig. 2). Equal velocity through the channels is achieved in Z-type. Multipass is selected to improve the flow velocity or heat transfer and also when there is considerable variance in the flow rates of streams. But generally not recommended due to low flexibility to increase or decrease the number of plates (refer to Fig. 3). The improper flow distribution in U-type can be handled by L-type channels (near the movable side) which also helps in lower pressure drop.

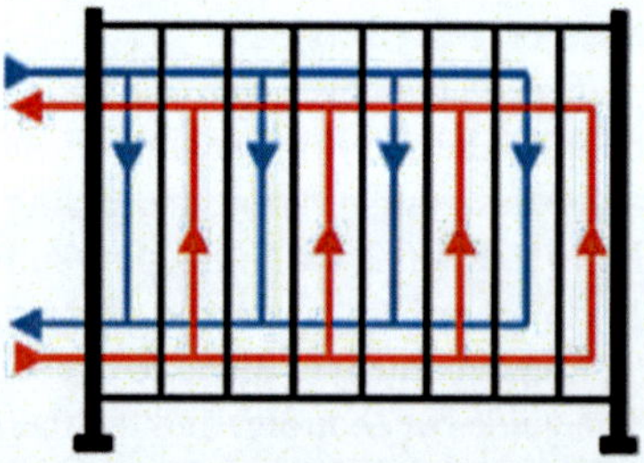

Fig. 1 U arrangement

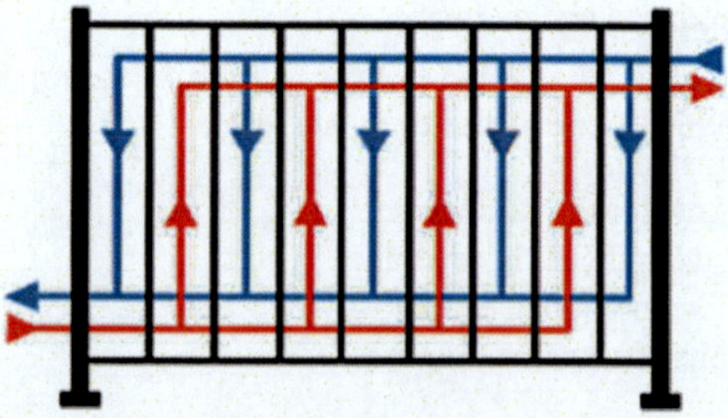

Fig. 2 Z arrangement

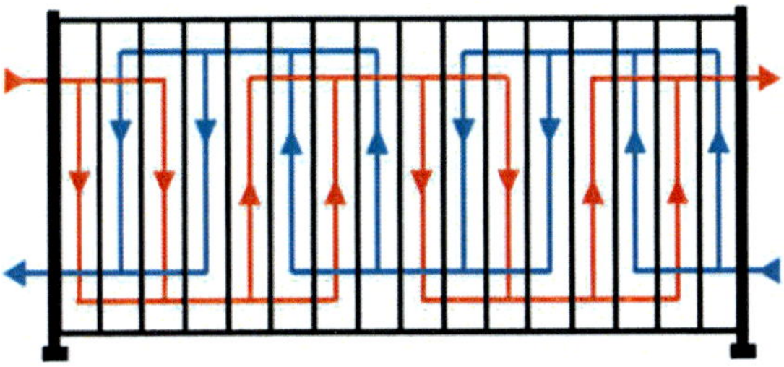

Fig. 3 Multipass arrangement [17]

Flow inside the channel is divided into two types. Crossed flow: It is tough to assemble but it avoids the stagnation areas. Straight flow: The straight flow sometimes led to the formation of stagnation areas, but they are easy to assemble (Fig. 7).

The flow distribution in a plate governs the port arrangement (refer to Fig. 8); a higher aspect ratio (length/width) helps in the improved distribution and pressure loss at entry. Pressure drop at entry, approx. 10% of total pressure drop, is considered a good design, however, it can be around 25% also in a single pass. It is to be noted that a higher pressure drop at entry in multipass will affect the heat transfer.

The main characteristic of PHE is its superior thermal performance (4–7 kW/m^2 °C (clean)), and very small approach temperature (as low as 2–3 °C).

1.4 Design Analysis

The stages involved in the design of PHE are establishing the fluid characteristics, defining plate geometry, calculating effective heat transfer area, establishing limited correction factor, heat transfer coefficient, and calculating pressure drop. These are explained in the plate design section.

There are two approaches to design PHE effectively, i.e., thermally controlled design and hydraulically controlled design. In thermally controlled design, if the design exceeds the allowable pressure drop for a given thermal duty, more plates are to be added resulting in a reduction in the pressure drop by reducing the velocity. In hydraulically controlled design, if the design pressure drop is lesser than the permissible pressure drop, it affects a larger temperature change across the plates than required, which leads over dimension. In such a case some plates are to be reduced and an increase in pressure drop is required by increasing the velocity.

Two important parameters which play the role in the designing of the heat exchanger are logarithmic mean temperature difference (LMTD) and pressure drop. Instead, the fact that in counter-current flow LMTD value is higher than co-current flow for the same specification means a lesser heat transfer area, leads to the smaller size of PHE, in several applications Co-current flow is chosen over counter-current, like in flooded evaporators or in offshore where space is a constraint to facilitate the piping route of bigger size inlet and outlet cold/hot medium.

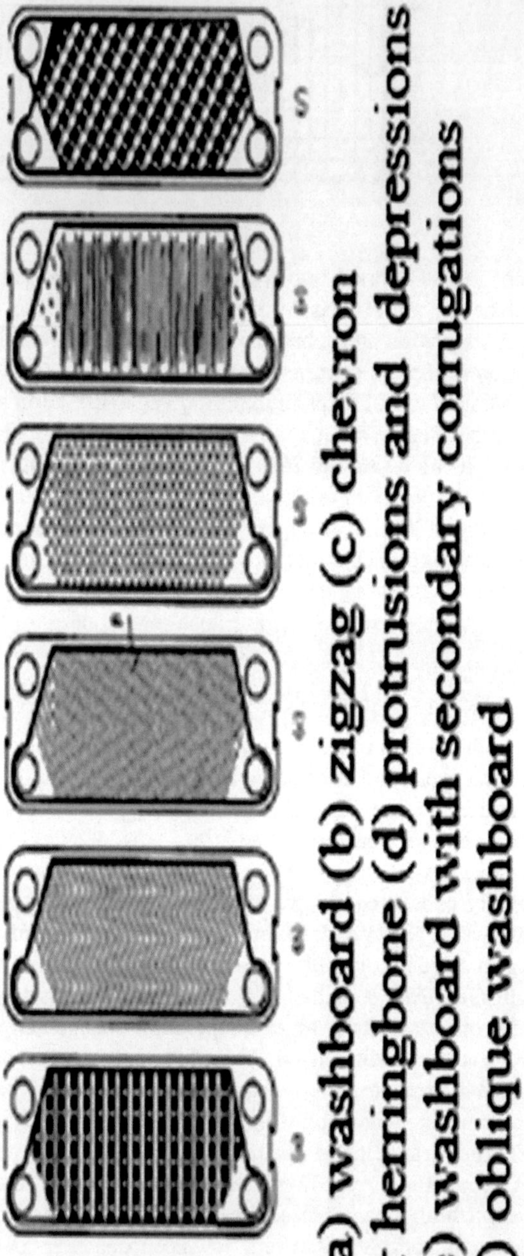

Fig. 4 Plate corrugations, Ref. [14]

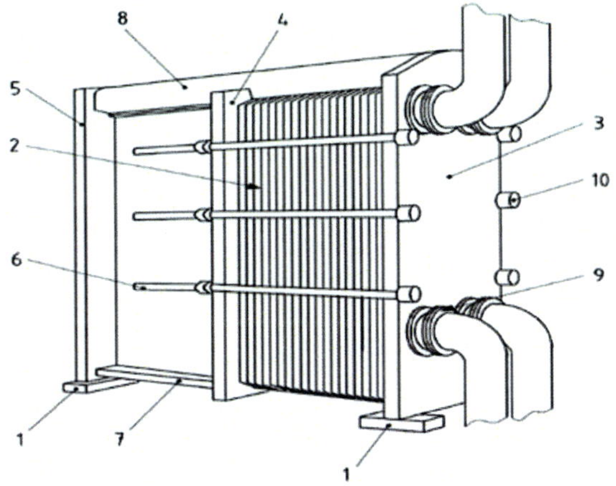

Fig. 5 PHE construction components [17]

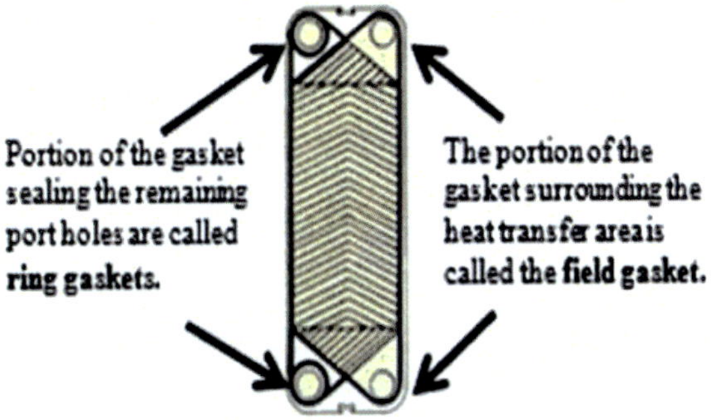

Fig. 6 Gasket demonstration [15]

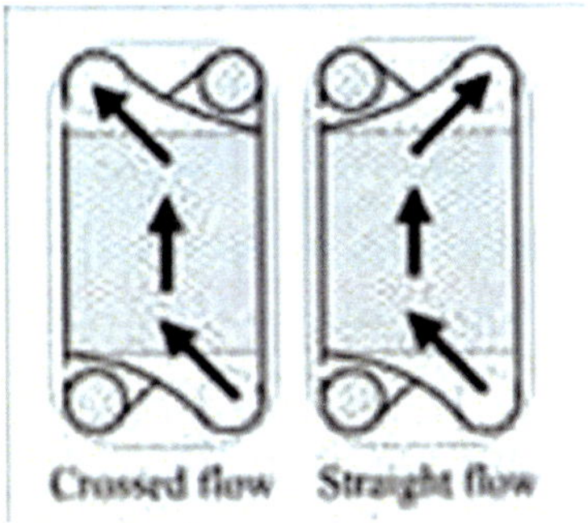

Fig. 7 Flow pattern inside the channel [15]

Pressure drops have both positive and negative implications on heat transfer. The negative aspect of pressure drop is the high pump power for upstream flow, thus expensive operation. However, the positive aspect of pressure drop is that it generates larger turbulence, thus improving heat transfer.

Fig. 8 Thermal mixing

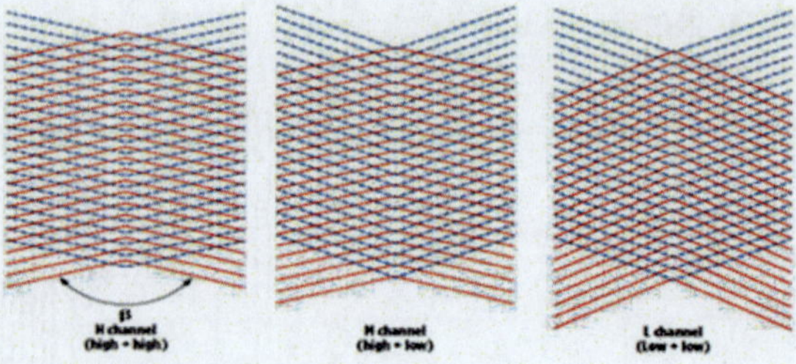

The PHE shall have an even flow distribution across the heat transfer surface. Velocity variations shall be minimal across the surface. The maximum allowable flow maldistribution shall be 10%. Many international end-users are recommending limiting the fluid velocity to 8.5 m/s (28 ft./s) for clean fluids and 4.6 m/s (15 ft./s) for fluids containing solids.

Fluid flow shall be arranged in the upward direction for vaporizing fluids or fluids containing entrained gas. Heat transfer calculations shall account for two-phase flow effects. Fluid flow direction shall be downward for condensing services such as steam.

Designers must keep the below relations while designing PHE.

$$v \sim 1/A;\ \delta P \sim v^2;\ U \sim (\delta P)^{1/3}$$

U = heat transfer coefficient, v = velocity through plate channel and A = area.

The pressure drop of a channel is summarized as proportional as below:

Pressing depth $\propto 1/(\delta P)^3$; pressing width $\propto 1/(\delta P)^2$; plate width $\propto \delta P$; plate length $\propto$ high theta means higher δP; port size $\propto$ small port size higher δP.

The number of plate requirements for PHE is a complex design and a function of N_C = number of channels, Φ = plate area enlargement factor, plate type, number of passes, and pressure drop.

Hence, different configurations of plates are to be selected keeping in view of all the above factors.

Thermal length (Θ) (*dimensionless number*) relates the performance of a channel geometry to a duty requirement.

$\Theta = dt/LMTD = (T_{in}-T_{out})/\ LMTD$; Θ depends on length of plate, diameter, corrugation angles, and pressure drop.

Therefore for effective design, it is important to select the channel geometry which satisfies each fluid's thermal length condition.

Plate Design

Understanding plate geometry is very important in order to attain optimized and economical design. Chevron plate is widely used. Therefore, plate design based on chevron angle is discussed in this paper.

The heat transfer and pressure drop are influenced by chevron angle (β). (β differs from β1 or β2, refer to Fig. 11, β2 = 180°–2β1, plate corrugation β1 generally varies from 25° to 65°.

Each plate size is hard-pressed in two dissimilar chevron angles, i.e., known as low theta (acute angle) and high theta plate (obtuse apex angle). The number of transfer units (NTU) range is the ratio of the temperature change of one medium to the LMTD across the plates. NTU correlation is as below: Ref. [18] as (NTU/L) ∝ (U/v) and (NTU/L)∝ (ΔP/v), where v is the velocity and L is the plate length. The above relations are dependent on β.

High-theta plate (pattern h)/high NTU plates > 1 and β approaches 25° signify higher turbulent flow resulting in increased high heat transfer, friction factor and Nusselt number increases, impeccable temperature approach, and high pressure drop.

Low-theta plate (pattern W)/low NTU plates < 1 and β approaches 65° signify lower turbulent flow resulting in low heat transfer, friction factor and Nusselt number decrease, high temperature approach, and low pressure drop.

To avoid multipass arrangement, mixing of high NTU plates and low NTU plates is proven now. Below mixing are possible, which are HH (H), HL (M), and LL (L). H channel: High NTU plates on both sides. M channel: High NTU plate on one side and low NTU plate on the other side. To address the issues arsies due to very different flow rates of the two fluids, M channels are used. L channel is used in low NTU plates on both sides.

Design Steps

Establishing the fluid characteristics: In this step, we obtain operating parameters for hot and cold fluid, respectively. The major parameters that will affect the heat transfer are flow rates, inlet and outlet temperature, density, viscosity, and specific heat values.

The heat load duty is calculated from the following two classic formulas: Q = U.A.LMTD; also Q = m.Cp.dt.

Therefore, the area required for the heat exchanger: A = Q/U.LMTD.

Overall heat transfer coefficient U is defined as: $1/U = (1/h1) + (1/h2) + t/k +$ Rfh + Rfc, where Q = heat load (kW); m = mass flow rate (kg/s); C_P = specific heat (kJ/kg °C); dt = inlet and outlet temperature difference on one side (°C); A = heat transfer area (m^2); h1 = heat transfer coefficient between hot medium and heat transfer surface (W/m^2 °C); h2 = heat transfer coefficient between heat transfer surface and cold medium (W/m^2 °C); t = thickness of heat transfer surface (m); R_{fh} = fouling factor (m^2 °C/W) hot side; R_{fc} = fouling factor (m^2 °C/W) cold side; k = thermal conductivity of metal (W/m °C); A_e = total effective area of plates.

If we calculate the area of a single plate, we can determine the number of plates required. (Effective area calculation of plate design is explained in the next section.)

Defining plate geometry: In the first step, we decide the plate thickness (t), chevron angle (β), effective width (L_w), and effective length (L_p). These values are obtained

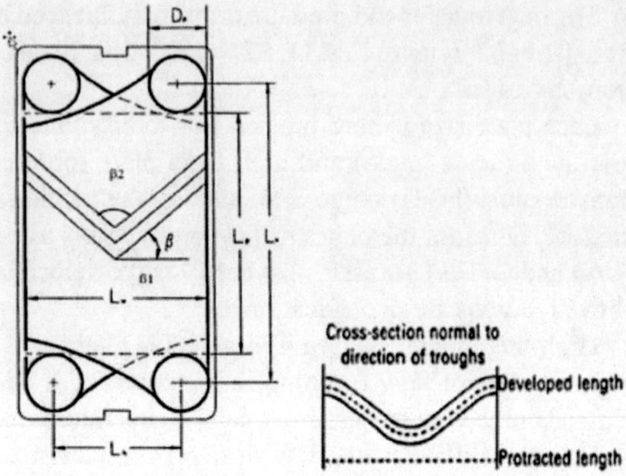

Fig. 9 Plate pattern [10]

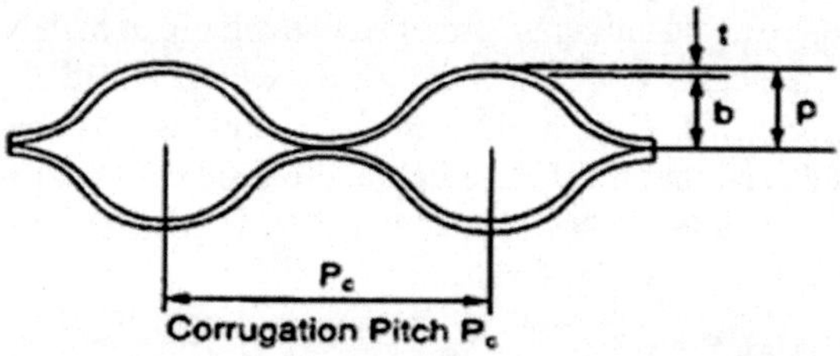

Fig. 10 Corrugation pitch [10]

from standard plate sizes as per manufacturer recommendations, or this can be further customized according to our requirements.

Effective plate length: The corrugations on plates increase the plate area depending on its pitch and depth (Figs. 9 and 10).

Φ = Developed length/protracted length. The enlargement factor (Φ) ranges from 1.1 to 1.25. The typical average is considered 1.17. L_p and L_w can be calculated from the port distance L_v, L_h, and D_p (port diameter) as below (refer to Fig. 11): $Lp \approx Lv - Dp$; $Lw \approx Lh + Dp$. So, L_p can be calculated as $L_p = L_v + L_h - L_w$.

Equivalent diameter, De (hydraulic diameter), $De = 2b/\Phi$, where b = channel average thickness (i.e., corrugation spacing around 2–4 mm); $b = p - t$, where p = plate pitch, t = plate thickness.

Calculating effective heat transfer area: Total effective heat transfer area: $Ae = Ap * Ne$, where N_e = effective number of plates, $Ne = Nt - 2$ and N_t = total number of plates (assumed initially).

Effective heat transfer area of plate: $Ap = \Phi.Lw.Lp$

Establishing LMTD correction factor: For single pass and multipass with equal passes, there is no requirement of LMTD correction as it is to be noted that the flows are properly distributed in counter-current. In case of unequal passes, a correction factor comes into the picture. It is to be noted that this correction factor is not as significant. Hence not discussed further.

Heat transfer coefficient: Heat transfer coefficient h (h1/h2 for respective fluid) depends on the correlation of Nusselt number (Nu) and Reynold number (Re), and Prandtl number (P_r) is required. Physically Nusselt number signifies the ratio of convective and conductive heat transfer, the Reynolds number signifies the turbulence in the operating liquids which directly affects the heat transfer, higher turbulence normally signifies better heat transfer and the Prandtl number also signifies the ratio of momentum and thermal diffusivity which signifies whether conductive heat transfer exceeds ($<< 1$) or convective heat transfer dominates ($>> 1$).

The following equations (proposed by Kumar [20]) are used to perform the calculation for heat transfer coefficient: $Nu = (hD_e/k) = J_h * (P_r)^{(1/3)} * (\mu_b/\mu_w)^{0.17}$; $h = [((J_h * (P_r)^{(1/3)} * (\mu_b/\mu_w)^{0.17}) * K)/D_e]$.

where $J_h = C_h * Re^y$ (also known as Kumar coefficient given by the formula), value of C_h, y depends on flow characteristics and chevron angles; μ_b = dynamic viscocity at bulk temperature; μ_w = dynamic viscosity at wall temperature; $P_r = (\mu C_p)/k$, Prandtl number; $Re = G_c * D_e/\mu$, where G_c is the channel mass velocity; $G_c = m_{ch}/(N_c * b * L_w)$, $N_c = (N_t - 1)/2N_P$.

where N_c is the number of channels per pass for one fluid and N_p is the number of pass.

For single-phase PHE (gasket type) kindly note that the constants used in the calculation for heat transfer and pressure loss are referred to as below (refer to Selection, Rating and Thermal Design, 2nd Edition, Page 395).

There are other correlations for obtaining the Nusselt number and friction factor which can be used alternatively. Muley and Manglik [19] proposed the following correlation for various values of β:

$$Nu = [0.2668 - 0.006967\beta + 7.244 \times 10^{-5}\beta^2]$$
$$Re^{0.728+0.0543\sin[(2\pi/90)+3.7]} \Pr^{1/3} (\mu_b/\mu_w)^{0.14}$$
$$f = [2.917 - 0.1277\beta + 2.016 \times 10^{-3}\beta^2]$$
$$Re^{[0.2+0.0577\sin(2\pi\beta/90+2.1)]}$$

Calculating pressure drop: The channel pressure drop is the addition of channel frictional pressure drop (P_c) and the port pressure drop (P_p): $\Delta P_c = 4f\,(L_{eff}N_p/D_e) \times (G_c^2/2p) \times (\mu_b/\mu_w)^{-0.17}$

where $f = K_p/Re^z$ (z is taken from Table 1); L_{eff} = effective length of the fluid flow path between inlet and outlet ports. $\Delta P_p = 1.4N_pG_p^2/2\rho$; $G_p = (m/\pi D_p^2)/4$, where m is the total flow rate in the port opening; D_p is the port diameter; G_p is the mass velocity of the port; N_p is the number of passes; ρ is the fluid density.

The total pressure drop can now be calculated by $\Delta P = \Delta P_c + \Delta P_p$.

Size of PHE: Based on the overall heat transfer requirement and pressure drop being within the allowable range, we decide the overall length, width, and height of the PHE.

A general rule for obtaining PHE dimensions currently practiced in the oil and gas industry is given by below relations.

Table 1 Heat transfer and pressure loss constants

Chevron angle (degrees)	Heat transfer			Pressure loss		
	Reynolds number	C_h	y	Reynolds number	K_p	z
<30	<10	0.718	0.349	<10	50	1
	>10	0.348	0.663	10–100	19.40	0.589
				>100	2.990	0.183
45	<10	0.718	0.349	<15	47	1
	10–100	0.400	0.598	15–300	18.29	0.652
	>100	0.300	0.663	>300	1.441	0.206
50	<20	0.630	0.333	<20	34	1
	20–300	0.291	0.591	20–300	11.25	0.631
	>300	0.130	0.732	>300	0.772	0.161
60	<20	0.562	0.326	<40	24	1
	20–400	0.306	0.529	40–400	3.24	0.457
	>400	0.108	0.703	>400	0.760	0.215
>65	<20	0.562	0.326	<50	24	1
	20–500	0.331	0.503	50–500	2.80	0.451
	>500	0.087	0.718	>500	0.639	0.213

Overall height of exchanger: Effective height of plates (L_p) + 2* Max port Dia (D_p) + 100 mm.

Overall width of exchanger: Effective width of plate (L_w) + 100 mm.

Overall length of exchanger: ((Plate thickness (t) + Mean gap between channels (b)) * number of plates (N_t) + 700 mm.

In special cases, PHEs are designed for condensation/evaporation operations, but for gases and vapors, this is not recommended due to the inadequate space inside channels and due to pressure limitations.

[11] If the application involves viscous fluids or we suspect chances of solids/slurries wide gap plates are most effective in application. Wide gaps are also excellent in steam services.

Maximum flow capacity also needs to be checked which determines the overall plate dimensions. The higher flow requires a higher port/nozzle diameter which decreases the net heat transfer area of the selected geometry. The bigger nozzle/port diameter requires more number of plates in parallel to meet the required heat duty. Therefore designer can use a wider plate that has more width than thermally required. The above necessities will lead to a higher chance of fouling and lower velocity. To addresses such concern larger diameter ports/nozzle can be fitted without increasing the width of the plate (max to approx. 450 mm) and can provide a high flow rate (max approx. 3,000 m^3/h) by using a "distance piece" which helps to stagger flange connection perpendicular to plate.

1.5 Material

In offshore facilities, material selection is critical due to the marine environment, wide operating ranges, and corrosive substances (CO_2, H_2S, and organic acid). Material selection for a PHE depends on several factors majorly including thermal transfer, operating temperature, corrosion resistance, and ease of fabrication.

Plate and gasket material selection is the most important consideration which affects the performance of the exchanger and their service life.

Plate materials: In plate material selection, the challenging aspect is that the thickness of plates cannot be increased beyond a certain extent (0.5 mm to 3 mm max), so the material selection should be immaculate.

Plates can be manufactured from several materials like high alloy or non-ferrous examples are SS316/316L, nickel alloy, titanium alloy, Monel, etc. Majorly PHEs in offshore industries are in seawater service, lube oil, and glycol application. Lean glycol/lube oil is non-corrosive, so SS316L is suitable; however, rich glycol with high temp duplex/super duplex or Incoloy is selected. In seawater service, copper/nickel alloys are normally considered basic material. ISO 21457 to be referred for material selection in seawater application.

If one of the heat exchanger's media seawater and titanium plates are used, then all of the port linings shall be in titanium.

Crevice corrosion due to plate design specifically in seawater application is one of the main concerns.

Titanium is widely used in offshore PHEs. The presence of chloride in liquid as well as in the environment causes limitation of the use of SS316/316L material if operating temp goes above 60 °C due to chloride stress cracking.

Gasket material: The most common cause of failure of gaskets is incompatible fluids or excessive pressure and temperature. Therefore the proper selection of gasket is paramount in the gasketed PHE. Vendor past project experiences details/records for use of such gaskets in similar services need to be checked thoroughly. Gaskets are either glued or clipped. For high fouling applications where cleaning of the plate is frequent glue gasket is preferred. In hydrocarbon services glued gasket is the best option. In the case of operating temperature above 115 °C, the low-pressure clip method is preferred due to having more gasket surface area. Clip-on-proven ranges are 170 °C at low pressure and 25 barg at a moderate temperature depending on the application.

For PHEs, the widely used gasket materials [16] are NBR (Nitrile rubber), HNBR (hydrogenated nitrile rubber), EPDM (ethylene propylene diene monomer), FPM (different grades), and FKM (fluorocarbon rubber).

It is well known that gasket stiffness increases and its elasticity decreases at low temperature. Operating below the minimum suitable temperature of the selected gasket will lead to reducing the sealing force of the gaskets. The material data sheet for the offered gasket is to be verified accordingly.

Table 2 Gasket material selection based on temperatures

Gasket material	Max operating temperature (°C)
NitrilezRubber (NBR)	100
EPDM	120
EPDM (peroxide cured)	150
Hydrogenated nitrile	160
FKM (fluoroelastomer)	170

Computability with fluid/chemical in use has to be checked properly. At high temperatures the oxidation process makes mostly the gasket harder (often) or sometimes softer (less likely) or some parts harder and some parts softer; however, it is rare. Therefore maximum permissible temperature limit with its duration is to be reviewed properly during vendor selection.

A general guide for the selection of gasket material based on the operating temperatures can be summarized in Table 2.

In case of lack of experience with proposed gaskets used by vendors for the applicable application, the acceptance of gasket is recommended subjected to witness an immersion test to measure gasket swelling, its hardness, and chemical suitability. Also, the witness glue testing where the final peel strength to be checked is recommended. The acceptable values are specified in API 662 [17].

Fixed cover plate and movable cover plate: As these plates seldom come in contact with liquid, hence carbon steels are sufficient to be used. Guide bar (bottom) and carrying bar (top) due to marine environment CS + SS316L Lined or solid SS 316L is preferred MOC.

2 Literature Review

There has been a lot of considerable and conclusive research work that has been presented to date on plate frame heat exchangers by evaluating different parameters in line to meet the industry requirements. Arsenyeva et al. (2011, pp. 4588–98) [1] grounded their optimization method on a mathematical model to define PHE thermal and hydraulic performance. Results showed an optimal solution with an appropriation of heat load and pressure drop of a particular stream. The model gives a substantial decrease in the heat transfer surface area of the PHE. However, for definite plate types, it is possible only in the narrow range of process conditions.

Jin and Hrnjak (2017, pp. 740–748) [12] in their paper illustrated that endplates in both frame and plate and brazed type can be assumed to work as the extended surface or as fins based on experimental data and numerical model.

The model used displayed trends that the fin efficiency reduces with an increase in heat transfer coefficient and efficiency increases with an increase with contact area. It was concluded that the endplate effect is stronger in the brazed plate heat exchanger because of the greater contact area and higher conductivity of the brazing material.

Kananeh et al. (2012, pp. 533–550) [3] concluded that there is a possibility to use nanocomposites as anti-fouling coatings to decrease fouling inside gasket type PHE. Test results showed that a few reduced fouling and details can be found in the respective paper, however, it is interesting to note that PTFE coating causes higher fouling than the standard SS plate.

Pinto et al. (2002, pp. 433–439) [2] in their research experimented by demonstrating the configuration of a PHE that can be denoted by a set of six distinctive parameters: 1. Number of channels; 2, 3. Number of passes on two sides; 4. Feed connection relative location; 5. Hot fluid allocation; 6. Type of flow. Thereafter, a methodology was designed to detect equivalent configurations for all of these parameters was stated. The constraints to this design were recognized as the number of channels, pressure drop, and velocities of channel flow. An algorithm was established, which helps to perform the screening. It successively eliminated infeasible and non-optimal elements by using subsets of the constraints from the set defined by the constraints on the channel numbers. Results showed that there could be multiple optimal solutions for a given configuration based on the reduced number of thermal simulations.

Tisekar Salman et al. (2016, pp. 56–62) [4] concluded in their research paper that convective heat transfer coefficient showed an increasing trend with an increase in both Reynolds number and mass flow rate for both parallel and counter flow arrangements. This result can be attributed to the fact that flow becomes more turbulent due to the corrugation of the plate and also due to high flow velocity. In the study, it was also established that the PHE effectiveness decreases with the increase in the mass flow rate of hot fluid.

Shah and Kandlikar [5] in their research paper developed a correlation between temperature effectiveness and LMTD correction factor as a function of NTU, heat capacity rate ratio, and the number of plates. This was also quoted in one of the papers published by Qiao, Hongtao, Vikrant Aute, Hoseong Lee, Khaled Saleh, and Reinhard Radermacher (2013, pp. 622–632) for 1, 2, and 3 pass plate and frame heat exchangers, respectively.

Kanaris et al. (2008) [6] in their research paper have tried to design an optimal design for PHE for sinusoidal herringbone plate configuration using a CFD model. Five dimensionless variables were selected as design parameters for the simulations, based on the characteristics which are the fraction of the entrance of the channel "blocked" with corrugations or interruptions, a measurement of the narrowness of the channel, a measure of the obtuseness of the corrugation, the sine of twice the angle of attack, and the Reynolds number.

The result showed that the optimal performance PHE can be achieved for the shortest distance between the plates and a less obtuse corrugated configuration. For increased Reynolds number the performance of the PHE can be bettered for wider channels and bigger angle of attack in low pumping operations.

Fábio et al. (2015) [13] developed two mathematical models for the design and optimization of PHE. The first model was based on the six distinct parameters as stated in the research of Pinto and Gut (2002, p. 433–439) under steady-state conditions. The second mathematical model was developed based on the assumption that any multi-pass PHE with a large number of plates is considered a combination of many concurrent and counter-current single-pass plate heat exchangers. Both the models showed the result values which were very close to each other.

Lines [7] suggested the use of non-conventional corrugation in the plate for plate and frame heat exchangers. In place of a conventional homogenous pattern for the plate, the l plate was divided into four quadrants of heat transfer section divided with two different angles for corrugations. As a result of using asymmetrical channels, area savings and cost reduction of about 16% and 10% can be achieved, respectively, which can be further optimized.

Dovic and Svaic (2007) [8] in their experiment concluded that higher chevron angle and higher pressing depth to wavelength ratio b/l resulted in higher heat transfer and high-intensity heat transfer and higher pressure drop due to more thorough intermixing. The exception to this is low Reynolds numbered flows (<10).

The too-high chevron angles also result in separation of the longitudinal flow component which will hamper the heat transfer. An increase in flow velocities and b/l in the case where the longitudinal flow component is large experimentally led to intensive heat transfer.

Raja et al. (2018, p.522–535) [9] performed a multi-objective optimization of the PHE-based adapting algorithm. Conflicting objectives considered were overall heat transfer coefficient and total pressure drop. It was observed a total of 98.2% decrease in heat transfer coefficient and a 72.1% decrease in total pressure drop between extreme design points.

3 Challenges and Mitigation

Various parameters which significantly improve PHE performance are discussed in this paper. Designing a plate heat exchanger for an offshore platform comes along with its own unique challenges. Therefore, a designer must focus on the below listed major challenges and their mitigation in the initial stage of the design itself and carefully evaluate the vendor's design.

- Inlet and outlet connection of hot/cold fluid plays important role in deciding the arrangement of piping for these exchangers, and several times, it has been seen that counter-current flow is not possible due to layout limitations in such cases co-current flow is recommended to manage the piping network.
- Proper investigation should be carried out like available pressure drop in the system as due to port size limitation pressure drop plays important role in the effectiveness of PHE.
- Avoid conservative value of fouling factors which consequences in oversized design with low velocities. It is well known that fouling is intensified in the case of low velocities. The velocity varies from 0.1 m/s to 1 m/s, the minimum

(which will increase fouling) is governed by the wall shear stress of 50 Pa, and the maximum is decided based on the permissible pressure drop across the equipment.

- Connection velocity and pressure drop in the fluid in the inlet connection pipework before it reaches the first plate also play a major role in the heat exchanger performance. On an offshore platform, the flexibility to reduce pressure using extended pipework is limited and hence it plays a significant impact in the design also.
- In the case of dirty seawater services, which are prevalent in offshore applications, a good measure to mitigate fouling is to keep sufficient wall shear stress value, which is basically the force exerted by working fluid to keep the particulate matter suspended in the fluid. It is directly proportional to pressure drop in the heat exchangers and for dirty seawater services, it could be as high as 75 Pa. The designer should as much as possible maintain a tradeoff between the wall shear stress and pressure drop.
- Manufacturer recommendation to be sought in case of system design is facing pressure and temperature variations, more than 10.34 barg/min and more than 11 °C/min.
- It is recommended to include a temporary strainer (conical strainer) on both service fluid inlets because in start-up cases dirt/abrasive may block/damage the plate corrugation. Regular cleaning is also recommended.
- Proper plate material and gasket material is paramount in the design selection of PHE. Material suitability is also discussed in this paper. The liner on the port must address erosion and corrosion in high flow, abrasive applications (SS316L is avoided in an abrasive application; duplex and nickel are recommended). It is also recommended to take precautions to minimize erosion of the port. Check with manufacturer recommendation for port velocity (generally around 20 ft/sec).
- The type of PHE can be selected on features based on Table 3.

Table 3 Selection of type of PHE based on features

Features	Gasketed	Brazed	Semi-welded*/welded
Design pressures	Up to 25 bar	Up to 30 bar	Greater than 60 bar
Design temperature	−35 to 220 °C	−195 to 220 °C	Based on the material used
Leakage susceptibility	More	Lesser	Less
Compactness	Slightly less compact	Compact	Highly compact
Flexibility	Higher can be modified for different heat duty	Less flexible	Difficult to cater to different heat duty
Cleaning	Mechanical and chemical possible	Chemical only	Chemical only
Toxic fluid services	Not preferred	Not preferred	Preferred
Corrosive Services	Less preferred	Better than gasket	Highly suitable
Efficiency	Less	More	Most

*Note that the semi-welded uses both gaskets and welded configuration and hence it contains features of both categories

4 Conclusions

- The most effective PHE selection is a major challenge in offshore oil and gas industries due to process parameters and environmental conditions. Increased thermal duties do not entitle to label as optimal design, and thus various aspects which need to be thoroughly checked in the design phase of PHE and its integration into the plant was discussed in this paper.
- There is a decent amount of research being done based on different analytical/mathematical models. Some key research works include using different corrugation patterns of plate type, studying the effect of endplates on the heat transfer, and larger differences in flow rate, velocity, fouling, and pressure drop were reviewed in this paper.
- The plate pattern requires significant research and data references of its performance, thus technical bid evaluation of each vendor may not be imperative. Thus each vendor design/selection shall be carefully reviewed during technical evaluation.
- Some of the important challenges in design and engineering in offshore industries were discussed. It is appropriate to conclude that PHEs can be used with few limitations and their mitigation as specified in this paper. Still, a lot of scope for future works like the effective design of PHE in vaporization and condensation service needs to be explored.

References

1. Arsenyeva OP, Tovazhnyansky LL, Kapustenko PO, Khavin GL (2011) Optimal design of plate-and-frame heat exchangers for efficient heat recovery in process industries. Energy 36:4588–4598
2. Pinto JM, Gut JAW (2002) A screening method for the optimal selection of plate heat exchanger configurations. Braz J Chem Eng 19(4):433–439
3. Kananeh AB, Peschel J (2012) Fouling in plate heat exchangers: some practical experience, heat exchangers-basics design applications. Jovan Mitrovic, IntechOpen. https://doi.org/10.5772/34026.533-550(2012)
4. Tisekar Salman W, Mukadam SA, Vedpathak HS, Rasal PK, Khandekar SB (2016) Performance analysis of corrugated plate heat exchanger with water as working fluid. Int J Res Eng Tech 5:56–62
5. Shah RK, Satish GK (1988) The influence of the number of thermal plates on plate heat exchanger performance. Current Research in Heat and Mass Transfer. New York: HemisTCPre PC, 227–241
6. Kanaris AG, Mouza AA, Paras SV (2006) Optimal design of a plate heat exchanger with undulated surfaces. Chem Eng Technol 29(8):923–930
7. Lines JR (1987) Asymmetric plate heat exchangers. Graham Manufacturing Company, Batavia
8. Dović D, Švaić S (2007) Influence of chevron plates geometry on performances of plate heat exchangers. Tehnički vjesnik 4(1, 2):37–45
9. Raja BD, Jhala RL, Patel V (2018) Thermal-hydraulic optimization of plate heat exchanger: a multi-objective approach. Int J Thermal Sci 124:522–535

10. Khan TS, Khan MS, Chyu M-C (2010) Experimental investigation of single phase conective heat transfer coefficient in a corrugated plate heat exchanger for multiple plate configurations. Appl Thermal Eng 30
11. Mitrovic J (2012) Heat exchangers: basic design applications. IntechOpen
12. Jin S, Hrnjak P (2017) Effect of end plates on heat transfer of plate heat exchanger. Int J Heat Mass Transfer 108:740–748
13. Fábio ASM, Carvalho EP, Mauro ASS (2015) Modeling and design of plate heat exchanger. Heat Transfer Studies and Applications Salim Newaz Kazi, IntechOpen. https://doi.org/10.5772/60885
14. Shah RK, Sekulic DP (2003) Fundamentals of heat exchanger design. Wiley & Sons Inc., New Jersey
15. Craston T. Plate heat exchanger: operational and maintenance manual. 202–005new.pdf (taco-hvac.com)
16. Shell DEP: Plate And Frame Heat Exchangers(Amendments/Supplements To ISO 15547–1:2005) Dep 31.21.01.32-Gen
17. American Petroleum Institute (API) Standard 662, plate heat exchangers for general refinery service; Part 1—plate-and-frame heat exchangers
18. Thulukkanam K (2013) Heat exchanger design handbook, 2nd edn. CRC Press. Boca Raton, Florida
19. Muley A, Manglik RM (1999) Experimental study of turbulent flow heat transfer and pressure drop in a plate heat exchanger with chevron plates. J Heat Transfer-Trans ASME 110–117
20. Kumar H (1984) The plate heat exchanger: construction and design. Proceedings of the First UK National Conference on Heat transfer, vol. 2, Leeds

Analysis of Cascade Thermal Energy Storage System

Hrishikesh Gupta, Priyanka Verma, Kirti Dhiman, Prince Chaudhary, and Neelam Khandelwal

Abstract In order to bridge the energy demand–supply gap, thermal energy storage units are high in demand. The thermal energy storage (TES) system stores thermal energy by heating or cooling phase change material (storage medium), and this whole process involves three steps: charge, storage and discharge. Generally, thermal energy can be stored in two forms, i.e., sensible heat and latent heat, and mostly latent heat storage is preferred over sensible heat because of its higher heat storage capacity than the sensible heat storage system. This paper analyses the designing and CFD simulation of a cascade heat storage unit by using ANSYS fluent as a modelling tool and selecting three PCMs, beeswax, Puretemp48 and Puretemp37, on the basis of design parameters. Calculation of parameters of cascade heat storage system at the time of charging and discharging is shown in this paper, i.e., the total time taken for charging at inlet temperature 363 K comes out 20,267 s (5.63 h), in which temperature drop for time intervals was observed, which was 25 °C after 30 min, 16 °C after 1 h 30 min, 10 °C after 2 h 30 min, and after 3.5 h 90% of PCM melted and temperature drop was only 4 °C. After 5 h 37 min, Puretemp48 and beeswax solidified completely. However, Puretemp37 was only 30% solidified and 70% was still liquid due to the very small amount of temperature difference between HTF and Puretemp37. At the time of discharging, the outlet temperature remained constant between 312 and 317 K.

Keywords C.F.D. · Cascade storage system · Thermal energy storage · Latent heat storage · Phase change material

H. Gupta (✉) · P. Verma · K. Dhiman · P. Chaudhary · N. Khandelwal
Department of Mechanical Engineering, J.S.S. Academy of Technical Education Noida, Noida, Uttar Pradesh, India
e-mail: hrishikesh.gu090@gmail.com

N. Khandelwal
e-mail: neelamkhandelwal@jssaten.ac.in

D. P. Mishra et al. (eds.), *Recent Trends in Thermal and Fluid Sciences*, Lecture Notes in Mechanical Engineering, https://doi.org/10.1007/978-981-19-3498-8_12

Nomenclature

U	Overall heat transfer coefficient (W/m^2 K)
HTF	Heat transfer fluid
m	Mass (kg)
PCM	Phase change material
A	Area (m^2)
NTU	Numerical transfer unit
ρ	Density (kg/m^3)
CSP	Concentrating solar plant
V	Volume (m^3)
CTES	Concentrated thermal energy storage
H	Height (m)
CFD	Computational fluid dynamics
c_p	Specific heat (J/kg K)
ε	Effectiveness
$\dot{m}$	Mass flow rate (kg/s)

1 Introduction

This paper describes a model for storage of thermal energy which is widely used in thermal power plants, especially CSP. The model proposed intends to increase the storage capacity of these plants where a large amount of heat energy gets wasted and may cause an environmental issue as well. Using this waste energy by using the cascade heat storage system will lead to a reduction in energy demand for running such power plants. A cascaded heat storage system can be used to optimize the use of fossil fuels, and it can also be used as a solution for heat transfer reduction by reducing temperature differences through the process of heat exchange. This work can be used as a reference for the further research on cascaded heat storage technology in the field of solar thermal storage.

In recent years, energy consumption has expanded dramatically because of global economic development. As a result, dependency on fossil fuels has increased, and owing to the extensive use of fossil fuels, greenhouse gas (GHG) emissions—carbon dioxide (CO_2) emissions in particular—have increased tremendously in the atmosphere and drove scientists to find solutions. Research on renewable energy resources provides a great solution for alternatives to fossil fuels.

Ismail et al. [4] presented many solutions to this problem, by making use of renewable energy sources and biofuel. But, some renewable energies (e.g. solar and wind energy) are only available during particular times. One of the solutions to this problem is to store energy in some devices, e.g., thermal energy storage devices. Fallahi et al. [2] discussed the enhancement of efficiency by conservation of energy, thus reducing the gap between supply and demand. Fayegh et al. [5] review

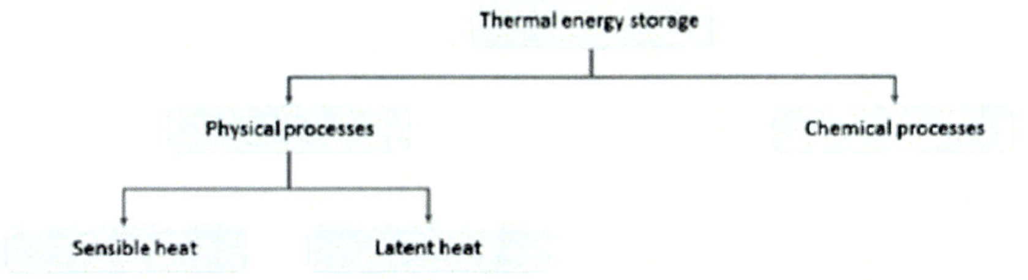

Fig. 1 Methods for thermal energy storage

different methods for storing heat energy. Figure 1 shows the representation of the classification of thermal energy storage systems based on physical and chemical processes. Gabriela [6] reviews on storage of thermal energy as latent heat as a captivating technique since it can supply energy having a high density which is stored over a relatively constant temperature range corresponding to the phase changing temperature of the storage material. Jane Diriken et al. [3] provide a review of shell-and-tube heat exchangers under specific boundary conditions during charging and discharging processes (HTF mass flow rates, HTF inlet temperatures, heat transfer enhancement techniques and type of PCM). During the charging of storage systems with single PCMs, heat from the heat transfer fluid (HTF) is transferred to the PCM very quickly. As a result, the temperature of the HTF decreases, resulting in a decrease in the temperature difference between the HTF and the PCM, resulting in a reduction in heat transfer at the storage end. As a result, the melting of PCM occurs quickly in those parts of the storage where HTF enters, but slowly at the storage end where HTF exits. The same issues were encountered during the discharging process: the PCM at the storage end which may not be used to raise the HTF's temperature.

2 Methodology

2.1 Proposed Design

Our storage system is a small-scale model proposed for thermal energy storage systems that are used in CSP plants. In practical CSP plants, the heat transfer fluid (HTF) is heated due to sunlight. This heated HTF is allowed to flow in the thermal energy storage system in order to store energy by means of heat transfer via convection.

The outer body of the heat storage system is made of a steel shell of diameter 20 cm and of length 50 cm and is separated into three vertical chambers with the help of steel separators. These chambers are filled with three different PCMs (Beeswax, Puretemp48 and Puretemp37). The chambers consist of four copper tubes with an

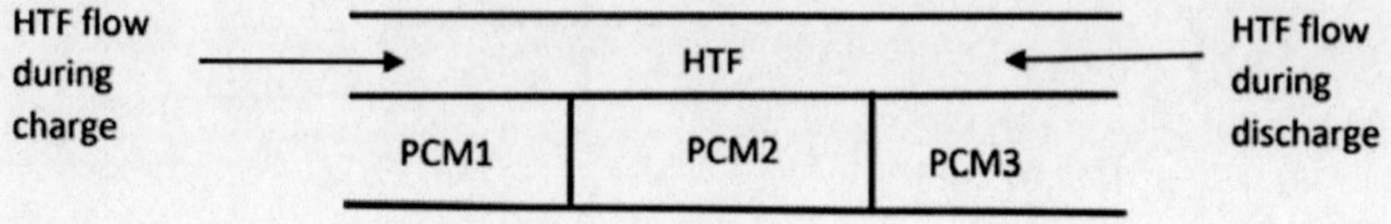

Fig. 2 M.P. PCM1 > PCM2 > PCM

Table 1 Thermo-physical properties of PCMs

Properties	Beeswax (PCM1)	Puretemp48 (PCM2)	Puretemp37 (PCM3)
Melting point (°C)	62	48	37
Pure solvent melting heat (J/g)	242	230	210
Thermal conductivity (W/m K)	0.25	0.15 (L) 0.25 (S)	0.15 (L) 0.25 (S)
Density (kg/m^3)	789.47	0.82 (L) 0.90 (S)	0.84 (L) 0.92 (S)
Specific heat (J/g K)	3.4	2.27 (L) 2.21 (S)	2.63 (L) S)

inner diameter of 2.5 cm. Inside these tubes, HTF flows at an inlet temperature of 90 °C (363 K). The ambient temperature is 27 °C (298 K) (Fig. 2; Table 1).

2.2 *Modelling*

According to Boda et al. [1], the overall heat transfer coefficient (U) range for heat transfer between light organics and water is 370–750 W/m^2 K and for medium organics and water is 240–650 W/m^2 K. The effectiveness of the system has been assumed as 0.9 and the mass flow rate has been taken as 0.1 kg/s during charging.

$$\varepsilon = 1 - e^{-NTU} \tag{1}$$

Thus, $NTU = 2.305$

$$NTU \cdot \dot{m}_{HTF} \cdot c_p = U.A_{tube} \cdot n \tag{2}$$

This gives $U = 490.71$ W/m^2 K which is within the permissible limit stated. The height of each shell has been calculated using density–volume relation.

$$\rho_{PCM} = \frac{m_{PCM}}{V_{PCM}} \quad (3)$$

$$V_{PCM} = \pi \cdot R^2{}_{tank} \cdot H_{PCM} \quad (4)$$

$$H_{PCM1} + H_{PCM2} + H_{PCM3} + H_{Seperator} \approx H_{Tank} \quad (5)$$

This gives $H_{PCM1} \approx H_{PCM2} \approx H_{PCM3} \approx$ 16cm where $PCM1$ is Beeswax, $PCM2$ is Puretemp48, and $PCM3$ is Puretemp37.

Relatively, the mass of each PCM is calculated using density–volume relation and is found to be 4 kg approximate for each PCM.

2.3 ANSYS Simulation

ANSYS 2020 R1 was used for the simulation and analysis of the storage system. Fluent, an ANSYS product, was used with the ANSYS Workbench for the simulation as the fluid was under motion inside the copper tubes. The simulation started by creating geometry, meshing, setting up in fluent and then furnishing the result. Details for these steps are as follows:

Geometry

The design modeler was used for creating the geometry of the model. Each of the chamber and tubes were filled with PCMs and fluid using proper commands such as Fill, Boolean, etc. (Fig. 3).

Mesh

Meshing was done with the help of ANSYS fluid flow (Fluent) in order to generate mesh at an appropriate size. Edge sizing was done on all the edges with hard biasing and the number of divisions was taken as 40. Later, named selection was done on the required parts. The number of nodes is 314,897 and the number of elements is 385,890 (Fig. 4).

Setup

General

1. The solver type is pressure and the velocity formulation is absolute.
2. Since the system has a variation with time, hence time has been chosen as transient.

Models

1. The energy equation was on.
2. Viscous was chosen as SST k-omega.
3. Solidification and melting were on for studying the phase change of PCMs.

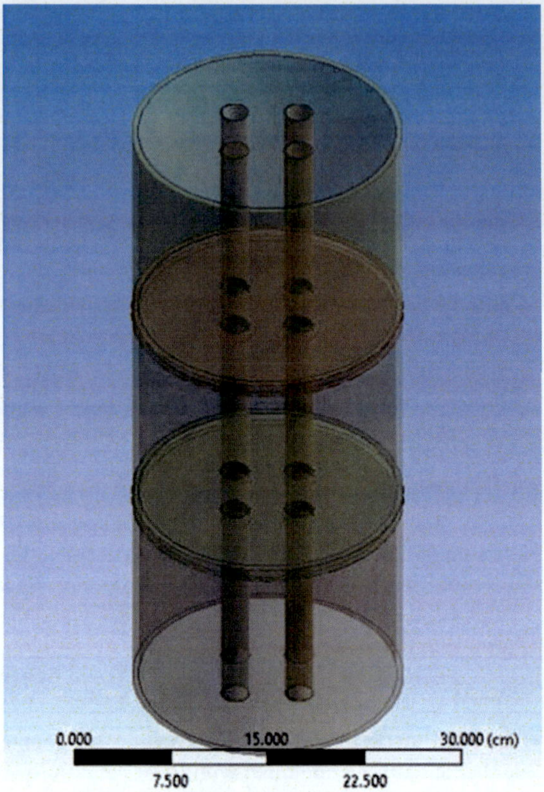

Fig. 3 Isometric view of the tank geometry on fluent design modeler

Materials

1. Beeswax, Puretemp48 and Puretemp37 were defined in the materials.
2. Their properties such as thermal conductivity, specific heat, density, pure solvent melting heat etc. were inputted as piecewise-linear at two different temperatures. Some properties which were constant were inputted as constant.
3. Steel and copper data were taken from a fluent database.

3 Results

3.1 Charging

Charging Result Discussion. From the above contours, one can easily observe that the HTF is losing heat to the surrounding causing the PCMs to melt and store energy. As time passes, more and more PCMs are getting melted and get charged completely

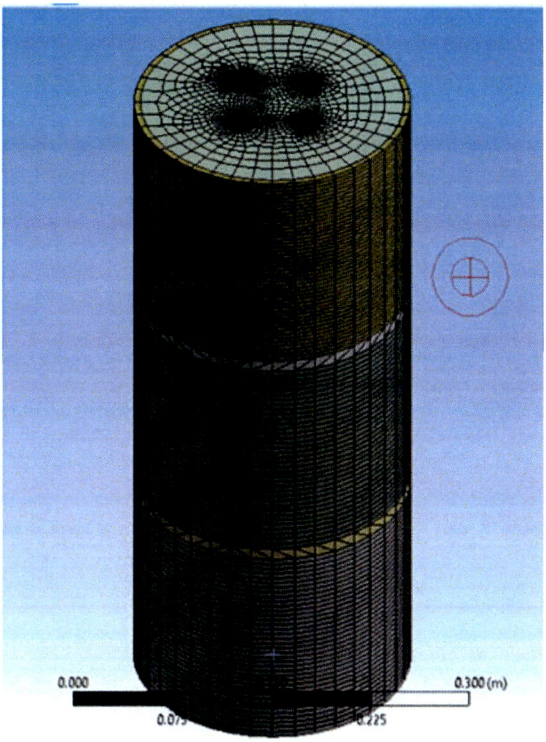

Fig. 4 Isometric view of the tank after meshing

after 20,267 s (5 h 37 min 47 s). The outlet temperature was getting dropped by 25 °C after 30 min, 16 °C after 1.5 h and only 4 °C after 3.5 h as most of the PCM was melted by now due to the rise in their temperature and hence it reduced the amount of heat absorption from HTF (Figs. 5, 6, 7, 8, 9, 10 and 11).

3.2 *Discharging*

Discharging Result Discussion

The same procedure was repeated as the charging, but here the inlet was provided from the bottom of the tank with a temperature of 27 °C (298 K). The simulation ran for 5 h 37 min, which was the charging time as well. During this duration of time, the Beeswax and Puretemp48 solidified by releasing heat energy which was absorbed by the HTF. The HTF temperature at the outlet was in the range of 39–44 °C which means water is getting heated by 12–17 °C. However, even after the solidification of these two PCMs, the third PCM Puretemp37 remains 70% melted. This is because the temperature difference between inlet water and this PCM was not

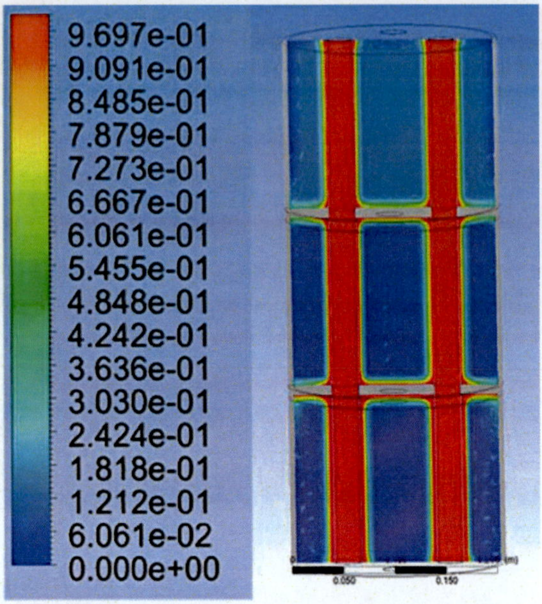

Fig. 5 Mass fraction contour after 30 min

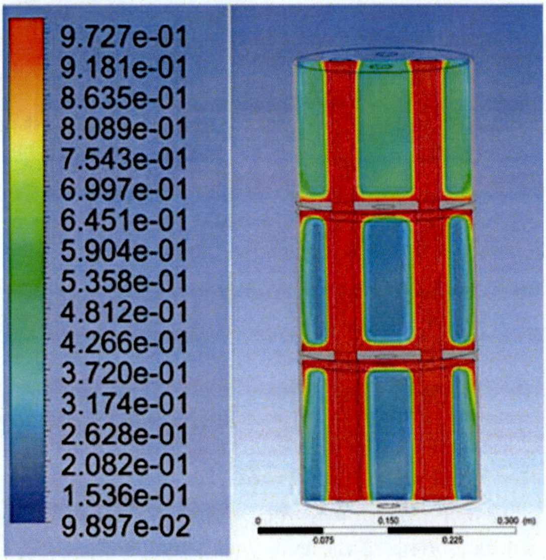

Fig. 6 Mass fraction contour after 90 min

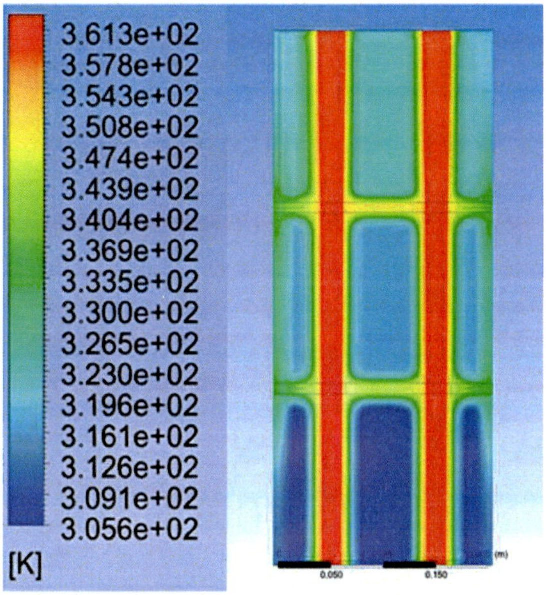

Fig. 7 Temperature contour after 90 min

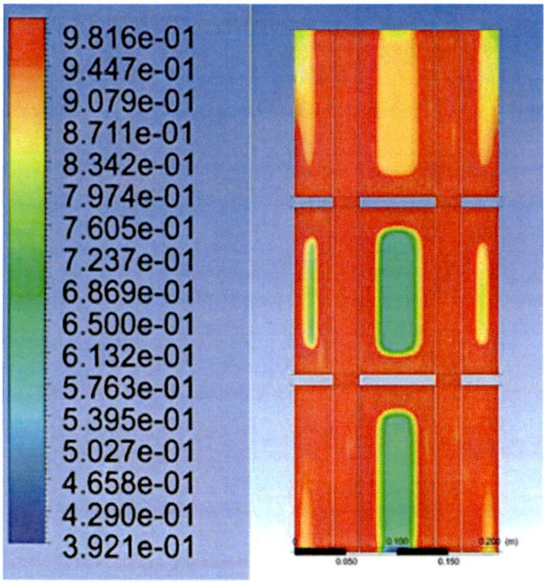

Fig. 8 Mass fraction contour after 3.5 h

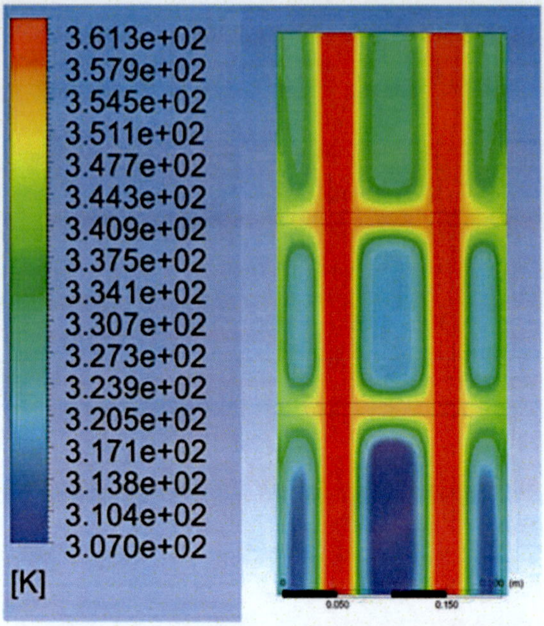

Fig. 9 Temperature contour after 3.5 h

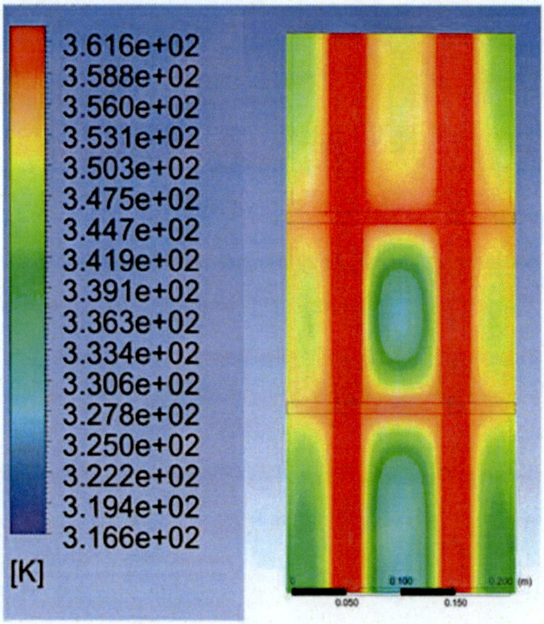

Fig. 10 Temperature contour after 5 h 37 min and 47 s (complete melting of PCM)

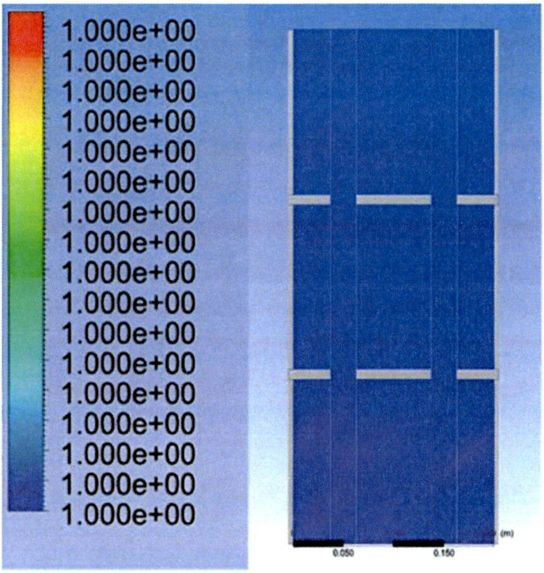

Fig. 11 Mass fraction contour after complete melting of PCM (5 h 37 min and 47 s)

very high for the transmission of heat energy effectively. Further simulation shows a rise in temperature of only 5 °C which is not very significant. Hence, any more simulation would not have provided worthy results (Figs. 12, 13, 14, 15, 16, 17, 18 and 19).

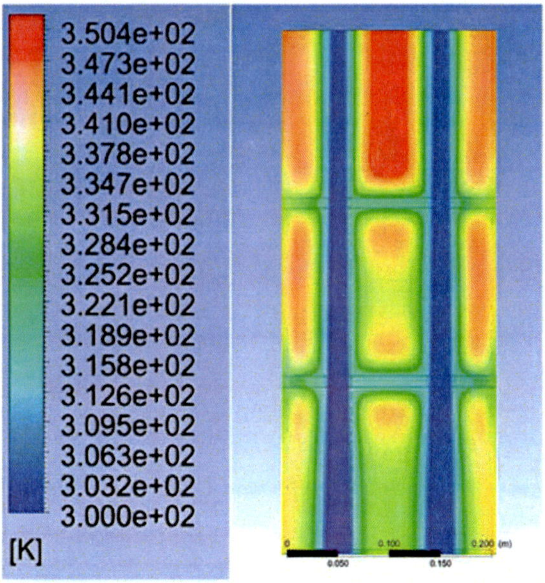

Fig. 12 Temperature contour after 30 min

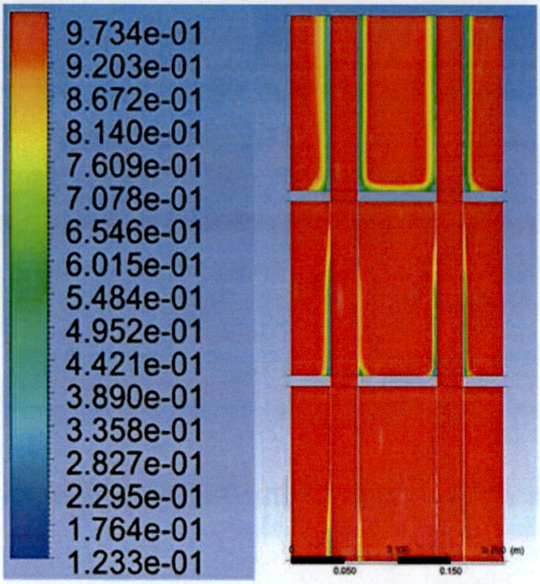

Fig. 13 Mass fraction contour after 30 min

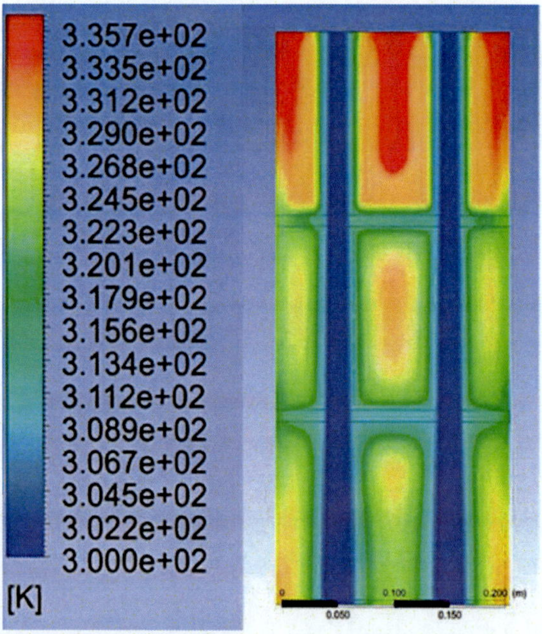

Fig. 14 Temperature contour after 1.5 h

Fig. 15 Mass fraction contour after 1.5 h

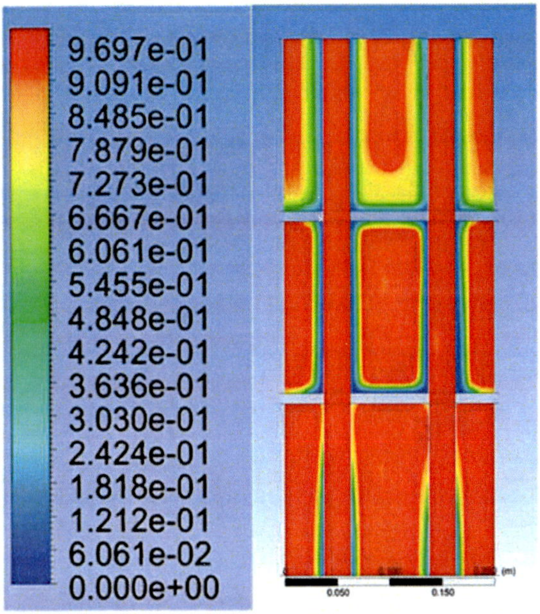

Fig. 16 Mass fraction contour after 3 h

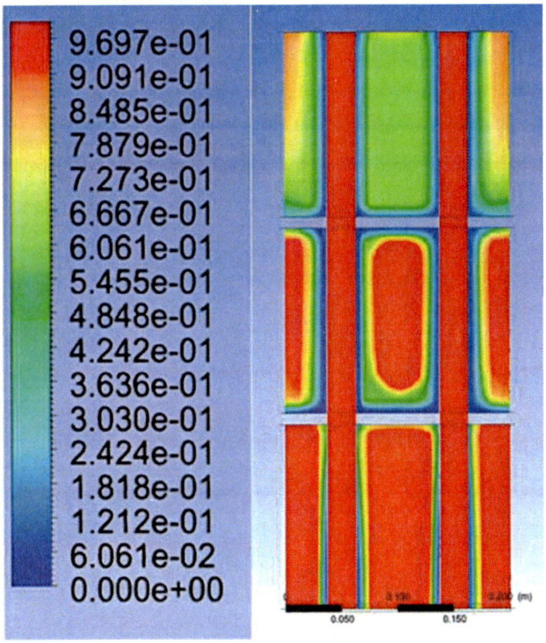

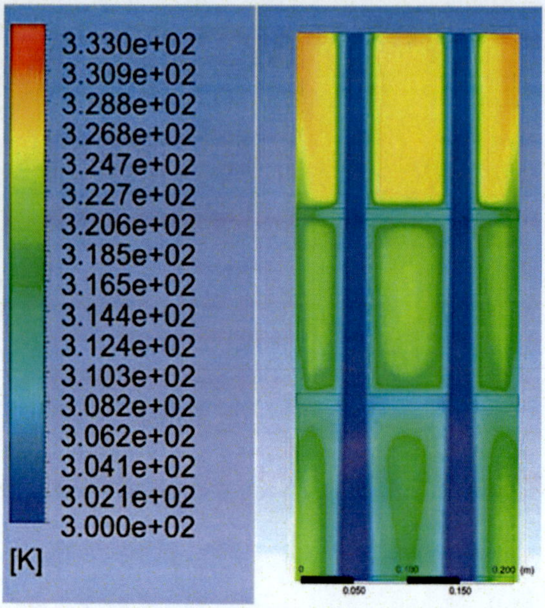

Fig. 17 Temperature contour after 3 h

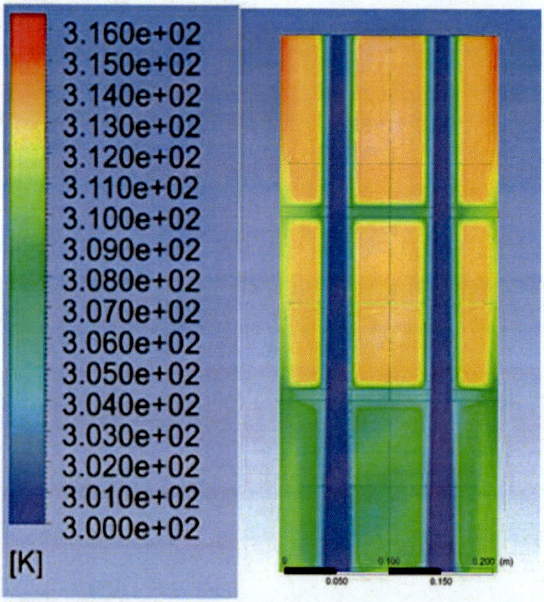

Fig. 18 Temperature contour after 5 h and 37 min

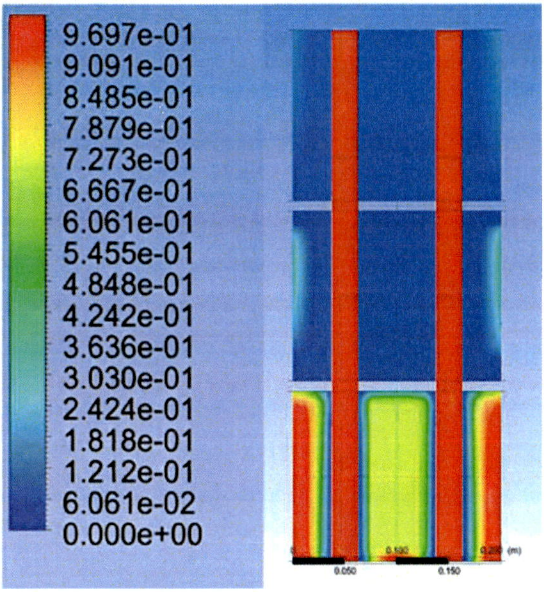

Fig. 19 Mass fraction contour after 5 h and 37 min

4 Conclusion

In this paper, CFD modelling of vertical axis cascaded thermal energy heat storage exchanger has been developed. The components, materials and type of storage system have been chosen based on a set of design criteria in order to achieve the maximum heat transfer rate possible, the longest storage duration and a cost-effective system. Also, constraints associated with PCM, HTF and heat exchanger materials have been studied and considered. Furthermore, simulation on ANSYS fluent was conducted and showed that the system stores thermal energy in three different PCMs in a reasonable time. After investigation of the entire process of charging and discharging, the mass fraction contour and temperature contour for each PCM at successive time intervals are observed and found that at the time of charging all three PCMs are completely melted and at the time of discharging Beeswax and Puretemp48 completely solidified. However, Puretemp37 was only 37% solidified and 70% was still liquid. It was clear from the observation that the energy was getting stored due to the melting of PCMs and can be reutilized by their solidification. Thus, it can be said that the utilization of such models with the CSP plant will lead to an increase in the amount of energy stored and reduce the energy demand for such operations.

References

1. Boda MA, Phand RV, Kotali AC (2017) Various applications of phase change materials: thermal

energy storing materials. Int J Emerg Res Manag Technol 6(4):167–171
2. Fallahi A, Guldentops G, Tao M, Granados-Focil S, Van Dessel S (2017) Review on solid-solid phase change materials for thermal energy storage: molecular structure and thermal properties. Appl Therm Eng 127:1427–1441
3. Gasia J, Diriken J, Bourke M, Van Bael J, Cabeza LF (2017) Comparative study of the thermal performance of four different shell-and-tube heat exchangers used as latent heat thermal energy storage systems. Renew Energy 114:934–944
4. Ismail MS, Moghavvemi M, Mahlia TMI (2014) Genetic algorithm based optimization on modeling and design of hybrid renewable energy systems. Energy Convers Manage 85:120–130
5. Koohi-Fayegh S, Rosen MA (2020) A review of energy storage types, applications and recent developments. J Energy Storage 27:101047
6. Socaciu LG (2012) Thermal energy storage with phase change material. Leonardo Electron J Pract Technol 20:75–98

Compressed Biogas Fuel Performance Enhancement Under Variable Compression Ratio Accompanied with Variable Ignition Location Spark Ignition Engine

Ashish J. Chaudhari, Vinay D. Patel, and Uday V. Aswalekar

Abstract Compression ratio of engines indicates the ability of that engine to harness the energy available in the fuel. Variation in compression ratio is a novel approach toward engine to work efficient at all different road conditions and provides an exceptional degree of control over engine performance. Conventional SI engine has a fixed compression ratio, a lower minimum value for starting the engine in cold start conditions or to avoid stalling of engine at full load condition. Conventional engines have maximum efficiency at full load condition and give exceptional self-ignition in cold start due to lower compression ratio, but while running on continuous fluctuating speed and torque requirement engine performances compromises between fluctuating requirement and losses its efficiency. In SI engines, automatic variable compression ratio improves cold start ability and low load operation, enabling the multi-fuel capability, increase of fuel economy and reduction of emissions. Automatic VCR differs from conventional VCR on the basis of changing the compression ratio according to requirement of speed and torque by using gear lever position as input signal to vary the compression ratio. The technique of variable spark location is added feature with variable compression ratio which enhance the charge combustion with maximum flame velocity post ignition. This automatic variable compression ratio with variable spark plug location under variable load and speed condition with compressed biogas fuel shows the substantial increase in efficiency with reduction of fuel consumption.

Keywords Compression ratio · SI engine · VCR engine · Automatic VCR engine · Compressed biogas

A. J. Chaudhari (✉) · V. D. Patel · U. V. Aswalekar
Faculty, Department of Mechanical Engineering, Vidyavardhini's College of Engineering and Technology, Vasai 401202, India
e-mail: ashish.chaudhari@vcet.edu.in

V. D. Patel
e-mail: vinay.patel@vcet.edu.in

U. V. Aswalekar
e-mail: uday.aswalekar@vcet.edu.in

D. P. Mishra et al. (eds.), *Recent Trends in Thermal and Fluid Sciences*, Lecture Notes in Mechanical Engineering, https://doi.org/10.1007/978-981-19-3498-8_13

1 Introduction

The world energy outlook reported the depletion of conventional energy sources by next 30 years [1]. The major consumers of conventional fuels are automobile industry, power plants, steel plant, forging plants are the most affected with this scenario. To cater this need, there are few viable sources of energy from renewables which could be explored to function those sectors. Many technologies were introduced and still under research for harnessing the maximum energy from renewable fuels. Some of the renewable source of energy from viewpoint of fuel sources are biodiesels, biogas, producer gas etc. Many authors were worked on the introduction of technologies such as variable valve timing, Direct Injection method, common rail direct injection, manifold injection, etc. using conventional fuel such as petrol or diesel. These techniques however may not as efficient as conventional fuels if fuel is from renewable category. Here it could be important to understand the contribution of fuel properties in the design of the engine in the tern combustion chamber. Two important factors in the design of the combustion chamber are compression ratio and spark plug location. The optimum values of those will depend on the various factors such as type of fuel, load on the engine, and speed as well The renewable fuel under consideration is the compressed biogas with higher percentage of Methane (CH_4). The fuel used herein is very low calorific value fuel and hence need more energy to cause combustion. Chaudhari et al. reported that the raw biogas fuel needs to optimize the engine CR at 10 with spark plug location at 2 mm inside the combustion chamber achieves best performance and lower emission [2]. Sahoo et al. emphasize the use of compression ratio technique in spark ignition engine and its effect using CNG fuel [3]. Yamin and Badran studied the effect of variation in spark plug location in cylinder head from center of cylinder toward wall radially on the performance of the engine using petrol fuel [4]. The author Srivastava and Agarwal mentioned about compression ratio of the prototype CNG engine and its effects on the engine performance, emission and combustion characteristics were investigated for laser ignition of CNG-air mixtures of varying mixture strength [5]. El-Adawy et al. describe the effect of air supercharging and different engine compression ratios on the performance of a biodiesel dual fuel engine. A direct injection naturally aspirated diesel engine was modified by fitting a supercharger and compression ratio control mechanism [6]. Cowans developed VCR charge mechanism which could be utilized efficiently in the engine [7]. Asthana et al. describe the performance of VCR engine with different mechanism [8]. The exhaustive literature work done by authors suggests that there is a scope to alter the engine structural specifications in order to accommodate alternative fuels efficiently. The most favorable component is the variable compression ratio. Very few authors reported the effect of VCR using compressed biogas fuel in spark ignition engine. Additionally, the spark plug location variation is a unique technique which was not found in literature.

2 Methodology

2.1 Engine with Auxiliary Piston Mechanism Using Petrol

The auxiliary mechanism consists of auxiliary piston and auxiliary cylinder as main components (Figs. 1 and 2). The auxiliary piston is hollow inside and provision is made to insert the spark plug concentric to the piston. It will have piston ring grooves to stop escaping of combustible gases from the chamber. Also an opening was kept on side wall of auxiliary cylinder for the wire connection of spark plug. This facilitates natural cooling of the mechanism as well due to direct exposure to the combustible gases inside combustion chamber. In order to automatic variation in compression ratio, one end of the piston is connected to ball screw (Fig. 3). With the help of ball screw precise and accurate motion will be achieved to the auxiliary

Fig. 1 Auxiliary piston accompanied with spark plug

Fig.2 Auxiliary cylinder with air vent for cooling accompanied with auxiliary piston

Fig. 3 Ball screw with nut assembly

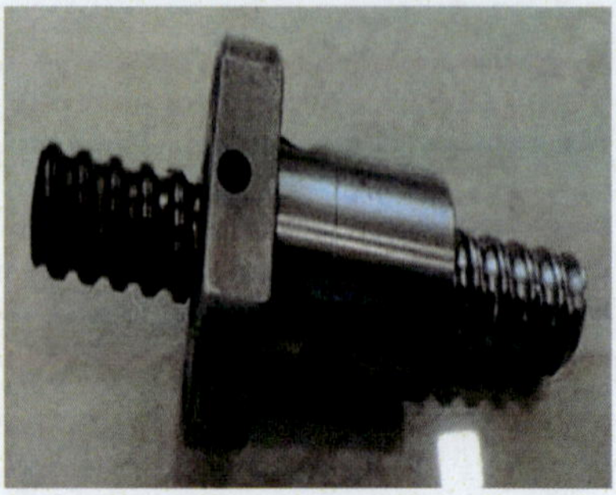

piston. Stepper motor is used to control the rotary motion of ball screw and achieve stepped regulation. IR sensors sensing the gear position acts as an input to Arduino and Stepper motor driver which controls the magnitude of angular rotation of ball screw. Ball screw is used for precise linear motion from rotational input. Using ball screw linear motion is provided to auxiliary piston due to which it precisely occupies space in combustion chamber.

The mechanism is installed on the cylinder head concentric to the main cylinder block of the engine. The auxiliary piston reciprocates inside and out of the main combustion chamber occupying the clearance volume of the engine. Thereby it could be able to change the compression ratio. New engine head will be mounted with same bolt and screw as previous. The additional feature of this mechanism is variable spark plug location (VSPL) in addition to alteration in compression ratio. This cylinder provides the path for sliding of auxiliary piston and also avoids ball screw shaft from buckling. Threads are provided on outer side in the bottom for clamping in cylinder head [9].

2.2 *Variable Compression Ratio Engine Set up Using Biogas Fuel*

As shown in Fig. 4, the high pressure compressed purified biogas stored at pressure 200–250 bar in cylinders at fueling station. High pressure Purified Biogas leaves the storage cylinders and flows toward engine unit where the vital conversion components are mounted and enters into one of them called the pressure reducer by means of Ni-stainless steel high pressure hose. The pressure reducer reduces its pressure from 200–250 bar to atmospheric pressure of 1.01325 bar. The properties of the fuels being used are tabulated in Table 1.

Then the low pressure Purified Biogas enters the engine by passing through the gas mixer where the Purified Biogas is mixed with air. The convergent-divergent section mixer for the CBG fuel supply system has been designed and developed. This mixer will be attached on the air intake manifold. The gas is supplied through the holes on the circumference of the mixer at interface of convergent-divergent section. While

Fig. 4 Experimental setup (Compressed purified Biogas)

Table 1 Comparisons of some physical properties of engine fuel [2]

Properties	Biogas	Petrol
Molecular formula	$CH_4 + CO_2$	C_8H_{12}
Density(gas) at 15 °C, kg/m^3	1.27(dry)	730
Lower heating value, MJ/kg	15.74	44.0
Heat of vaporization, MJ/kg	0.5	0.33
Stoichiometric A/F ratio	5.36	14.7
Research octane number	110	91

air is supplied through this mixer, at the inlet air manifold, low pressure could be reached and hence gas from the container could be sucked inside mixer. Appropriate Purified Biogas and air mixture enter the engine's combustion chamber where this mixture is combusted to develop power which is utilized to drive the vehicle.

The experimental procedure consists of starting the engine normally using 100% petrol fuel. Initially set speed on 2400 rpm at no load on the brake dynamometer. Once engine output stabilized, operates flow regulating valve on biogas cylinders allowed the flow of compressed biogas to the engine. The quantity of biogas supplied could be measured by flow measuring device. This will cause a rise in speed as fuel supplied is both petrol and biogas. To stop the petrol supply to engine slowly turns of the regulating valve on petrol tank. Once engine output gets stabilized, it will be running on 100% compressed biogas. Now set the required compression ratio by operating the auxiliary piston, using flat belt dynamometer apply load to the engine output shaft in increment of 0.5 kg in the range of 0–3 kg and record the observations. Same procedure is carried out for all compression ratios.

3 Result and Discussion

The experimentations are carried out on the engine set up with novel VCR mechanism for varying compression ratio and spark plug location using compressed biogas and petrol as fuel. Lot of studies were reported by different researchers in this area [10–12].

The experimentation using petrol fuel is as discussed in the following graphs shows the variation of brake power with break mean effective pressure.

In this Fig. 5, the brake power is found increasing with increase in BMEP for all the compression ratios under consideration. At no load condition, there could not be any variation in output power from CR4.8 to CR16.33. However at 3.21 bar BMEP, there is substantial rise by 16.86% has been recorded. However running engine at CR16.33 observes knocking during combustion. CR10.4 also demonstrate rise of 24.31% with respect to CR4.8. This is happened due to rise in cylinder pressure during compression stroke owing to reduction in combustion chamber volume. This causes the complete combustion of the fuel air charge.

The BSFC is a function of mass of fuel consumed by the engine which was found decreasing with increase in pressure as shown in Fig. 6.

At no load condition, the engine is cold and hence consumes more fuel to initiate the combustion. this is the reason, the SFC is much higher at no load for all CR. With increase in BMEP from 0.1 bar to 1.09 bar the slope of decrement is much larger which further reduced for higher pressure. This is owing to the rise in cylinder temperature, combustion gets faster and complete combustion occurs reducing the consumption of fuel.

The Brake thermal efficiency (BTE) variation with the BMEP is drawn in Fig. 7. As shown in figure, the BTE is found increasing with increasing in pressure for all CR under consideration. At original CR4.8, the BTE rises from 0.5 to 9.19% with respect to BMEP. At 3.21 bar the BTE was recorded to be 12.89% which is maximum at CR16.69 than that of CR4.8.

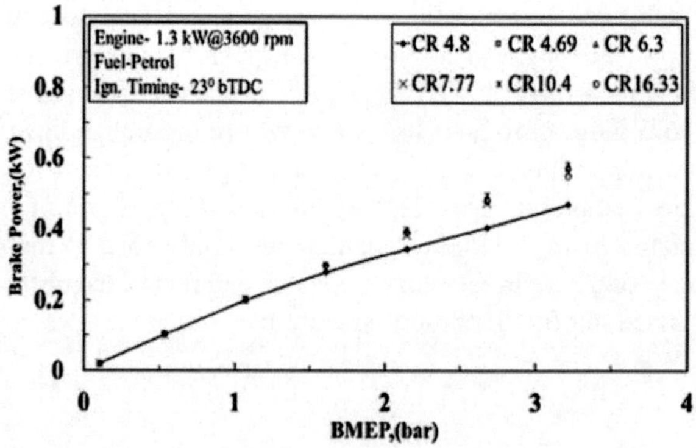

Fig.5 Variation of brake power with BMEP for variable CR using petrol fuel

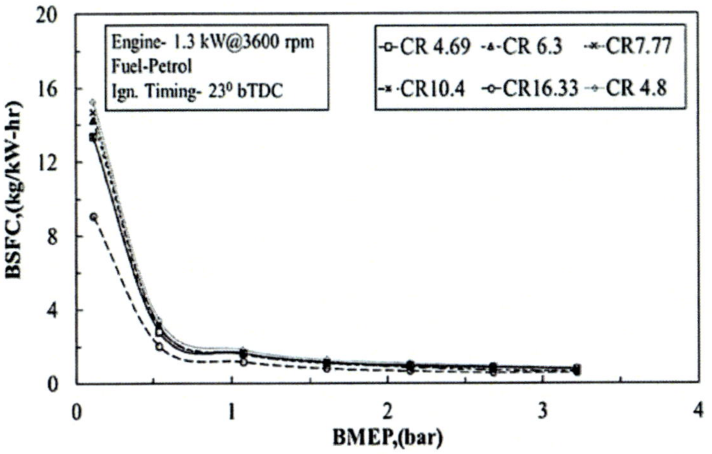

Fig.6 Variation of BSFC with BMEP for variable CR using petrol fuel

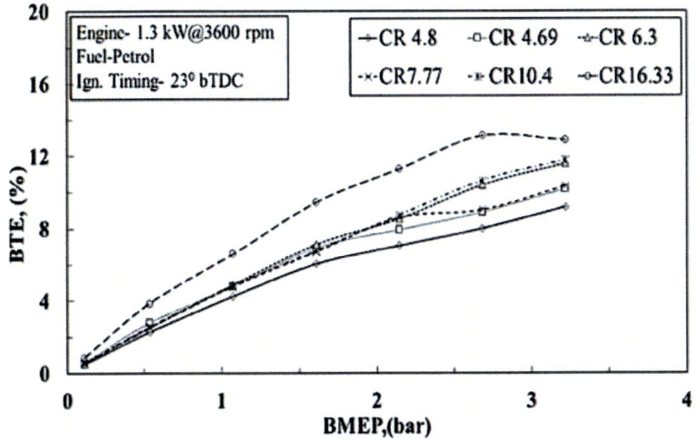

Fig.7 Variation of BTE with BMEP for variable CR using petrol fuel

The brake power variation with BMEP for compressed biogas fuel is as shown in Fig. 8. It is found that the BP increases with increase in BMEP for all compression ratios under consideration.

At CR10.48, the rise in BP is found to be 96.34% with rise in BMEP. CR16.33 shows lower power output 0.57 kW compared to 0.58 kW at CR10.48. At BMEP 3.21 bar, there could be 17% rise in brake power was reported [13–15].

The BSFC variation with BMEP for compressed biogas is shown in Fig. 9 where the trend follows same for all CRs. Initially BSFC found reduced with greater slope from no load condition. The decrement of 8%, 11%, 10%, 8%, 13%, and 11%, respectively, at CR4.8, CR4.69, CR6.3, CR7.77, C R 10.4, and CR16.33. With the

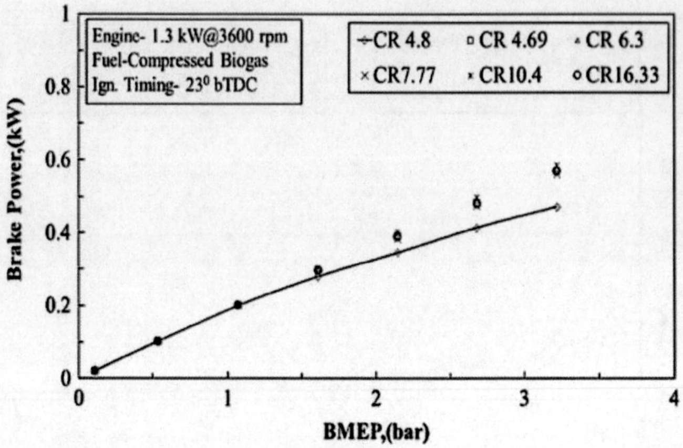

Fig. 8 Variation of BP with BMEP for variable CR using CBG fuel

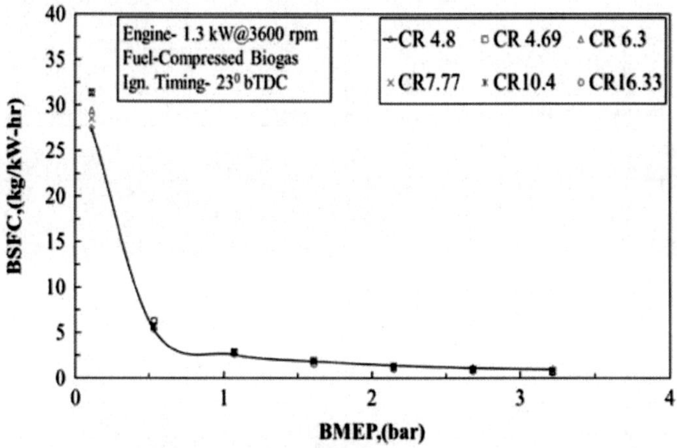

Fig. 9 Variation of BSFC with BMEP for variable CR using CBG fuel

increase in CR there is reduction in fuel consumption ultimately lowers BSFC. From CR4.8 to CR16.33, overall reduction of 66% is recorded using CBG fuel.

The performance of CBG fuel in variable compression ratio spark ignition engine in terms of brake thermal efficiency (BTE) is as shown in Fig. 10. The rate of increase of BTE with BMEP found increment with constant slope for all CR. The increment of 96–98% has been recorded for CR from no load to maximum load condition. For original CR, at maximum load the BTE was 7.79% which was 12.99% at CR16.33 showing increment of 40%. This may be for the reason of increase in compression pressure with reduction in CR causing higher cylinder temperature. This intern reduces the combustion duration achieving faster combustion. There was

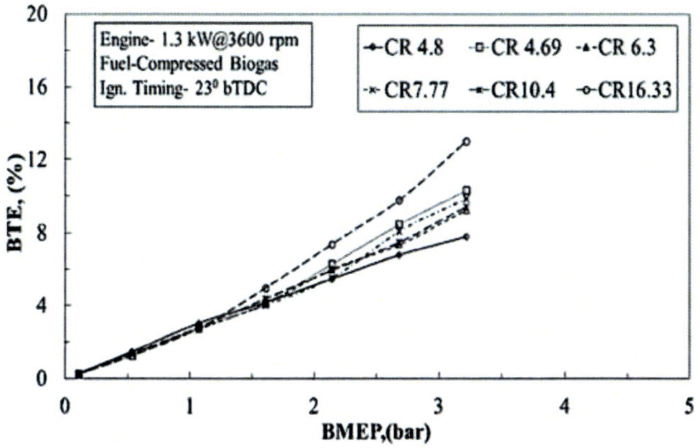

Fig. 10 Variation of BTE with BMEP for variable CR using CBG fuel

an increment of 32%,19%, 26%, and 20%.was achieved with increment in CR from 4.69, 6.3, 7.77, and 10.4, respectively.

The comparative nature of BSFC and BTE for both the fuels under consideration were analyzed in Fig. 11. As shown in the figure, the BSFC for petrol fuel is lower with respect to that of CBG for all BMEP from no load to full load condition. At zero load, the BSFC was 31 kg/kW hr for CBG which was 13 kg/kW hr for petrol fuel.

The higher BSFC represent more fuel consumption of CBG. This may be for the reason of lower calorific value of the CBG as compared to petrol fuel. However with increase in BMEP the BSFC was in close agreement for both the fuels. At 3.21 bar, the

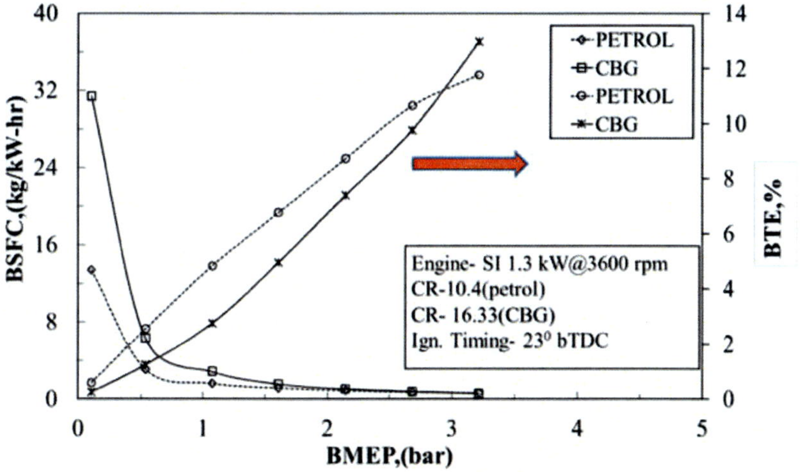

Fig. 11 Variation of BSFC and BTE with BMEP for optimum CR using petrol and CBG fuel

BSFC recorded was 0.60 and 0.66 kg/kW hr for CBG and petrol fuel. This could be happened due to the higher compression pressure increases the combustion chamber temperature and hence increased the combustion rate.

The BTE variation shows the trend of increasing efficiency with increase in BMEP for both the fuels CBG and petrol. At lower load BTE found to be 3% and 5% for CBG and petrol fuel. The optimum CR for petrol fuel is 10.4 while it is 16.33 for CBG fuel. The performance of the engine with petrol fuel shows higher efficiency by 12% at maximum BMEP or maximum load which was 13% with CBG at same maximum BMEP. This could be happened due to higher compression ratio for CBG fuel whereas for petrol CR is lower. It was seen from BTE curve of petrol fuel (Fig. 7), the efficiency is higher for CR16.33, but the engine combustion was subjected to knock and uneven combustion. Owing to this optimum CR for petrol fuel is set at 10.4. CBG is higher octane fuel and hence it will perform better with higher CR and happened here as well.

4 Conclusion

The experimental study of the variable compression ratio with the renewable fuel such as compressed biogas was carried out successfully. In addition to this the effect of variable spark plug location could be observed for those fuels. The engine under consideration could be found working more efficiently with the gaseous fuel. The important outcomes of the experimental study are as follows:

1. The petrol fuel performs better if the compression ratio is increased above original compression ratio. The optimum CR is found to be 10.4. However CR16.3 also attains higher efficiency but subjected to unavoidable knock causing uneven combustion.
2. The optimum spark plug location with respect to CR10.4 is 7.16 mm inside combustion chamber. It may attain the exact location of combustion center, and the flame initiates close to combustion center and develops faster across opposite end of combustion chamber.
3. The compressed biogas fuel is another renewable fuel tested in the engine. The fuel supply system needs major modification as compared to basic petrol fuel. The engine needs pilot fuel as petrol and then it could be switch over to 100% CBG fuel.
4. The performance of the engine using higher octane CBG fuel is found to improve with an increase in CR from original 4.8 to CR16.3. Even engine is found running smoot with higher CR16.3. The efficiency is found increasing continuously with increase in CR. The optimum CR for the CBG fuel is 16.33 with HONDA GK 100 engine.
5. As the original engine is designed considering petrol as fuel, the compression ratio and spark plug location were fixed for optimum performance. However, if the compression ratio is increased as per load on engine, it must alter the spark

plug location. Ahead of this, even the working fuel is other than petrol such as CBG, it becomes compulsory to rethink about optimizing the compression ratio and spark plug location in SI engine. The optimum sparkplug location in CBG is 9.1 mm protruded inside the combustion chamber. This causes the faster combustion of fuel air charge and hence attain maximum efficiency.

6. The novel mechanism designed in the study makes the engine versatile to accommodate any octane fuel to achieve maximum efficiency. This mechanism could be operated by driver in vehicle simply from the dashboard achieving the optimum compression ratio as well as spark plug location for the particular fuel available.

Acknowledgements The authors are thankful to the University of Mumbai for research funding to this project.

Appendix A

The more precise method of estimating uncertainty in experimental results has been presented by Kline and McClintock as well as Moffat by sequential perturbation [16, 17]. Accordingly, the uncertainty in the result has been evaluated as shown in Table 2.

Table 2 The uncertainty associated with computed parameters for HONDA Engine and measuring instruments

Computed parameters	Uncertainty (%)
Brake power	±17.90
Brake thermal efficiency	±11.25
Brake specific fuel consumption	±11.23

References

1. International Energy Agency (2018) World energy outlook report
2. Chaudhari A, Kulkarni V, Sahoo N (2017) Effect of variable compression ratio using an innovative mechanism on multi fuelled spark ignition engine. Sadhana Acad Proceed Eng Sci 43:211
3. Sahoo S, Srivastava DK (2021) Effect of compression ratio on engine knock, performance, combustion and emission characteristics of a bi-fuel CNG engine. Energy 233(15):121144
4. Yamin JA, Badran OO (2002) Analytical study to minimise the heat losses from a propane powered 4-stroke spark ignition engine. Renew Energy 27:463–478
5. Srivastava D, Agarwal A (2013) Laser ignition of single cylinder engine and effects of ignition location. SAE Technical Paper, paper no. 2013-01-1631. https://doi.org/10.4271/2013-01-1631

6. El-Adawy M, El-Kasaby M, Eldrainy YA (2018) Performance characteristics of a supercharged variable compression ratio diesel engine fueled by biodiesel blends. Alexandria Eng J 57(4):3473–3482
7. Cowans KW (2002) High efficiency engine with variable compression ratio and charge. US Patent No-0104492 A1
8. Asthana S, Bansal S, Jaggi S, Kumar N (2016) A comparative study of recent advancements in the field of variable compression ratio engine technology. SAE Tech Pap 2016-01-0669
9. Heywood JB (1988) Internal combustion engine fundamentals. McGraw Hill international editions, New York
10. Sridhar G, Paul PJ, Mukunda HS (2001) Biomass derived producer gas as a reciprocating engine fuel—an experimental analysis. Biomass Bioenergy 21:61–72
11. Choi GH, Han SB, Chung YJ (2003) The effect of hydrogen enrichment on exhaust emissions and thermal efficiency in a LPG fuelled engine. KSME Int J 17(8):1196–1202
12. Gumus M (2011) Effects of volumetric efficiency on the performance and emissions characteristics of a dual fuelled gasoline and LPG spark ignition engine. Fuel Process Technol 92:1862–1867
13. Porpatham E, Ramesh A, Nagalingam B (2007) Effect of hydrogen addition on the performance of a biogas fuelled spark ignition engine. Int J Hydrogen Energy 32:2057–2065
14. Park C, Park S, Lee Y, Kim C, Moriyoshi LS, Y., (2011) Performance and emission characteristics of a SI engine fueled by low calorific biogas blended with hydrogen. Int J Hydrogen Energy 36:10080–10088
15. Hu E, Huang Z, Liu B, Zheng J, Gu X, Huang B (2009) Experimental investigation on performance and emission of a spark-ignition engine fueled with natural gas-hydrogen blends combined with EGR. Int J Hydrogen Energy 34:528–539
16. Kline SJ, McClintock FA (1953) Describing uncertainties in single-sample experiments. Mech Eng 75:3–12
17. Moffat RJ (1982) Contributions to the theory of single-sample uncertainty analysis. ASME J Fluids Eng 104(2):250–260

Numerical Simulation of a Lid-Driven Cavity Problem by Solving General Pressure Equation Using OPENFOAM: A Solver Validation

Vivek Kumar and Abhishek Kundu

Abstract Using general pressure equations, a CFD method was employed to simulate the unsteady viscous incompressible flow. OpenFOAM-v2012 solvers have gotten a lot of attention from academicians, researchers, and professionals in recent years. Access to the code in its entirety, quick solver development, customization, and a large and open user community are only a few of the software's main features. The lid-driven cavity is a basic problem in which the algorithm can be successfully tested. In this case, the lid-driven cavity problem was solved using the OpenFOAM-v2012 platform by solving the general pressure equation. One of the configurations is the square cavity. The results of Ghia et al. (J Comput Phys 48:387–411, 1982) are compared to the patterns of constant flow produced by OpenFOAM-v2012 for various *Re* ranging from 100 to 3200. For various *Re*, the square cavity's isothermal flow is explored in detail. Stream function and vorticity discretized equations are computed using an implicit approach. "Streamline pattern and vorticity contours at different Reynolds number $Re = 100$, 400, 1000, and 3200 are analyzed. The lid-driven cavity problem was solved using the CFD tool OpenFOAM-v2012 and a high-resolution scheme, that is, second order centered schemes in space and a third order Runge–Kutta scheme in time. The computational results show that as *Re* increases from 100 to 3200, secondary vortices begin to develop and grow larger in size". For lower Reynolds numbers, the primary vortex's center is toward the corner on the top right. With increasing *Re*, it travels closer to the geometric center of the cavity.

Keywords Square cavity · OpenFOAM-v2012 · Computational fluid dynamics · General pressure equation

V. Kumar (✉) · A. Kundu
Applied Mechanics Department, MNNIT Allahabad, Prayagraj 211004, UP, India
e-mail: vivekmay2013@gmail.com

A. Kundu
e-mail: abhishekkunduamd@mnnit.ac.in

D. P. Mishra et al. (eds.), *Recent Trends in Thermal and Fluid Sciences*, Lecture Notes in Mechanical Engineering, https://doi.org/10.1007/978-981-19-3498-8_14

1 Introduction

The understanding of fluid pressure is a critical subject in fluid dynamics. Another method for studying incompressible fluid flow has recently been proposed. It is based on the use of a "general pressure equation" (GPE) (1) obtained from compressible "Navier–Stokes's equation" (NSE). "The Incompressible Navier–Stokes (INS) equations are a combination of hyperbolic and elliptic partial differential equations that are incompressible [1]. The incompressible flow problem is solved using a traditional finite volume method with a Pressure Poisson equation" (PPE) [2].

$$\frac{\partial q}{\partial t} + \frac{\partial f}{\partial x} + \frac{\partial g}{\partial y} = \frac{1}{Re}\nabla^2 q \tag{1}$$

where q-vector of primitive variables, f- and g-flux functions are

$$\mathrm{q} = \begin{pmatrix} G \\ u \\ v \end{pmatrix}, \mathrm{f} = \begin{pmatrix} \frac{1u}{Ma^2} \\ u^2 + p \\ uv \end{pmatrix}, \mathrm{g} = \begin{pmatrix} \frac{1v}{Ma^2} \\ uv \\ v^2 + p \end{pmatrix}$$

$$\mathrm{p} = \mathrm{G} + \frac{1}{2}(u^2 + v^2)$$

where u- and v-fluid velocity components are in the direction of x and y, respectively, and p—pressure, Ma—Mach number, and G—Grand potential.

Actually, the "incompressible flow problem can also be solved by GPE [2]. The people who used to solve the incompressible flow problem used to do it with the help of Pressure Poisson Equation [2]. The pressure Poisson equation takes a longer time to solve the incompressible flow problem, that's why people shifted from Pressure Poisson Equation (PPE) to General Pressure Equation (GPE) [2, 3]. One can also use a high-resolution scheme [4] to solve the convective part of the general pressure equation".

Lid-driven cavities, for example, have been utilized as a sample problem for repeatedly analyzing and improving numerical approaches on repeated occasions. "Lid-driven cavity flows are used in a wide range of industrial processing applications, including short-dwell and flexible blade coaters". Cavity flows include a wide flow-rate range characteristics, from pure rotating at the recirculating region's center to vigorous expansion near the lid's edges. For a wide range of Reynolds values, the 2-D lid-driven cavity flow has been numerically simulated. In many cases, accurate benchmark findings are supplied for both constant and periodic solutions around the essential *Re*.

The main aim of this project is to use the CFD tool OpenFOAM-v2012 to test the lid-driven cavity problem using the "general pressure equation". For Reynolds numbers between 100 and 3200, this session investigates the isothermal, incompressible, laminar flow in a square cavity, the top wall (lid) of which travels at a steady speed. Ghia et al. [5] have published results for this problem using a number of solution strategies. The topic of this research is a lid-driven cavity with a single geometric configuration.

2 Numerical Methodology

Figure 1 depicts the one geometric arrangement [6] that was examined for simulation in this study. All of the square's boundaries are walls in a square cavity. The top wall moves at a constant speed in the x-direction, The other three, on the other hand, remain unmoving. The flow is considered to be laminar, and the icoFoam solver for the isothermal, laminar, incompressible flow is used to solve the problem on a uniform mesh.

The entire procedure is divided into three stages:

(1) Pre-processing
(2) Processing
(3) Post-processing.

2.1 Pre-processing

In the main directory of OpenFOAM-v2012, a directory is created and named square cavity. Three subdirectories are required: initialize, constant, and system. The initial conditions for fluid characteristics are specified in initialize. There are two subdirectories in the constant directory: "PolyMesh and transport properties". "BlockMesh and boundary case files" are included in PolyMesh. Now, in OpenFOAM-v2012, the geometry of a square lid-driven cavity must be created and stored in a "BlockMesh file". OpenFOAM-v2012 allows you to edit your files.

All geometries in OpenFOAM-v2012 are created in three dimensions and operate in a 3-D Cartesian coordinate system. By default, OpenFOAM-v2012 solves the problem in 3-D, but it can be programmed to solve it in 2-D by giving a "special

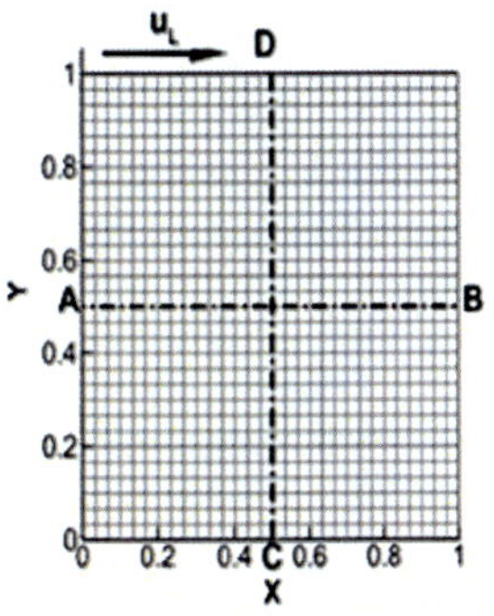

Fig. 1 Geometry of square cavity

empty boundary condition on boundaries normal to the third dimension for which no solution is required". Instead of a single case file, this data is kept in a group of files, a case file in OpenFOAM-v2012. Simulated data includes mesh, fields, attributes, control settings, and so on.

2.2 *Processing*

OpenFOAM-v2012 programs can be executed in one of two different ways: as a foreground operation, in which the shell waits for the command to finish before prompting the user, or as a background process, in which the shell accepts future commands without waiting for the command to finish. It is necessary to run icoFoam in the foreground in this case. To launch the icoFoam solver, navigate to the case directory and execute icoFoam in the terminal window. It shows the actual time, highest Courant number, starting and concluding all the fields of residuals in the terminal panel. It should simultaneously compute the vorticity data and stream data at each time step.

2.3 *Post-processing*

ParaView 5.9.0 can be used to view findings quickly; they are recorded to time directories. Return to the ParaView 5.9.0 window and pick the cavity in the case module for OpenFOAM-v2012 Properties panel. In ParaView 5.9.0, you may observe pressure, velocity, stream function, and vorticity contours. In ParaView 5.9.0, the meshing of a square cavity is shown in Fig. 2.

3 Results and Discussion

Based on the outcomes, provided in Ghia et al. [5], a detailed contrast was made. Reynolds numbers have increased from a low to a high level for the lid-driven cavity problem. Both qualitative and quantitative comparisons have been made. The plots of both streamline and vorticity are utilized for qualitative contrast, and the component of u-velocity is displayed at various distances from the cavity's base for quantitative comparison. Some of the photos are shown below in Figs. 3 and 4 for $Re = 100$ and Figs. 6 and 7 for $Re = 3200$. For $Re = 100, 400, 1000$, and 3200, OpenFOAM-v2012 created streamline and vorticity contours for a square cavity with a grid size of 129×129. The benchmark results of Ghia et al. [5] are compared to these results. OpenFOAM-v2012 outcomes are to the right, while Ghia et al. [5] are on the left. The results are similar with Ghia et al. [5].'s findings. The current solution is useful for solving general pressure equations rather than pressure Poisson equations, which

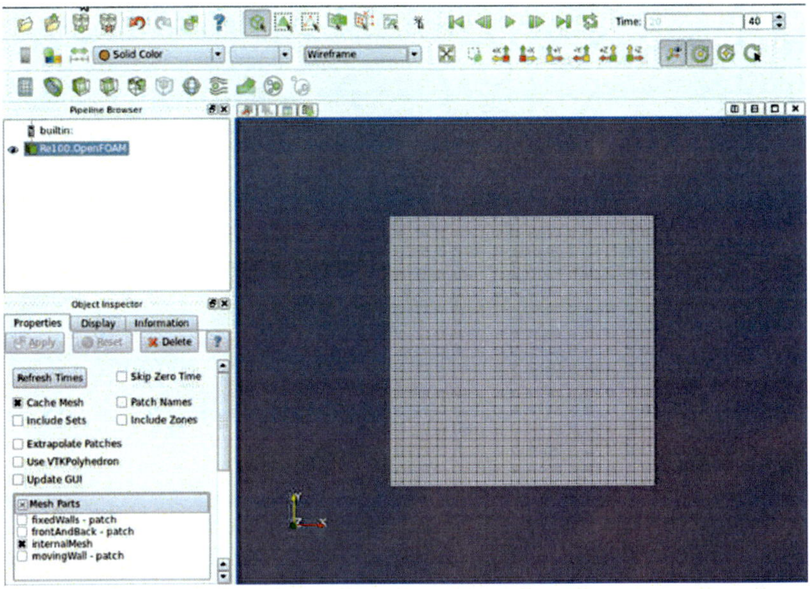

Fig. 2 Mesh creation in ParaView 5.9.0

Fig. 3 Comparison of $Re = 100$ streamline contours with Ghia et al. [5] results (129 × 129)

are easier to formulate and solve in a shorter amount of time. The computer results also demonstrate that as *Re* goes from 100 to 3200, secondary vortices tend to appear and grow in magnitude. For lower Reynolds numbers, the major vortex's center is toward the top right corner. With increasing *Re*, it moves closer to the geometric center of the cavity. Secondary vortices can be seen even in the upper left corner at high Reynolds numbers (see Fig. 6). For $Re = 3200$, the streamlines and vorticity contours are shown. Secondary vortices may be seen developing from the bottom

Fig. 4 Comparison of $Re = 100$ vorticity values with Ghia et al. [5] results (129 × 129)

of both corners, and the number of secondary vortices increases as the Reynolds number increases. If one uses a lower resolution approach [7] for convective fluxes, it is exceedingly at higher Reynolds numbers; it's harder to capture these micro-scale vortical features. At maximum Reynolds numbers, when capturing small-scale corner vortices is difficult, the requirement for using a high-resolution method plays a key role (Fig. 5).

The non-dimensional values of u are shown along vertical bars passing through the centroid of the cavity. A plot of vorticity contours, streamlines for solution domain and at various Reynolds numbers, the u-velocity along the cavity's Center line is obtained. For $Re = 100$, 400, 1000, and 3200, Fig. 9 depicts the fluctuation in u-velocity along a vertical bar going across the cavity's centroid. The maximum magnitude of the velocity component has been shown to grow as the Reynolds number

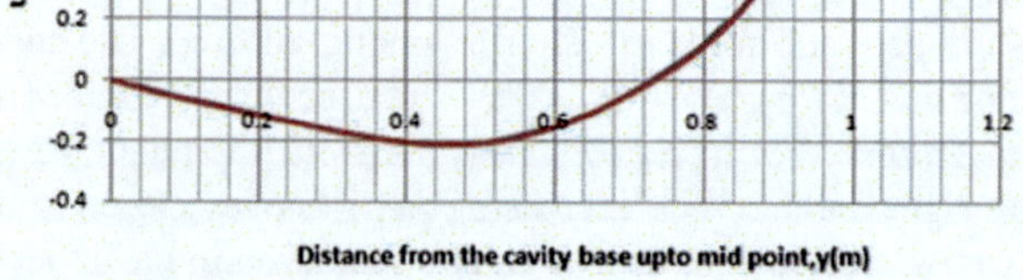

Fig. 5 Quantitative validations of $Re = 100$ using Ghia et al. [5] results (129 × 129)

Fig. 6 Ghia et al. [5] compared the results of streamline contour for $Re = 3200$ (129 × 129)

Fig. 7 Vorticity contour for $Re = 3200$ compared to Ghia et al. [5] results (129 × 129)

increases. It's also worth noting that the maximum value shifts toward the cavity's walls. These depict how, as Re grows, the boundary layers flatten out (Fig. 8).

4 Conclusions

- The OpenFOAM-v2012 platform was used to solve a "2-D incompressible viscous lid-driven cavity problem using the general pressure equation".
- This general pressure equation has the simplest form and state forward to use.
- It also has the potential to reduce the computational effort.
- Lid-driven cavity is the simplest problem to check the ability of the general pressure equation.

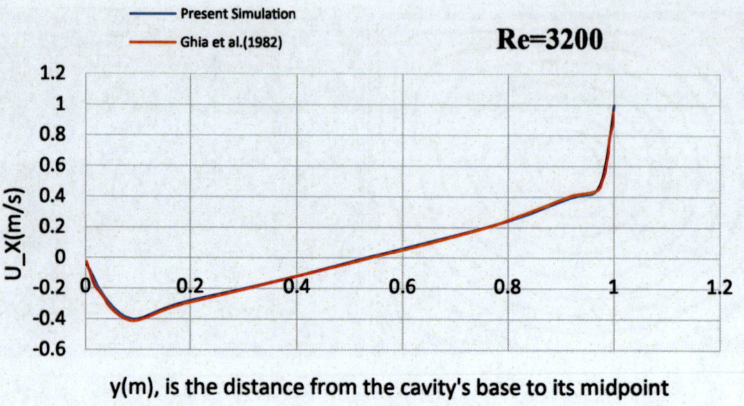

Fig. 8 At $Re = 3200$, quantitative validations based on Ghia et al. [5] findings (129 × 129)

Fig. 9 At different Reynolds Numbers (*Re*), u-velocity component comparison

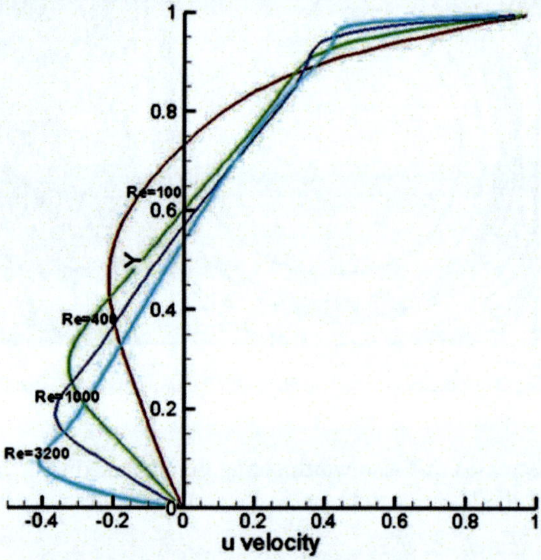

- The vorticity and streamline plot indicate the presence of the small-scale vortices due to the use of high-resolution scheme at a maximum Reynolds number.
- Increase in Reynolds number increases the magnitude of the velocity field.
- Future scope lies in the use of high-resolution solver using general pressure equation.

References

1. Landau LD, Lifshitz EM (1987) Fluid mechanics. Pergamon, Oxford
2. Toutant A (2018) Numerical simulations of unsteady viscous incompressible flows using general pressure equation. J Comput Phys 374:822–842
3. Hashimoto T, Tanno I, Yasuda T, Tanaka T, Morinishi K, Satofuka N (2019) Optimized finite difference method with artificial dissipation for under-resolved un steady incompressible flow computations using kinetically reduced local Navier-Stokes equations. Comput Fluids 184:21–28
4. Greenshields CJ (2010) OpenFOAM, OpenFOAM user guide
5. Ghia U, Ghia KN, Shin CT (1982) High-Re solutions for incompressible flow using the Navier-Stokes equations and a multi-grid method. J Comput Phys 48:387–411
6. Paramane SB, Sharma A (2008) Consistent implementation and comparison of FOU, CD, SOU, and QUICK convection schemes on square, skew, trapezoidal, and triangular lid-driven cavity flow. Numer Heat Transf Part B 54:84–102
7. Kundu A, De S (2017) Navier-Stokes simulation of shock-heavy bubble interaction: comparison of upwind and WENO schemes. Comput Fluids 157:131–145

Design Analysis of Solar Water Distillation System by Using Flat Plate Collector

Neelam Khandelwal, Jatin Singh, Ishita Singh, Manikant Singh, and Raj Kumar Kesharwani

Abstract Among all the sources of renewable energy, solar energy has the potential. There are different applications where solar energy can be used in place of conventional energy resources. The present work analyzes the performance of solar distilling plants where a flat plate collector is used as a solar energy source. Solar distilling plants are mainly used where the need for small plants exists and it is available at less cost. It is widely used to convert saline water into pure water for drinking purposes without using conventional sources of energy. Radiation, convection, and conduction are the basic modes of heat transfer. The parameters, i.e. Volume of Water Required per day, temperature of hot water required, temperature of Inlet water, Wind speed, and Solar Radiation have been used for calculating the efficiency of flat plate collector, and parameters, i.e. Solar Radiation and different condensing cover material have been used for the Performance Analysis of Solar Still alone and Solar Still with FPC at a different time of a day. It has been found that coupling a solar flat plate collector with a still has increased the volumetric efficiency by 32.78% theoretically. This project aims to produce around 6–7 L of distilled water from 10 L of saline water during 8 h using solar energy by the use of a flat plate collector.

Keywords Solar still · Flat plate collector · Efficiency of collector · Solar energy

1 Introduction

The present scenario in many parts of the world is that there is no availability of a suitable source of clean drinking water available in lakes, rivers, sea, etc. which contains a lot of impurities and carries diseases, and hence, it is not fit for drinking purposes. Solar energy is a renewable energy resource on earth, and with the help of

N. Khandelwal · J. Singh (✉) · I. Singh · M. Singh · R. K. Kesharwani
Mechanical Engineering Department, JSS Academy of Technical Education, Uttar Pradesh, C20/1 Sector 62, NoidaNoida, India
e-mail: Jatinsingh1801@gmail.com

N. Khandelwal
e-mail: Neelamkhandelwal@jssaten.ac.in

D. P. Mishra et al. (eds.), *Recent Trends in Thermal and Fluid Sciences*, Lecture Notes in Mechanical Engineering, https://doi.org/10.1007/978-981-19-3498-8_15

this energy resource, the process of converting saline water into pure water is known as solar distillation. In our project, we have used a flat plate collector to increase the heat input of solar still to get more condensed water vapor as a result. Solar distillation is a cost-saving phenomenon over other types of distillation. A flat plate collector is coupled to the simple solar still to increase its distilled water output, thereby producing pure water with renewable resources (i.e. solar energy). In order to enhance the performance of solar still, there have been many improved designs and modifications made by several researchers. As Valsaraj (2002) demonstrated experimentally, floating aluminum sheets are capable of increasing a still's output in comparison to a conventional still as a result of floating aluminum sheets. Tiwari et al. (2003) also included a condensing cover inclination of 30° of an active solar distillation. Tripathi and Tiwari (2004) studied the effect of changing the length of the north wall of a solar still for a given height on the daily output. However, the change in the height of the north wall did not affect the daily output for a given height. A symmetrical greenhouse with no mirrors and an asymmetrical greenhouse with mirrors, using ASGHT technology, were experimentally compared, and the result indicated that the ASGHT still produced a 25% better yield than the SGHT still by utilizing mirrors inside the wall. Having performed their experiment, [3] proved that not only saline water but also tap water yields an essentially higher yield. A simple still can always be made more productive by adding FPCs to it, as [3] proved. Tiwari (2006) indicated that the water depth is a parameter for evaporative heat transfer. Ayber (2006) to be mathematically modeled and stimulated has demonstrated an inclined solar water distillation system. The productivity of the system was also shown to be affected by solar intensity. In addition, Badran and Abu-Khader (2007) pointed out that the thickness and other effective parameters such as solar intensity can also affect the system's productivity. Sahoo et al. (2008) tested the efficiency of solar-powered stills with blackened surface and thermocol insulation in order to remove fluoride contaminants from drinking water. Using water circulation to the sidewalls of the still, Kumar and Bai (2008) improved the performance of a solar still by concentrating condensate on the condensation surface. Abdallah and Badran (2008) proved that there is an increase in productivity by around 22% by the use of sun tracking and the reason behind that is the increase of overall efficiency by 2%. Abdullah et al. (2009) tested solar stills using a variety of absorbing materials. The study of Sayur et al. (2011) showed that there is heat generation when solar rays pass into a still basin that is coated with black materials. These materials absorb more solar energy. Heat is absorbed sufficiently in the lamp, resulting in an increase in the temperature of the water, leading to evaporation and then condensing on the inclined cover glass and being collected in the collection chamber. Based on Tiwari's (2013) research, solar energy is most effectively utilized in those countries where sunlight is abundant. In his study, El-Maghlany (2015) indicated that an optimal glass cover tilt angle depends on the direction of each surface when considering double slope solar stills.

The purpose of this study is to develop a theoretical model. Although several still configurations have been numerically studied, there are no theoretical models that incorporate the flat plate collector, and designing high-efficient stills could be greatly enhanced with the development of such a model.

2 Methodology

The methodology of our project involves designing solar still as per the output requirements. Based on our application requirements, we estimated the output of a solar still at around 3 to 4 L. Moreover, we are using an external flat plate collector, which helps to preheat the input water resulting in increasing the output efficiency.

2.1 Design Calculations

2.1.1 Modeling of Flat Plate Collector

Flat plate collectors are the cheapest and easiest to fabricate among other non-concentrating solar collectors except when the absorber itself is used as a collector. These solar collectors are used to intercept and absorb solar radiation with a concentration ratio of 1 and a temperature range of 40–60 °C.

The present model consists of a good casing and used a very low thermal conductivity material in it that prevents the energy losses from the top, bottom, and sidewalls which is having a huge contribution in optimizing the useful heat required to elevate heat transfer fluids temperature up to the desired value which further helps to improve the efficiency of the flat plate collector.

The useful heat energy in W to increase the temperature of water from room temperature to desired output temperature can be calculated as

$$\mathrm{Q_u} = \mathrm{m} * \mathrm{C_p} * \Delta T \tag{1}$$

The mass flow rate of the water present in the collector tubes is calculated by

$$m = \rho V \tag{2}$$

Total incident heat energy in the month of June is given as

$$\mathrm{Q_i} = \mathrm{DNI} * \mathrm{A_c} * \eta_{\mathrm{coll}} \tag{3}$$

Total useful heat energy with the consideration of total heat losses is given by

$$\mathrm{Q_u} = \mathrm{A_c}\left[\mathrm{Q_s} - \mathrm{U_L}\left(\mathrm{T_{pm}} - \mathrm{T_a}\right)\right] \tag{4}$$

Top loss heat coefficient is calculated as

$$U_t = \left[\frac{M}{\left(\frac{c}{T_p}\right)\left(\frac{T_p - T_a}{M+f}\right)} + \frac{1}{hw} \right]^{-1} + \left[\frac{\sigma (T_p^2 - T_a^2)(T_p - T_a)}{\frac{1}{\varepsilon p + 0.005(1-\varepsilon p)} + \frac{(2M+f-1)}{\varepsilon_c} - M} \right] \quad (5)$$

Bottom loss coefficient is calculated as

$$U_b = \frac{k_i}{\delta_b} \quad (6)$$

Side loss coefficient is calculated as

$$U_s = \frac{(L_1 + L_2)hk_i}{(L_1 L_2 \delta s)} \quad (7)$$

Total heat loss-transfer coefficient is calculated as

$$\mathrm{U_L} = \mathrm{U_t} + \mathrm{U_b} + \mathrm{U_s} \quad (8)$$

Efficiency of the collector is calculated as (Tables 1, 2 and 3)

$$\eta = (\mathrm{Q_u}/\mathrm{Q_i}) * 100 \quad (9)$$

Table 1 Specifications of flat plate collector

Components	Size
Length of collector (mm)	1000
Area of collector (m^2)	0.8
Width of collector (mm)	800
Thickness of collector (cm)	150
Thickness of absorber plate (cm)	1
Copper tube diameter (cm)	1
Width of plate (mm)	720
Length of absorber plate (mm)	940
Copper tube thickness (cm)	0.1
Tube overall length (cm)	1440
HTF	Water
Insulation thickness at bottom (cm)	4
Insulation thickness at side (cm)	3
Base	Wood
Wood glass cover thickness (cm)	0.4
Glazing area (cm^2)	8000
Optimum tilt angle (degree)	30°

Table 2 Meteorological data of solar radiation of 20th June

SI. No	Time (in hours)	Solar radiation (W/m^2)
1	08:00	579
2	09:00	740
3	10:00	759
4	11:00	772
5	12:00	781
6	13:00	898
7	14:00	836
8	15:00	704
9	16:00	535

Table 3 Hourly variation of temperature of hot water, mass flow rate, and Efficiency of flat plate collector with respect to Solar Intensity data of 20th June

SI.No	Time (in hour)	Solar radiation (W/m^2)	Q_i (W)	Q_u (W)	Mass flow rate (kg/s)	Outlet temperature (°C)	Efficiency of FPC
1	08:00	579	440.04	124.03	0.00142	46.15	28.16
2	09:00	740	562.4	220.75	0.00181	54.54	39.25
3	10:00	759	576.84	232.30	0.00186	55.25	40.27
4	11:00	772	586.72	240.20	0.00189	55.78	40.94
5	12:00	781	593.56	245.67	0.0019	56.32	41.38
6	13:00	898	682.48	317.98	0.0022	60.01	46.59
7	14:00	836	635.36	279.12	0.0020	58.80	43.93
8	15:00	704	535.04	198.86	0.0017	53.33	37.16
9	16:00	535	406.6	96.11	0.00131	42.77	23.63

2.1.2 Modeling of Solar Still

Base Area

In solar stills, the base area is of utmost importance for design purposes, as it gives us the area necessary for the requisite amount of solar radiation to be incident on the still to yield the expected output. The amount of solar energy needed to produce 3–4 L of water can be calculated by using the following formula:

$$M_w * L_v = Q_{req} \tag{10}$$

Since the value of Latent heat of vaporization of water has to be taken as 2260 kJ/Kg.

Table 4 Specifications of solar still

Components	Size
Length of solar still (mm)	900
Breadth of solar still (mm)	400
Base area of solar still (m^2)	0.36
Height of solar still (excluded water stored in it) (mm)	231
Glass cover length (mm)	462
Total height of solar still (excluded plywood thickness) (mm)	270
Glass cover inclination (degrees)	30°

During June, we have analyzed averaged data of the amount of solar energy which is incident in Noida, for the purpose of calculating the amount of incident solar energy (Q_{inc}). We have assumed the period of 8 h (8:00 A.M. to 4:00 P.M) per day for the incident solar energy and we are considering commercial glass as a glass cover in our solar still, which transmits a minimum of 80% light rays (Table 4).

Area of Base required is given by

$$A = Q_{req}/Q_{inc} \quad (11)$$

External Flat Plate Collector

By using an External Flat Plate Collector, more sunlight is incident on the Solar still, resulting in an increased rate of evaporation.

A flat plate collector on the solar still can increase the amount of solar energy incident on it by an amount that can be calculated as follows:

$$Q = Q_{inc} * Ac * \eta \quad (12)$$

Therefore, the total incident solar energy in the solar still is Q1 is given by

$$Q1 = Q_{inc} + Q \quad (13)$$

Therefore, the amount of distilled water that can be obtained by using FPC can be calculated by (Tables 5 and 6)

$$Mw = Q1/L_v \quad (14)$$

Table 5 Hourly variation of productivity of solar still alone and productivity of solar still with FPC with respect to solar radiation data of 20th June

SI. No	Time (in hours)	Inlet water Still temperature (in degrees)	Solar radiation (W/m^2)	Productivity of solar still alone (in liters)	Productivity of solar still with FPC (in liters)
1	08:00	46.15	579	0.265	0.472
2	09:00	54.54	740	0.339	0.709
3	10:00	55.25	759	0.348	0.737
4	11:00	55.78	772	0.354	0.757
5	12:00	56.32	781	0.358	0.770
6	13:00	60.01	898	0.412	0.945
7	14:00	58.80	836	0.383	0.852
8	15:00	53.33	704	0.323	0.656
9	16:00	42.77	535	0.245	0.406

Table 6 Hourly variation of productivity of solar still by using different condensing cover material with respect to solar radiation data of 20th June

SI. No	Solar radiation (W/m^2)	Productivity of Solar still by using glass sheet condensing cover	Productivity of solar still by using acrylic sheet condensing cover
1	08:00	0.266	0.156
2	09:00	0.339	0.199
3	10:00	0.348	0.205
4	11:00	0.354	0.208
5	12:00	0.358	0.211
6	13:00	0.412	0.242
7	14:00	0.383	0.225
8	15:00	0.323	0.190
9	16:00	0.245	0.144

3 Results & Discussions

Solar Radiation Flux Variation with Time can be seen in Fig. 1.

Figure 2 shows the variation of distillate output by using different condensing cover materials with Solar Radiation at a particular time. From the theoretical analysis, it is observed that the productivity of solar still when glass is used as the condensing cover gets more than the daily productivity of solar still having acrylic sheet condensing cover. A maximum of 3.028 L/day amount of water is obtained with the solar still having the glass cover in comparison to 1.78 L/day amount of water when an acrylic sheet is used (Fig. 3).

From the above comparison, it can be concluded that there are variations in productivity by using still alone and by using still with the use of FPC with the variation in

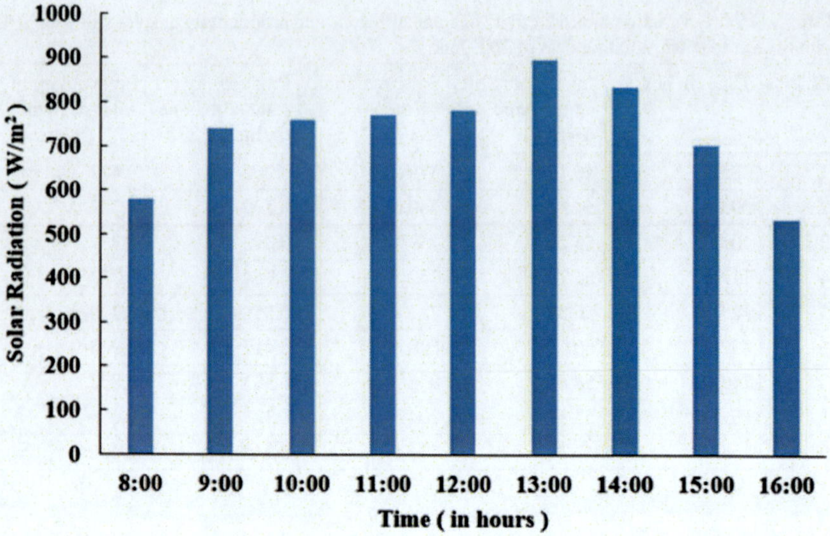

Fig. 1 Variation of Solar Radiation with respect to time

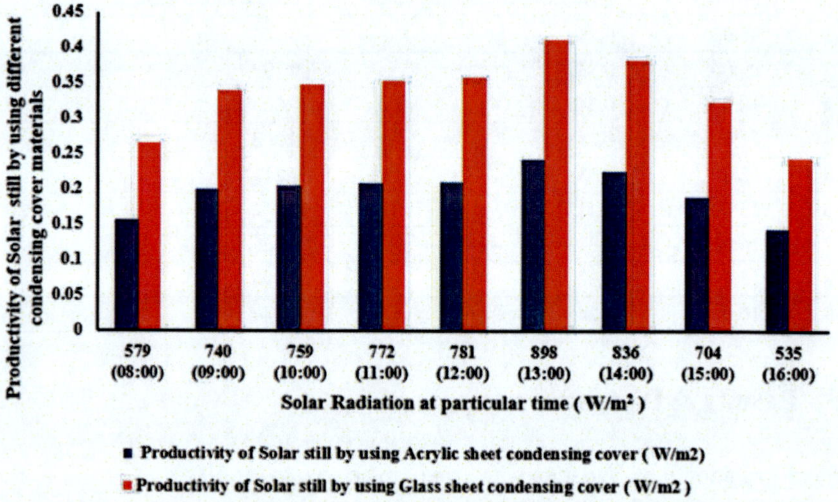

Fig. 2 Variation of productivity of solar still by using condensing cover materials with solar radiation at a particular time

solar radiation at a particular time. The graph presented above clearly shows that by coupling the FPC with Solar Still, the amount of distilled water has increased by up to 3.28 L as compared to still without using a flat plate collector for 10 L of input water.

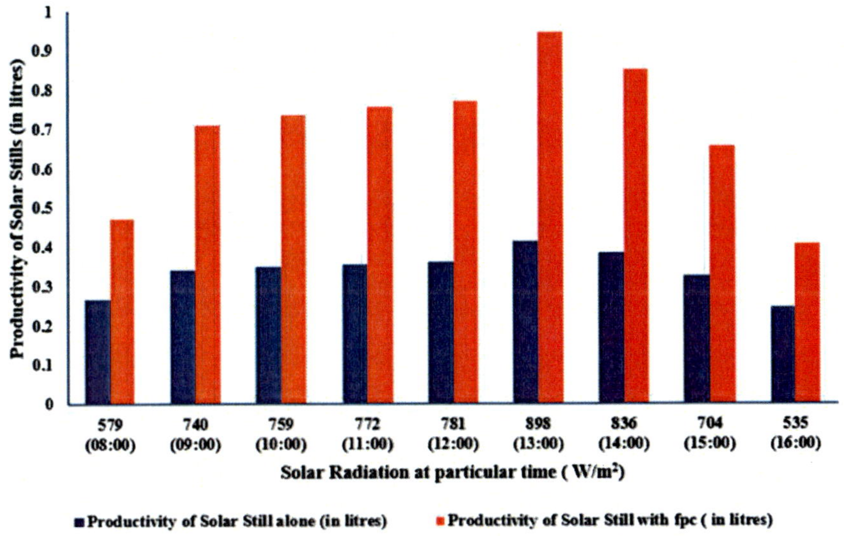

Fig. 3 Variation of Productivity with respect to Solar Radiation at a particular time

4 Conclusion

In this paper, we have done the Mathematical modeling of Solar Still with the use of a flat plate collector. To compare the performance of solar stills alone and solar stills coupled with FPC, several parameters, such as solar radiation and different condensing cover material, were assessed. From our investigation, several points can be concluded, which are as follows:

- From Fig. 2, We have observed that the Volumetric efficiency of solar still when glass is used as the condensing cover material has increased by 12.48% in comparison to having acrylic sheet condensing cover material.
- Our results and comparison with other researchers show that the solar still using a flat plate collector resulted in an increase of 32.78% in the volumetric efficiency of the system than by using still alone for an input of 10 L of water per day.

References

1. Aybar HS, Egelioglu F, Atikol U (2005) An experimental study on an inclined solar water distillation system. Desalination 180:285–28
2. Singh AK, Yadav RK, Mishra D, Prasad R, Gupta LK, Kumar P (2020) Active solar distillation technology: a wide overview. Desalination 493:114652
3. Badran AA, Al-Hallaq AA, Salman IAE, Odat MZ (2005) A solar still augmented with a flat-plate collector. Desalination 172:227–234

4. Sahoo U, Singh SK, Barbarate I, Kumar R, Pant PC (2016) Experimental study of an inclined flat plate type solar water distillation system, Sahoo et al. Renewables
5. Rajagopal T, Aranganathan K, Arun kumar RV, Boobalan M (2020) Performance enhancement in solar flat plate collector. Int J 9(02)
6. Mathur K, Thomas M, Lineswala P, Nayar S (2015) Solar distillation of water. Int Res J Eng Technol (IRJET) 02(07)
7. Duffie JA, Beckman WA (2013) Solar engineering of thermal processes, fourth edition
8. Kumar SN, Shantharaman PP (2015) Fabrication of solar water distillation system. Int Res J Eng Technol (IRJET) 2 (1)
9. Journal of King Saud university—Engineering Sciences 30(1): 78–85, January 2018
10. www.google.com/amp/s/www.tutiempo.net/amp-en/solar-radiation/noida.html
11. www.fdocuments.in/document/chapter-ii-literature-review-2018-12-24-chapter-sssssssii-literature-review-21-general.html

Evacuated Tube Solar Collector Performance with Combined Effect of Triple Integrated Helix and Thermal Energy Storage

Vinay D. Patel, Anurag Churi, Shreedhar Davane, Ashwin Dhangar, Aashirwad Patil, and Ashish J. Chaudhari

Abstract Evacuated Tube Solar Collectors (ETSC) are increasingly in use worldwide because of their high thermal efficiency and high working temperature. Being a passive system, its usage in rural areas is increasing. So, in order to improve the performance of ETSC, the heat pipe evacuated tube solar collector is being utilized for experimentation. The insulated reservoir tank accommodating the manifold is provided with twisted tape, and the outside of the manifold is completely filled with phase change material. Additionally, the parabolic reflectors are added to improve the absorption of incident radiation on the evacuated tube over 360°. The experimentation will be performed with different flow rates of 0.3, 0.5, 1 and 2 LPM of water. The performance with twisted tape shows an improvement in temperature rise by 10 °C with respect to normal ETSC. Due to the addition of a reflector, the performance could be further improved showing an efficiency of 24.5% at 2 LPM. The improvement in the efficiency could be quantified and plotted. Phase change material plays a crucial role owing to which the fall in temperature in post noon session is in range of 5–8° only.

Keywords Evacuated tube solar collector · Parabolic reflector · Solar water heater · Twist tape · Phase change material

V. D. Patel (✉) · A. Churi · S. Davane · A. Dhangar · A. Patil · A. J. Chaudhari
Department of Mechanical Engineering, Vidyavardhini's College of Engineering and Technology, Vasai (West), Palghar 401202, Maharashtra, India
e-mail: vinay.patel@vcet.edu.in

A. Churi
e-mail: anuragchuri696@gmail.com

S. Davane
e-mail: shreedhardavane001@gmail.com

A. Dhangar
e-mail: ashwindhangar1999@gmail.com

A. Patil
e-mail: aashirwadpatil.pp00@gmail.com

A. J. Chaudhari
e-mail: ashish.chaudhari@vcet.edu.in

D. P. Mishra et al. (eds.), *Recent Trends in Thermal and Fluid Sciences*, Lecture Notes in Mechanical Engineering, https://doi.org/10.1007/978-981-19-3498-8_16

1 Introduction

A solar water heater acts as an alternate and renewable energy source for water heating applications. Evacuated tube solar water heating system is one of the most proven and effective methods of solar thermal energy collection systems. The main constructional feature of ETSC is the heat pipe. The material used for heat pipe is copper, owing to its better thermo-physical properties. The function of ETSC consists of absorbing the incident rays falling on the evacuated tube and heating the primary working fluid in the evacuated tube. This heated fluid raises against gravity and reaches the transmission zone to distribute the heat to the secondary fluid. In order to increase the rate of heat transmission from the primary working fluid to the secondary working fluid, the heat pipe has a bulb-like protruded part which is highly sensitive to rapid changes in temperatures. In order to achieve maximum heat transfer, different techniques were adopted by researchers. Thianpong et al. explain the procedure of inserting twisted tape as a swirling flow generation for more heat transfer enhancement and obtained 86.7% heat transfer with respect to plain tube [1]. Chai et al. performed experiments on ETSC with inner concentrating reflective coating on the evacuated tube in order to increase the absorption of maximum solar radiation. They achieve the efficiency of 0.65–0.72 at solar radiation 950 W/m^2 which is around 10% higher than the conventional one [2]. Ismail et al. performed a study of ETSC using rectangular and circular absorbers in an evacuated tube. They found enhancement in efficiency for circular section by 1.1–8.8% and for rectangular section 0.5–4.5% [3]. A new ETSC introducing nano-phase change material (PCM) along with fins technology for thermal efficiency improvement of the ETSC was studied by Elarem et al. They added copper nanoparticles to paraffin wax and investigated the heat transfer behavior of PCM using ANSYS CFD [4]. The effect of optimized fins in addition to PCM accelerates the melting of it for effective heat transfer in ETSC. Pawar and Sobhansarbandi performed the CFD analysis of heat pipe in solar collector with and without PCM tritriacontane paraffin ($C_{33}H_{68}$) material with a melting point of 72 °C [5]. The average deviation of 2.04 and 4.80% was observed with respect to experimental results with and without PCM, respectively. Sadeghi et al. experimented with the ETSC for thermal characteristics of the nanofluid. To improve the heat transfer, the authors use a parabolic concentrator beneath each tube and a spiral aluminum coil inside the heat exchanger. Three different artificial techniques were implemented to analyze the records [6]. The literature did reflect the contribution of researchers for improving the efficiency of the ETSC system. However, the literature on experimental performance of heat pipe based ETSC system combined with twisting tape, PCM and parabolic reflector was not noticed.

This paper emphasizes this aspect of performance improvement of the ETSC system. The radiations from the sun incident on the tubes cover only the upper half portion. The bottom half portion of the tube is not exposed to sunlight causing less heat absorbed by the evacuated tube. To overcome this, parabolic reflectors are used that cover the whole length of the evacuated tube. The advantage of this is the incident rays falling on the parabolic reflector will concentrate in form of the beam of radiation over the bottom portion of the evacuated tube which causes heating of the bottom half portion as well. The sheet is curved into the desired shape to reflect maximum radiation (90–95%). Each tube is provided with the reflector, and tubes are placed at the focal point of the respective reflector. Here, the reflective aluminum mirror sheets have been used which have a reflectivity of more than 85%. The water flowing in the heat exchanger is streamlined through the manifold and does not get enough heat from convection because it does not cover enough internal surface area of the manifold. This is where twisted tape sometime referred to as 'helix inserts' come into play [7]. There are various types of twisted tape inserts generally used for heat transfer enhancement. It is a completely passive system and the major advantage is the smaller size of the heat exchanger will suffice for the application. In the experimental setup, analyses are done for twist tapes of aluminum material having a width of 6 cm and twist angle of 90°. They play a significant role in increasing the surface contact of water in the tube and reducing the velocity of water in the tubes resulting in an increase in the temperature of the water. The insulated reservoir tank around the heat exchanger is filled up with the PCM (paraffin wax) [8]. The experiments are performed on the setup with different volume flow rates and plotted the results in the next sections.

2 Experimental Evaluation

The experimental setup consists of an evacuated tube made of borosilicate glass. The tube consists of two glass tubes concentric with a vacuum between the two layers of the glass tube. A vacuum between the two layers of the evacuated tube insulates against the heat loss. The tube absorbs solar energy and converts it to usable heat which is carried forward by the working fluid inside the evacuated tube. The tube is 1800 mm in length. The evacuated tube is subjected to heavy thermal stress and abrupt change in temperature.

To enhance thermal efficiency, the heat pipe is located inside the evacuated tubes. This consists of a copper pipe filled with acetone solution that is to be heated. The copper tubes gain the heat from the evacuated tube and transfer it to the acetone solution. The acetone solution inside the tubes gets heated and converted into vapor and lifted upon the top of the tube in the bulb of the heat pipe. As the heat pipe is completely inside the evacuated tube, no external thermal losses take place. Figure 1 shows the complete view of the evacuated solar tube collector. Initially, there is cold water in the tank. This heat is transmitted to the water flowing through the manifold. The helix inserts inside the manifold create turbulence and drop the velocity of the flow. This allows water to collect more heat from the water pipe, as it covers more surface area. This heated water is then collected in a water collector where the temperature is monitored. The experimentation was performed for variable flow rates through the manifold [9, 10]. The water flows from overhead tanks to the manifold through flow control valve and flow measuring device. There is a copper pipe inside the manifold through which the water flows. The primary function of it is to transmit

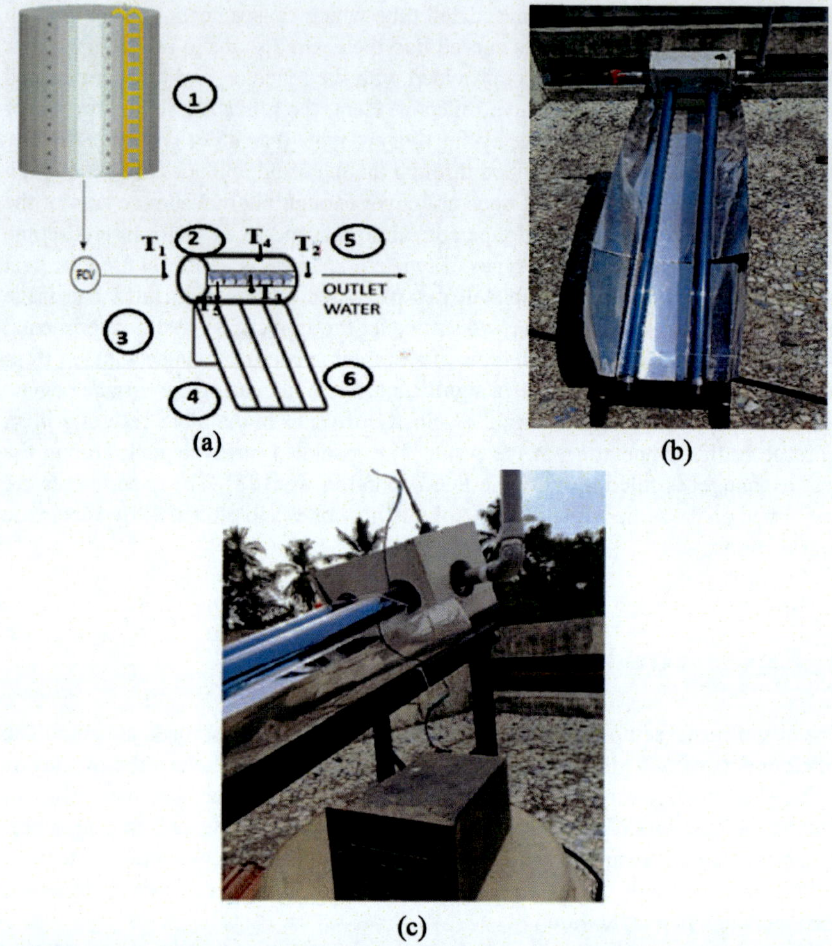

1. Water Tank, 2.Heat exchanger with PCM and tube, 3. Triplex twisted tape, 4. Parabolic trough collector, 5. Thermocouples, 6. Evacuated tube with heat pipe.T_1-Temperature at inlet to heat exchanger, T_2-Temperature at outlet of heat exchanger, T_3.Temperature of PCM at tube wall, T_4-Temperature of PCM at Shell side,T_5-Temperature of working fluid in evacuated tubes.

Fig. 1 **a** Schematic of experimental setup, **b** Experimental setup of evacuated solar tube collector with helix inserts and **c** Measuring instruments for recording temperature

heat from the heat pipe to the water flowing through itself. The job of the water pipe is not only to transmit the heat but also to retain it, and hence, the use of copper as a material is most suitable as the PCM material is surrounded from outside the copper tube which helps to retain the heat. The shell accommodates the copper water pipe and helix insert. The material used for the shell is stainless steel to offer rigidity and strength to the overall structure. It is axially covered with caps from both sides and allows the connection of pipes for the inward and outward flow of water. The wax is solid at first but melts as it reaches 60 °C. It has excellent capacity to store heat and hence is useful as an insulation. These temperatures will be used to calculate efficiency at each point using Eq. 1.

$$\eta = \left(\dot{\text{m}} \times C_p \times (T_{\text{max}} - T_{\text{min}})/I \times A \times N\right) \tag{1}$$

where N = number of tubes; C_p = specific heat of water, J/kg K; I = solar radiance, W/m^2; A = area of single evacuated tube, m^2 and $\dot{\text{m}}$ = mass flow rate, kg/m^3.

3 Result and Discussion

The experimental setup as shown in Fig. 1 will be tested further on the terrace of Vidyavardhini's College of Engineering and Technology (19°28′N–19°47′N latitude and 72°48′E–72°8′E longitude) on clear sunlight day with maximum incident radiation of 950 W/m^2 measured using a solar lux meter. Three repeated trials have been recorded from 9 a.m. to 5 p.m. with variable mass flow rates (0.3, 0.5, 1 and 2 LPM). Temperatures are recorded at all points as shown in Fig. 1 using the K-type thermocouple and digital indicator. The variation of temperature with time with and without twist tape is as shown in Fig. 2 for the mass flow rates 0.3, 0.5, 1 and 2 LPM. At the mass flow rate 2 LPM achieves a maximum temperature of 75 °C at 3 p.m. with twisted tape over 62 °C without it causing a 20.96% increment. This may be happened due to increases in the turbulence effect in the tube causing more heat to be absorbed by water. At other flow rates, the percent increment of 34%, 13.11% and 12.90% is recorded with the use of twisting tape for 1, 0.5 and 0.3 LPM, respectively. The maximum temperature variations are obtained during a day period from 1 to 3 p.m. when the solar intensity of radiation is at its peak. The effect of the parabolic reflector on the performance of the twisted tape-based ETSC system is as shown in Fig. 3 for varying mass flow rates of 0.3, 0.5, 1 and 2 LPM. The temperature of the water at the outlet was found to be 75 °C showing a percent increment of 5.6% with the provision of the reflector on the ETSC.

The attachment of the reflector shows an increment of 6.3, 7.8 and 11.1% at other flow rates of 1, 0.5 and 0.3 LPM also with that of no reflector.

This may be happened due to an increment in the area for absorbing the incident rays over the complete circumference of the tube. This causes more heat transferred to the water in the manifold.

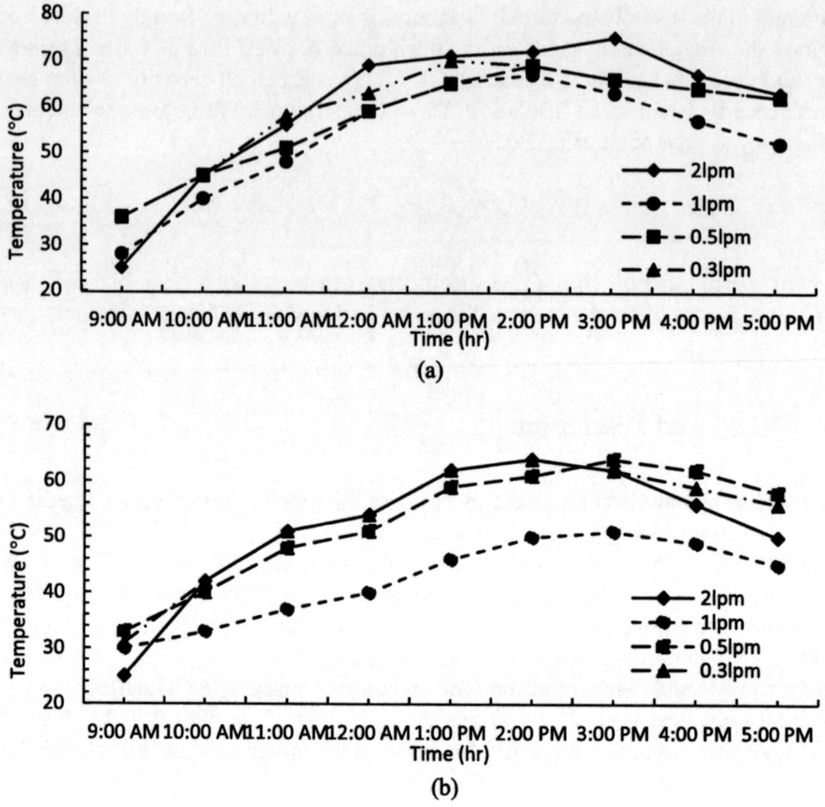

Fig. 2 Temperature–time. **a** with twist tape; **b** without twist tape

The efficiency variation with and without twist tape at different mass flow rates of fluid is as shown in Fig. 4a. From the figure, it is clear that when the fluid passes at 2 LPM through the manifold, it achieves maximum efficiency of 43.7% using twisted tape over 34.1% without twisted tape. The efficiency of 27.6, 26.9 and 24.5% is recorded for the fluid flowing at rates of 1, 0.5 and 0.3 LPM which is reported higher than the one without twist tape. The rise in efficiency may be the combined effect of PCM along with the twisted tape. Similarly, the effect on the efficiency of twisted tape ETSC with and without a reflector is plotted in Fig. 4b.

As seen in the figure, the efficiency of 24.5% is evaluated at the 2 LPM flow rate through the manifold which is 14.1% without reflector use. A similar trend was also observed for other flow rates where efficiency of 19.6, 16.9 and 15.1% is evaluated for 1, 0.5 and 0.3 LPM. This may be the combined effect of the twist tape and PCM for higher efficiency.

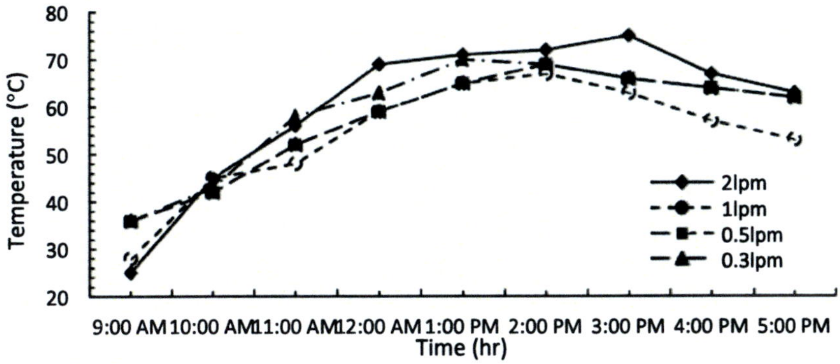

(a)

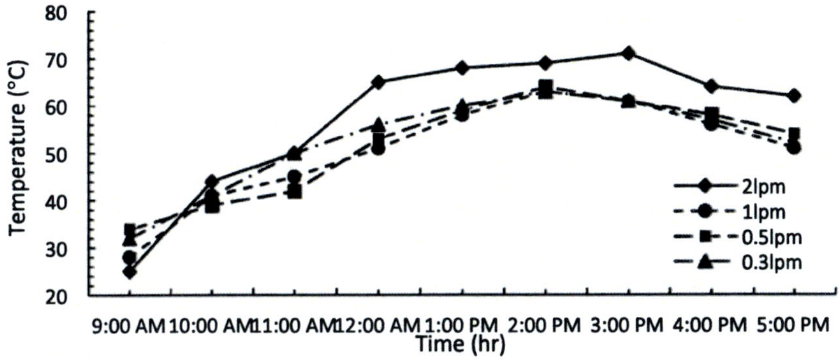

(b)

Fig. 3 Temperature–time. **a** with twist tape and reflector; **b** with twist tape but without reflector

The PCM temperature (T_3) as shown in Fig. 1 is plotted in Fig. 5 with time. The paraffin wax temperature obtained in the manifold of the evacuated solar tube collector reaches 61 °C at 1.00 p.m. and it remains steady with a percent variation of 8.92% till 5.00 p.m. This will be the major portion of the rise in efficiency of the ETSC system.

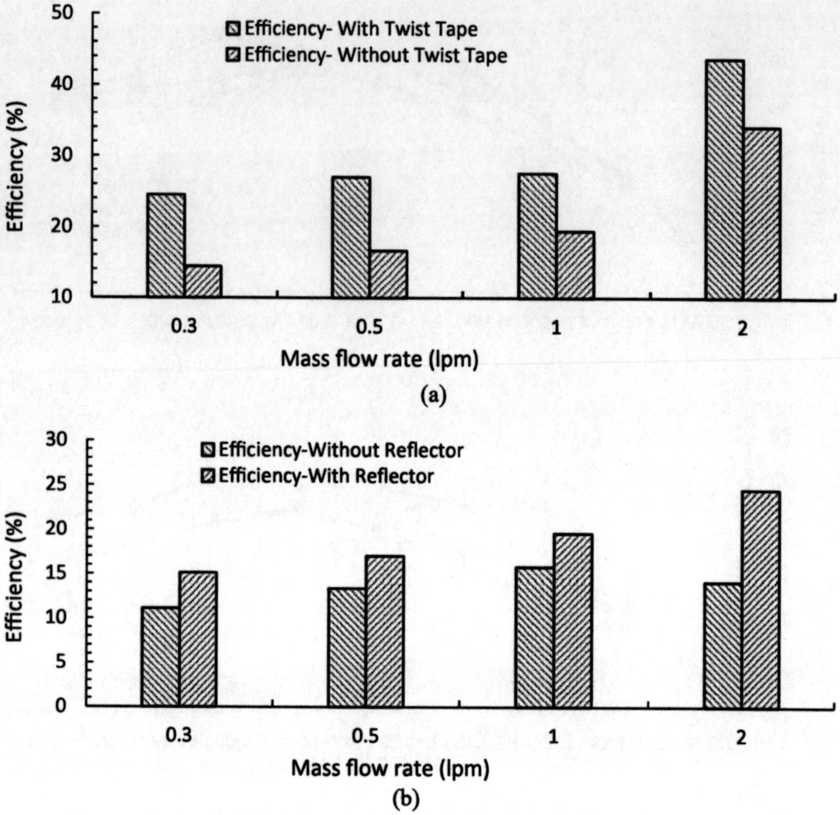

Fig. 4 Efficiency comparisons. **a** With and without twist tape; **b** With and without reflector

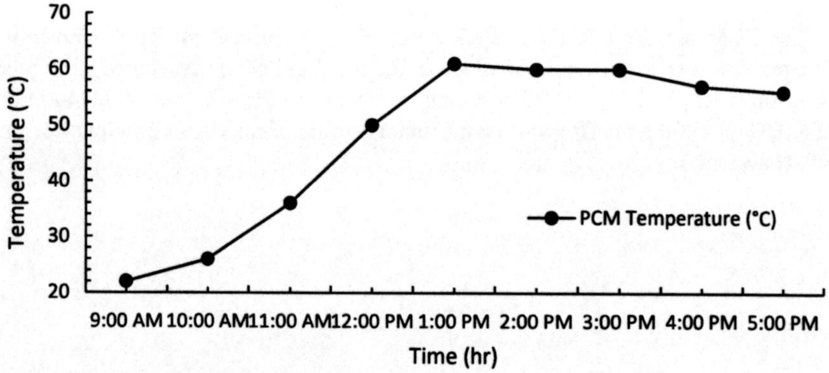

Fig. 5 Temperature of PCM over daytime

4 Conclusion

The experimental analysis of the ETSC with twisted tape and PCM is carried out successfully. The ETSC with twisted tape is efficient when the flow rate is higher especially. The efficiency of the ETSC with twisted tape and parabolic reflector will further improve the performance of the system. In the experimental readings taken for comparison of setup with reflector and without reflector, the major difference is observed between time ranges of 11 a.m. to 1 p.m. This is because during the peak hours of high-intensity sunlight, the downside of the reflector doesn't get enough heat as it does while having the reflector on the setup. The phase change material plays an important role in the performance. Despite the addition of twist tape and reflector as accessories, the PCM stays as one of the most important and integral parts of the setup as the heat retention and redistribution capacity of the PCM is the reason the working fluid (water) stays warm when the heat addition is low or absent.

Acknowledgements The authors are thankful to Vidyavardhini's Trust, Vasai, for their support in Evacuated Tube Solar Collector Performance with Combined Effect of Triple Integrated Helix and Thermal Energy Storage experimental setup.

References

1. Thianpong C, Eiamsa-ard P, Eiamsa-ard S (2012) Heat transfer and thermal performance characteristics of heat exchanger tube fitted with perforated twisted-tapes. Heat Mass Transfer 48:881–892
2. Chai S, Yao J, Liang JD, Chiang YC, Zhao Y, Chen SL, Dai Y (2021) Heat transfer analysis and thermal performance investigation on an evacuated tube solar collector with inner concentrating by reflective coating. Solar Energy 220:175–186
3. Ismail KAR, Teles MPR, Lino FAM (2021) Comparative analysis of eccentric evacuated tube solar collector with circular and rectangular absorber working with nanofluid, Cleaner Eng Technol 3:1–13, 100105
4. Elarem R, Alqahtani T, Mellouli S, Aich W, Khedher N, Kolsi L, Jemni A (2021) Numerical study of an evacuated tube solar collector incorporating a nano-PCM as a latent heat storage system. Case Stud Thermal Eng 24:100859
5. Pawar VR, Sobhansarbandi S (2020) CFD modeling of a thermal energy storage based heat pipe evacuated tube solar collector. J Energy Storage 30:101528
6. Sadeghia G, Najafzadeh M, Americ M (2020) Thermal characteristics of evacuated tube solar collectors with coil inside: an experimental study and evolutionary algorithms. Renew Energy 151:575–588
7. Singh BP (2019) A review paper of twisted tape inserts. Int J Appl Eng Res 14(9). ISSN 0973-4562
8. Shukla A, Buddhi D, Sawhney RL (2019) Solar water heaters with phase change material thermal energy storage medium: a review. Renew Sustain Energy Rev 13:2119–2125
9. Zambolin E, Del Col D (2010) Experimental analysis of thermal performance of flat plate and evacuated tube solar collectors in stationary standard and daily conditions. Sol Energy 84:1382–1396
10. Michaelidis IM, Eleftheriou PC (2011) An experimental investigation of the performance boundaries of a solar water heating system. Exp Thermal Fluid Sci 35(6):1002–1009

Low-Speed Wind Power Generation System: An Overview

B. K. Srivastava, Vijay Shankar, and A. K. Singh

Abstract This work aims to accomplish a wind-powered turbine's substitute marshaling for powering a generator utilizing low-speed wind and using the easy mechanics of wind circulation inside and out the closed spaces like auditoriums, theaters, warehouses, and industrial plants. Turbo ventilators are another option for a motor-powered ventilation arrangement. The rooftop turbo ventilator is now comprehensively accepted in industrial ventilation and furthermore has salient ventilation properties that can be employed to cool economic, residential, and plenty of institutional offices. It works efficiently and successfully at even lower wind speeds; hence, it always performs accordingly throughout its service life. This provides a gateway toward clean energy using the resources in our hands. Current work represents the overall development and generalized modeling of the system with a relevant gear mechanism system that holds a better approach to harvest the electrical energy from low-speed roof ventilators.

Keywords Turbo ventilator · Renewable energy · Wind energy · Green energy

1 Introduction

The energy from natural resources is renewable energy that is also mentioned as a clean energy source that is utilized in various utilities with the help of different solar collectors [1] and associated technology [2, 3] such as solar distillation [4, 5], steam generation [6], and power generation [7], by considering the environmental safety

B. K. Srivastava
Department of Applied Sciences, Galgotia College of Engineering and Technology, Greater Noida, India
e-mail: bipin.srivastava@galgotiacollege.edu

V. Shankar
Department of Chemistry, B.S.N.V.P.G. College, Lucknow 226001, India

A. K. Singh (✉)
Galgotia College of Engineering and Technology, Greater Noida, India
e-mail: agashok26@gmail.com

D. P. Mishra et al. (eds.), *Recent Trends in Thermal and Fluid Sciences*, Lecture Notes in Mechanical Engineering, https://doi.org/10.1007/978-981-19-3498-8_17

perspectives [8, 9] under the energy matrix [10, 11] and economic viabilities [12, 13]. These produce electricity without harming the environment. Because of the energy requirements and changes within the political and social environment, wind turbines began to roll exponentially across the planet within the last 30 years. Renewable energy reports for more than 13% of the global energy supply and more than 22% of the worldwide electricity needs. As of February 2021, India's total installed wind energy capacity was slightly more than 38 G.W., the fourth largest installed capacity globally [14]. The project deals with craft design and implementing the system using the turbo exhaust ventilator set off in low and high wind speed conditions. This model can work efficiently at different wind velocities contrasted with classic horizontal axis wind turbine, which might contribute to its increasing popularity for mass utilization of wind energy in electricity production [15].

Within this proposed system, it is coupled with a free-running turbo ventilator to the generator with the assistance of a gear mechanism, thereby producing electricity and using this for lighting the industry. Since only a few regions within the globe have windy conditions throughout the year, this method isn't restricted to a few chosen areas, as shown in Fig. 1. Here, the ventilator rotates with the wind energy and the heat is generated inside the industry. The Turbo ventilator sucks the heated air, and it is transferred to the outside by its rotation. Both the energies are consumed here for the rotation. The identical concept is utilized during this study, but the interior space temperature and exterior atmospheric wind are also used along with the instantaneous observations. So, the continuous development of ventilators is often required to achieve targeted energy in many countries around the globe. The wind from any direction is sufficient to rotate the ventilator in the clockwise direction. The

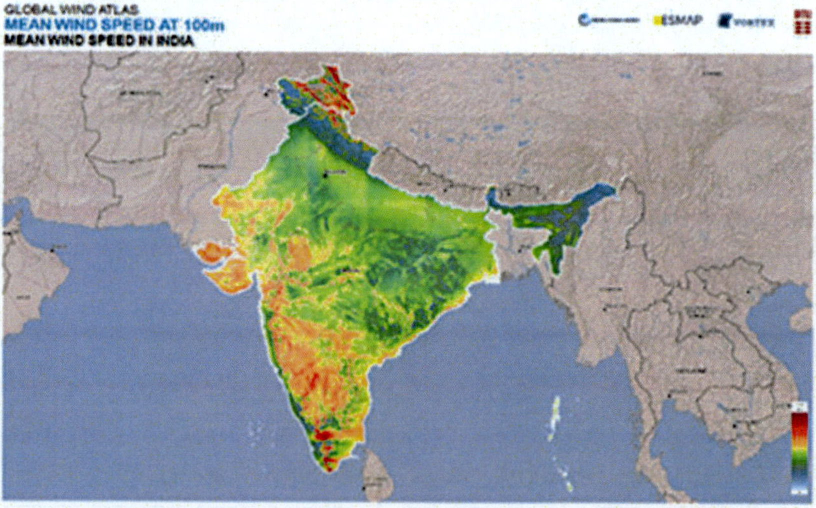

Fig. 1 Mean wind speed in India

ventilator blade is so designed that the wind from any direction and even a negligible amount of wind are merely needed to generate optimum electricity [16].

The wind-powered Turbo ventilators make the system much more effective for ventilation of various buildings and warehouses. This ventilator works with the energy absorbed from the wind. Turbo ventilators are spherical metallic vents with vanes in them. A tiny amount of wind is enough for the rotation of the turbo ventilator. As the wind velocity increases, the rotation of the turbine will also increase and dissipate the heated air. The heated air inside the chamber rises. Whenever the turbo ventilator rotates, the siphon actuates to push the hot air out through the vanes in the replacement of fresh air that comes into the workstation [17].

The main objectives for the present work are as follows:

- To erect opportunities in areas having lower wind velocities, with starting wind velocity of about 1.01 m/s with the Omnidirectional capabilities of Turbo ventilator fan.
- To transform wind energy into electric power with remarkably cheap running and maintenance costs.
- To mark the possible governing elements that are responsible for the performance of the system.
- The use of a gear mechanism increases the rotational speed of the system.

The global energy system in 2017, which was more than 75% powered by fossil fuel, and more than 60% powered by fossil fuels, as shown in Fig. 2. In 2018, further addition of energy by non-renewable fuels performed by renewable for 3rd successive year, and that's why the increase in electric power by fossil fuels was more than 8% become substantially more than each renewable summed, as shown in Fig. 3.

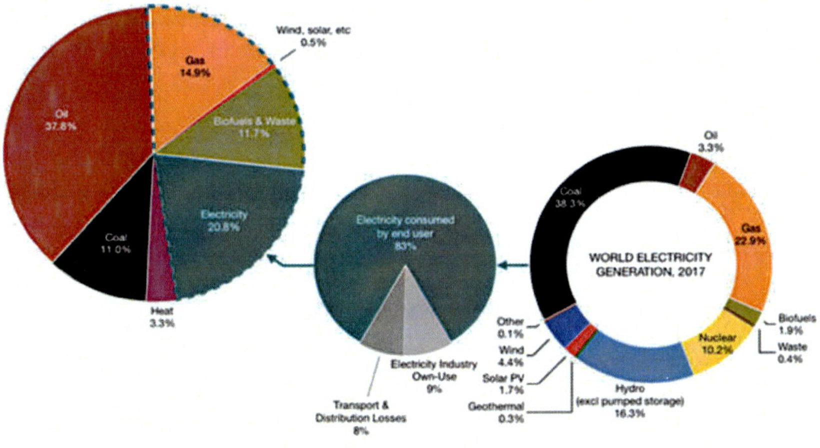

Fig. 2 World energy data [16]

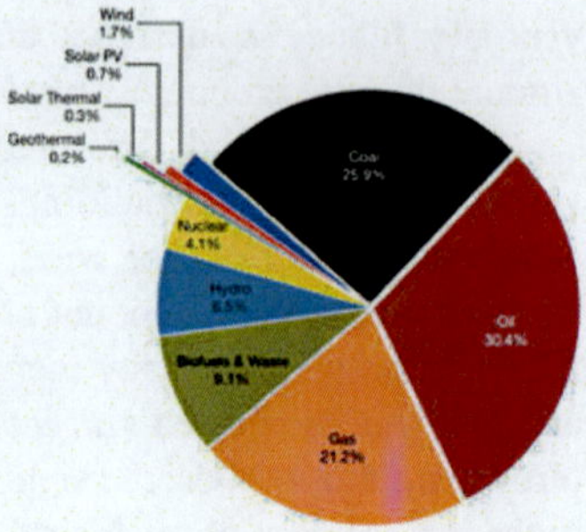

Fig. 3 Division of Energy Resources [16]

2 Construction

Turbo ventilator fan: Turbo ventilator is an apparatus used to regulate heated air and humidity by throwing interior heated air through its vents to the outer atmosphere, thus bringing fresh air inside the chamber or shed or industry as shown in Fig. 4a, b. When heated air arises upward, it starts to create low pressure at the mid of the turbo ventilator, due to which heated air is thrown outwards, and the turbine starts to rotate. This phenomenon is often termed the stack effect. The stack effect makes the turbo ventilator keep spinning due to air circulation. A computational fluid dynamic analysis has been done to conclude that the ventilator has a throat diameter of 300 mm with a higher rotational speed (rpm) at any given wind velocity [18]. The primary reason to focus on this size is that it needs higher rpm to produce more power. Axial-Flux Permanent Magnet (AFPM).

Generator: Utilizing the permanent magnet in a coreless axial-flux producer makes the assembly easier and acceptable for locally fabricated open hardware wind power purposes. The rotor component is installed by utilizing a permanent magnet. The turbo ventilator rotational speed drives the rotor and field windings that are coupled with the DC generator. Magnetic flux over the gap and slashed through armature

Fig. 4 **a** Turbo ventilator. **b** Turbo ventilator

windings to induce a potential difference in armature windings, hence, induced electromotive force generated accordingly. After all, the AFPM generator's total output potential difference would be reformed, and selecting a peculiar minimal frequency became more pertinent to the total cost of the AFPM generator than its functional aspects. This happens because the minimal frequency dictates the count of pairs of poles and hence the total counts of the magnet. Permanent magnets employed in this system are rectangular in shape neodymium magnet, which is believed to be one of the most excessive and rare-earth materials used in the fabrication procedure of the smaller wind turbine [19].

Rectifier: The generator produces alternating current output, which shall be stored as DC power for further uses. This action of converting alternating current into the direct current is labeled as rectification. With the help of a full-wave bridge rectifier, our goals will be achieved to convert AC into DC for which the diode rectifier is used on a large scale and widely accepted [20].

Gear: The primary reason to employ a gear mechanism is to ramp up the rotary speed of the turbo ventilator. With the aim to obtain twice rotational velocity, one can get more power out from the AFPM generator (Fig. 5). It has been designed a gear system for this particular system as shown in Fig. 6, which can ramp up as shown in Table 1.

Assembly: The rotating turbo ventilator has various modules. One of the components is typically known as the propeller shaft. In this particular case, propeller shaft has a diameter of 16 mm. It varies from product to product. The customized pinion for this project will be mounted on the bottom of the propeller shaft, and adjacent pinion gear will be placed so that rotary motion can be carried out. With gear, an AFPM generator is directly attached, as shown in Fig. 5. That is covered under the assembly part as represented in Fig. 7.

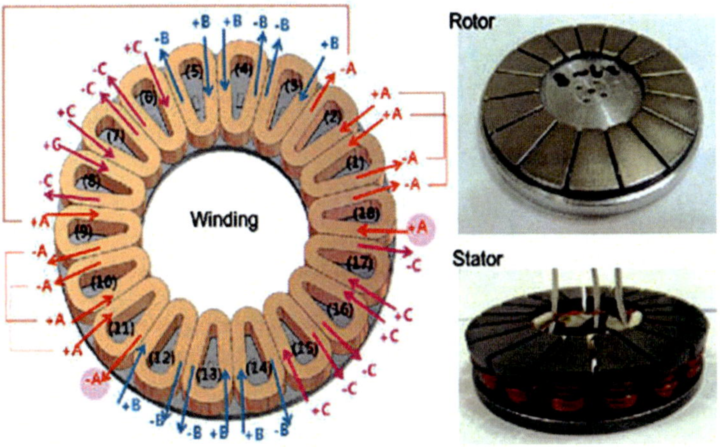

Fig. 5 AFPM generator rectifier [21]

Fig. 6. 3-D model of gear assembly

Table 1 Variation of rotational speed against the wind speed with and without AFPM gears [21]

Wind velocity (m/s)	Rotation (rpm)	AFPM speed (no gear) (rpm)	AFPM rotational speed with gear (rpm)
1.0	30	30	60
2.0	45	45	90
2.75	56	56	112
3.44	62	62	124
4.55	107	107	214
5.25	120	120	240
7.60	132	132	264
8.87	151	151	302
9.5	195	195	390

3 Working

Turbo Ventilator is powered by wind to create economical and adequate ventilation which is quite helpful for techno-green impact on the environment [22–24]. This ventilation is helpful in natural convection and is mainly based on temperature and pressure differences. The mounting of Turbo Ventilators is on rooftops of industries, warehouses, auditoriums, etc. The blades on the rotors are exposed to space outside the hall and are rotated along with wind movement. It comes in passive ventilation as no external force is used to circulate the air inside and outside the building. The lower density and light-weighted warm airflow move toward the roof by natural forces and thermal buoyancy. When the turbines rotate, they suck the warm air out through the vent and transfer them from the indoor to outdoor. Thus, a pressure drop is created all over the turbo ventilator. On account of this pressure drop, a denser fresh breeze flows inside through the rooftop ventilator. A pattern is being assembled and causes the rotation of blades. Hence, the turbo ventilator functions on natural convection

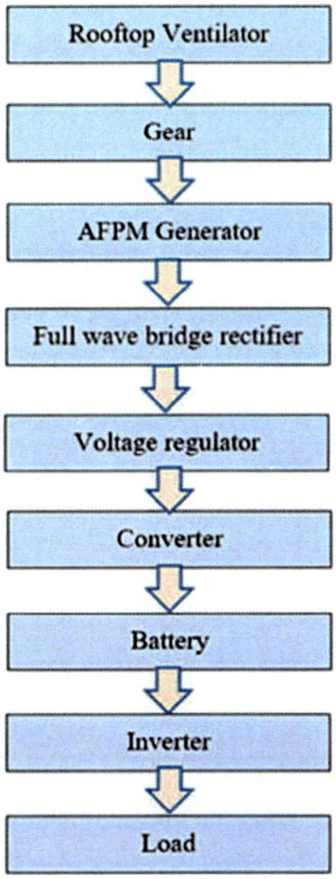

Fig. 7 Probable layout of the assembly

and decreases the indoor temperature without external power usage. An electric generator may be attached to the turbo ventilator for electricity production. The primary function of electric generators or alternators is the conversion of mechanical energy into electrical energy. The power generation is based on Faraday's law, i.e., the induction of electromotive force by the conductor's movement in a magnetic field.

A spur gear mechanism is employed with a gear ratio of 2:1 and module, m (2 mm), to ramp up the rotational speed of the AFPM generator with twice the rate of turbo ventilator as represented in Table 1. One singular rotation of turbo ventilator leads to two cycles of AFPM generator. In coreless AFPMs, the stator coil winding is sandwiched between two Neodymium permanent magnet discs with opposite polarity of S–N and N–S arrangement. The elector motive force produced by the AFPM is alternating and unsteady. A full-wave bridge rectifier is attached to the generator. This contains four diodes that transform the AC into DC. On the DC output of the rectifier, a capacitor is placed, which can handle up to 50 V. Placing a capacitor on the DC output of the rectifier is like putting a large room for electrons to travel, resulting in a steadier flow toward the load. A Linear Voltage Regulator is attached to the capacitor, which does not allow voltages of more than 24 V. When

Table 2 Designed gear system for a particular system

Design parameters	Values
Module (m)	2 mm
Pinion teeth (TP)	16
Gear teeth (TG)	32
Gear ratio (G)	2
Pinion-gear center distance	48 mm
Circular pitch dia. of pinion	32 mm
Circular pitch dia. of gear	64 mm
Addendum	2 mm
Dedendum	2.5 mm
Tooth thickness	3.1416 mm
Bottom clearance	0.5 mm
Whole depth	4.5 mm

the voltage becomes higher than 24 V, the Voltage regulator wastes its power as heat within the associated circuitries and terminals. When the input voltage increases further, it shuts itself down for safety. Then, a DC-to-DC converter is placed whose function is to take any voltage up to 30 V as input (regulated by a voltage regulator) and bring it to a specific voltage level. This converter gives a couple of options like (1) to set a minimum input voltage, (2) to set the output voltage, and (3) to limit the output current. The obtained DC output is now steady and safe to use. This can be directly used to supply power to light-emitting diode bulbs for lighting. An inverter is an electronic device used to convert DC electricity to AC electricity. And, the DC loads like chargers or light-emitting diode bulbs for lightning can be attached here. This output power can also be stored in battery storage in DC for future use and applications. It provides a preview for the future evolution of the setup and stimulates the application of clean energy at smaller sputum. This locally manufactured turbo ventilator has been demonstrated to have robust design and fabrication proceeding. These procedures can be effortlessly understood narrowly from pivotal scientific understanding. Also, they can be altered by the consumer and technocrats who establish them to suit to cater to local requirements by employing locally sourced components (Table 2).

4 Conclusion

This paper briefly illustrates the idea of electric power using turbo ventilators. This study demonstrates the output potential difference and current commensurate to wind speed and turbo ventilators' rotary speed.

The overall work refers to and recommends utilizing turbo ventilators, especially working under the low wind velocities and suitable for most low arid regions and small workplaces such as domestic purposes and kitchen applications.

The presented assembly is the key to the genesis of open-sourced hardware for smaller wind turbines that could yield electricity for the industrial use case at a lesser price and with the technological know-how on the fabrication of renewable power resources.

With the employment of a gear mechanism, this system generated more electricity than it could without any modification. This paper has an excellent view on future betterment and evolution within the industrial and the domestic or residential applications.

References

1. Singh AK (2020) Samsher: analytical study of evacuated annulus tube collector assisted solar desaltification system: a review. Sol Energy 207:1404–1426
2. Singh AK, Singh DB, Mallick A, Harender, Sharma SK, Kumar N, Dwivedi VK (2019) Performance analysis of specially designed single basin passive solar distillers incorporated with novel solar desalting stills: a review. Sol Energy 185:146–164
3. Singh AK, Singh DB, Dwivedi VK, Kumar N, Yadav JK (2018) A review of performance enhancement in solar desalination systems with the application of nanofluids. In: ICACCCN, pp 814–819
4. Singh AK, Singh DB, Dwivedi VK, Tiwari GN, Gupta A (2020) Water purification using solar still with/without nano-fluid: a review. Mater Today Proc 21:1700–1706
5. Singh AK (2021) Samsher: a review study of solar desalting units with evacuated tube collectors. J Clean Prod 279:123542
6. Singh AK (2020) Singh: an inclusive study on new conceptual designs of passive solar desalting systems. Heliyon 7:e05793
7. Singh AK, Yadav RK, Mishra D, Prasad R, Gupta LK, Kumar P (2020) Active solar distillation technology: a wide overview. Desalination 493:114652
8. Singh AK, Srivastava BK, Shankar V, Prasad R, Yadav RK, Shukla AK, Nandan G, Kumar P (2020) Energy matrices and enviroeconomical perspective of solar decontamination system. Mater Today Proc. https://doi.org/10.1016/j.matpr.2020.09.198
9. Singh AK, Singh D, Lohumi MK, Srivastava BK, Gupta HP, Prasad R (2021) Potable water production by single slope active solar distillation unit—a review. Adv Energy Environ. Lecture notes in civil engineering, Springer. 142:31–42
10. Singh AK, Singh DB, Mallick A, Kumar N (2018) Energy matrices and efficiency analyses of solar distiller units: a review. Sol Energy 173:53–75
11. Singh AK (2021) Samsher: material conscious energy matrix and enviro-economic analysis of passive ETC solar still. Mater Today Proc 38:1–5
12. Singh AK, Dwivedi G, Srivastava BK, Tiwari BK, Kumar P, Yadav RK, Singh M, Shukla AK, Nandan G (2021) Study of analytical observations on energy matrix for solar stills. Mater Today Proc. https://doi.org/10.1016/j.matpr.2021.02.109
13. Singh AK, Samsher (2021) Tech-en-econ-energy-exergy-matrix (T4EM) observations of evacuated solar tube collector augmented solar desaltification unit: a modified design loom. Mater Today Proc. https://doi.org/10.1016/j.matpr.2021.09.088
14. Kumar A, Kumar K, Kaushik N, Sharma S, Mishra S (2010) Renewable energy in India: current status and future potentials. Renew Sustain Energy Rev 14(8):2434–2442

15. Shinde R, Lavhale V, Nair A, Pawar S, Mahajan R (2016) Generation of electric power using turbo ventilators. Int J Curr Eng Technol 6:89–93
16. Jadhav GK, Ghanegaonkar PM (2015) Performance characteristics of turbo ventilator: a review. Int J Innov Technol Res 3(4):2276–2282
17. Dongre K, Meshram AA, Bhoyar AR, Tamgade RU, Kapse AR, Jethani S (2020) Electricity generation by using rooftop ventilator. Int Res J Eng Technol 7(2):3006–3008
18. Jadhav GK, Ghanegaonkar PM, Garg S (2016) Experimental and CFD analysis of turbo ventilator. J Build Eng 6:196–202
19. Dangeam S (2011) An electric generator is driven by a roof ventilator. Energy Procedia 9:147–158
20. Shatri DC, Irwanto M, Syafawati AN, Shema SS (2011) Power generation roof ventilator. In: International conference on environment and industrial innovation, vol 12, pp 183–187
21. Minaz MR, Celebi M (2017) Design and analysis of a new coreless PMSG with three rotor and double stator. Results Phys 7:183–188
22. Singh AK, Kumar P, Srivastava BK, Yadav RK, Singh M, Kumar A, Tyagi A, Nirala A, Prasad R (2021) Comparative observation of energy matrix and environ-economy for solar still systems. Mater Today Proc 47(13):3730–3734
23. Singh AK (2022) Samsher: techno-environ-economic-energy-exergy-matrices performance analysis of evacuated annulus tube with modified parabolic concentrator assisted single slope solar desalination system. J Clean Prod 332:129996
24. Singh AK, Kumar P, Kishore R, Singh P, Kant N, Singh P, Nandan G, Jaiswal A, Prasad R (2021) Stumpy-paced green power generation system. In: RDCAPE. IEEE Explore, pp 351–355

Highly Confined Flow Past a Circular Cylinder

Dheeraj Kumar Das, Achhaibar Singh, and D. K. Singh

Abstract A computational fluid dynamics method is used to study the highly confined flow around a circular cylinder. Numerical solutions are obtained by solving time-independent incompressible Navier–Stokes equations in 2D. The study of the effect of highly confined flows using the finite-element method explores the steady and unsteady flow regimes in a laminar and incompressible flow around the circular cylinder. Effect of high blockage on flow characteristics, such as streamlines, vorticity field, surface pressure, eddy length, time series of C_d and C_l, Strouhal number, phase plots and variation of drag and wake length with Re is presented.

Keywords Finite-element method · Circular cylinder · Blockage · Strouhal number · GMRES

1 Introduction

This study explores the highly confined flow around a circular cylinder. Effects of blockage are investigated after varying Re from $40 \leq Re \leq 200$. The blockage is defined as the ratio of characteristic length (perpendicular to the flow) of the domain to the width of the channel. In the current work, the blockage is used as 0.9. The effects of blockage and boundary conditions are employed on the problem statement and flow characteristics are investigated.

Dennis et al. [1] investigated the semi-analytical method for non-rotational cylinder from Re = 5 to 100. Badr et al. [2] considered both steady and unsteady regime at low Re around circular cylinder but Ingham et al. [3] and Tang et al. [4]

D. K. Das (✉) · A. Singh · D. K. Singh
Mechanical Engineering Department, Netaji Subhas University of Technology, New Delhi 110078, India
e-mail: dheerajkumardas@gmail.com

A. Singh
e-mail: asingh@nsut.ac.in

D. K. Singh
e-mail: dksingh@nsut.ac.in

D. P. Mishra et al. (eds.), *Recent Trends in Thermal and Fluid Sciences*, Lecture Notes in Mechanical Engineering, https://doi.org/10.1007/978-981-19-3498-8_18

Table 1 Earlier studies concerning flow around a highly confined circular cylinder

Study	Re	Blockage	Flow regime	Method
Dennis et al. [1]	5–100	–	Steady	Semi-analytical
Sahin et al. [7]	0–280	0.1–0.9	Steady and unsteady	Finite-volume
Sen et al. [8]	6–40	0.000125–0.8	Steady	Finite-element

investigate only steady flow regime. Kang et al. [5] investigate laminar flow around a circular cylinder and Mittal et al. [6] investigate wide range of blockage. Sahin et al. [7] used finite-volume method for Re = 0–280 and B = 0.1–0.9. They considered the study in both the regime, steady and unsteady. Numerical solutions are obtained by solving 2D Navier–Stokes equations. Present work aims to present the flow characteristics at very high blockage which were not included in earlier studies (Table 1).

2 The Governing Equations and Problem Set-Up

The incompressible Navier–Stokes equation in two-dimensional flow field is

$$\rho(\mathrm{u}\cdot\nabla)\mathrm{u} = -\nabla \mathrm{p} + \mu\nabla^2\mathrm{u}$$

$$\nabla\cdot\mathrm{u} = 0$$

In a circular cylinder, a no-slip boundary condition is used at the lateral walls of the computational domain. At both lower and upper boundaries, the velocity is set as zero. At the inlet of the domain, the flow is considered a fully developed flow and the velocity profile is parabolic. On the downstream side, the Neumann boundary condition is specified for the velocity. No-slip boundary condition means zero relative tangential velocity between the fluid and the cylinder and it is applied on the surface of the cylinder (Fig. 1).

3 The Finite-Element Mesh for Bounded Flow

The mesh used for bounded flow consists of three numbers of blocks. The central block consists of a cylinder and two other neighboring blocks as left and right blocks. For highly confined flow around a circular cylinder, there is no upper and lower as shown in Table 2. Effects of blockage have influence over the flow parameters.

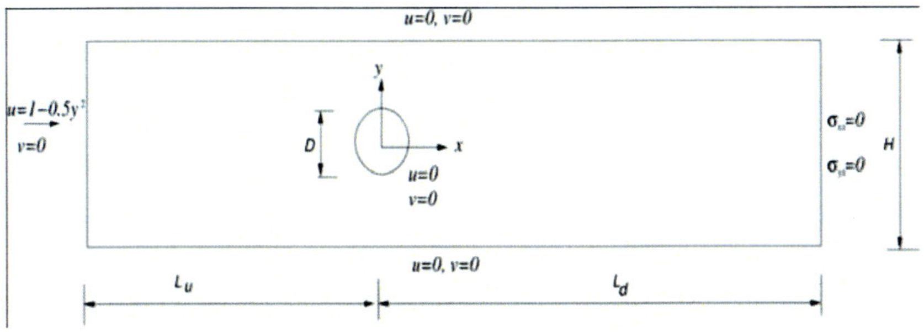

Fig. 1 Problem description of highly confined fluid flow around a circular cylinder

Table 2 The structured finite-element mesh consists of three blocks, namely the left block, the right block and the central block for bounded flow

Left block	Central block	Right block

4 Results

Results are presented for $40 \leq \text{Re} \leq 200$, and the effects of blockage are investigated. The Gauss–Legendre quadrature formula is used for calculating the element level matrix. The ODE/PDE is converted into the linearized algebraic through finite-element discretization of fluid flow equations. These linear algebraic equations are solved by matrix-free implementation of GMRES (Generalized Minimal Residual Method).

5 Comparison of Unbounded and Bounded Flow

The comparison of drag with Re for unbounded and bounded flow is shown in Fig. 2. The drag force decreases with an increase in Re and results are presented at Re =

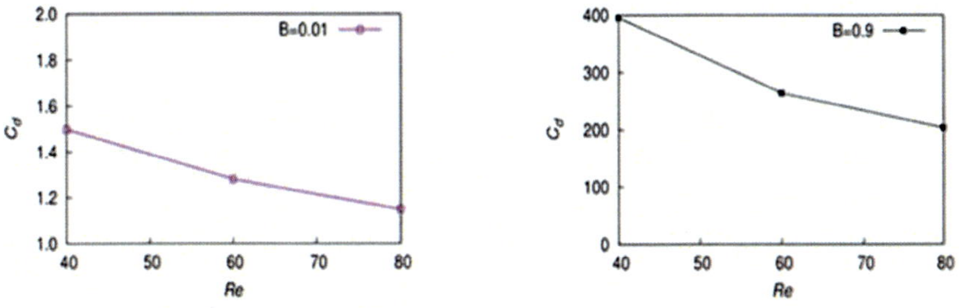

Fig. 2 Variation of drag with Re for unbounded and bounded flow past a circular cylinder

40, 60 and 80 for B = 0.01 and 0.9. For high blockage, there is no pressure recovery and this causes a very large value of drag.

6 Steady Flow Results at Re = 40, 60 and 80

6.1 Streamlines at Re = 40, 60 and 80

The streamlines are symmetrical at each Re = 40, 60 and 80 as shown in Fig. 3 and eddies are increasing with an increase in Re. The bubble elongates approximately linearly with Re for B = 0.9.

6.2 Distribution of Drag and Eddy Length with Re

The Re is varied from 40 to 200 and drag and eddy length variations are shown in Fig. 4. Eddy length is the stream-wise distance between the wake stagnation point and rear stagnation point.

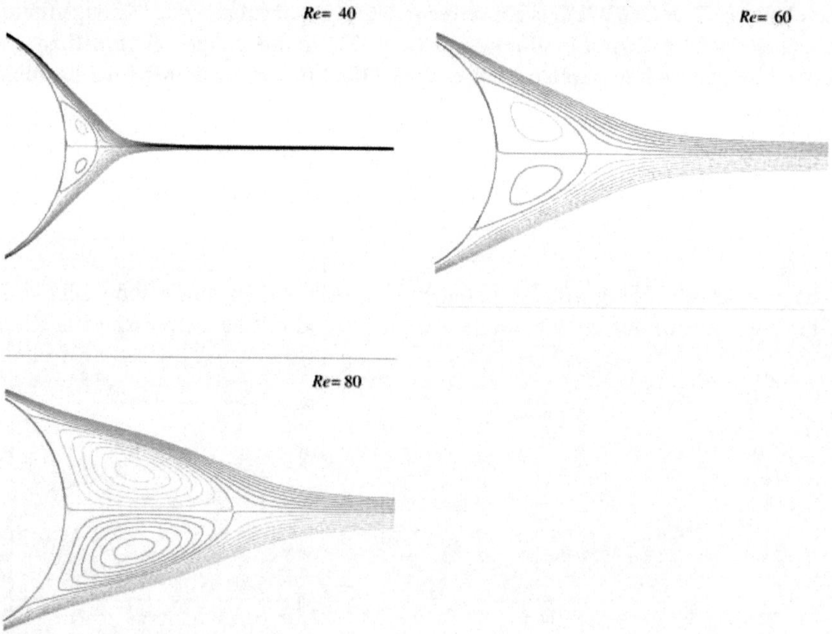

Fig. 3 Streamlines at Re = 40, 60 and 80

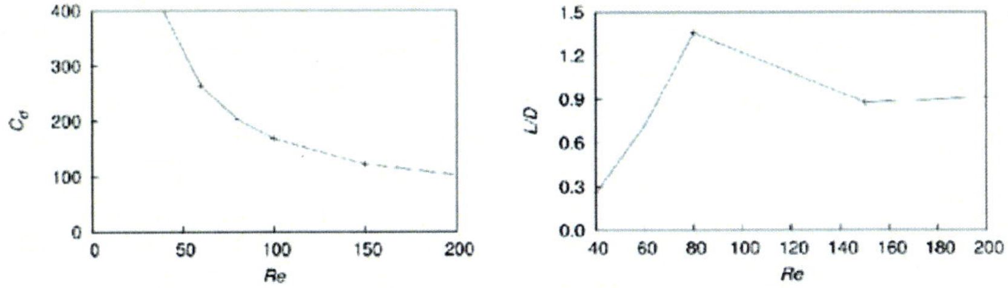

Fig. 4 Distribution of drag and eddy length (L/D) with Re

6.3 Surface Pressure Distribution

Figure 5 shows the distribution of pressure at the surface of the cylinder for Re = 40, 60 and 80 under steady-state conditions. It is found that the maximum pressure occurs at the forward stagnation point (0°) and its value is more in the case of low Re. Minimum pressure after $\theta = 90°$ over the half cycle. There is a sharp decrease in surface pressure just before $\theta = 90°$ and the minimum pressure is near about $\theta = 100°$. After that, there is not much change in surface pressure for different Re. Favorable pressure gradient is found between forward stagnation and cylinder shoulders.

Base point pressure is almost same for all Re as shown in Fig. 6. Forward stagnation point pressure is more in case of low Re and it decreases with increasing Re.

7 Unsteady Flow at Re = 150, 200 and B = 0.9

Vorticity contours, time series of $C_d - C_l$ and phase plots are presented for unsteady flow at Re = 150, 200 and B = 0.9

Fig. 5 Distribution of pressure along the surface of cylinder at Re = 40, 60 and 80

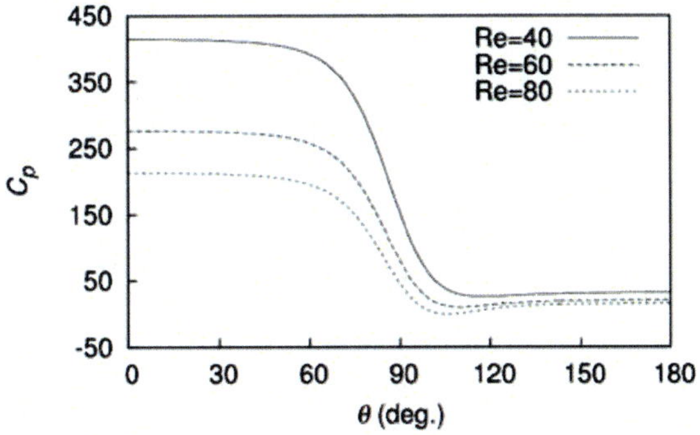

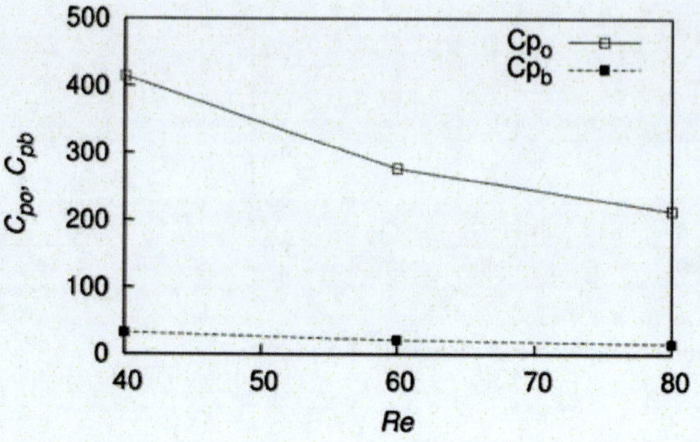

Fig. 6 Distribution of pressure at forward stagnation point and base pressure for Re = 40, 60 and 80

7.1 Vorticity at Re = 150 and 200

The vorticity contours at Re = 150 and 200 are shown in Fig. 7a, b. Both the contours represent the 2S mode of frequency because two vortices are shed per cycle. One vortice is shed from the top and another vortice is shed from the bottom of the cylinder. Instantaneous vortices shed in the downstream characterize the Karman vortex street.

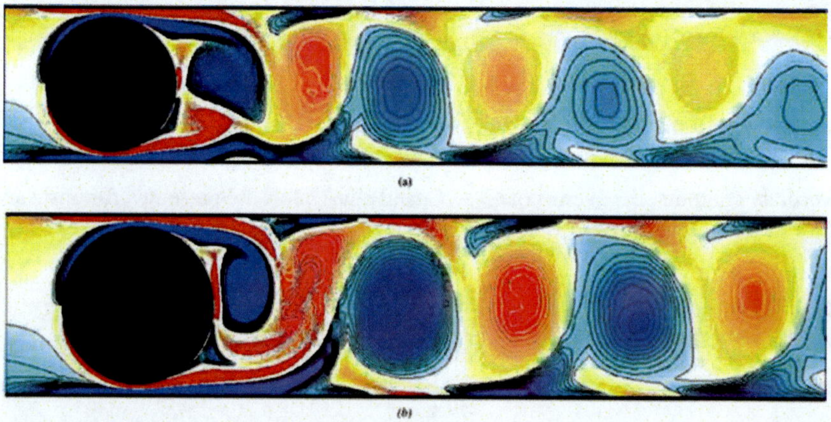

Fig. 7 Vorticity contours at Re = (**a**) 150 and (**b**) 200

7.2 Time Series of C_d and C_l at Re = 150 and 200

Time histories of drag and lift coefficients are plotted for Re = 150 and 200 after the convergence state is reached. Drag oscillates four times of St at Re = 150 and two times of St at Re = 200 (Figs. 8, 9, 10 and 11).

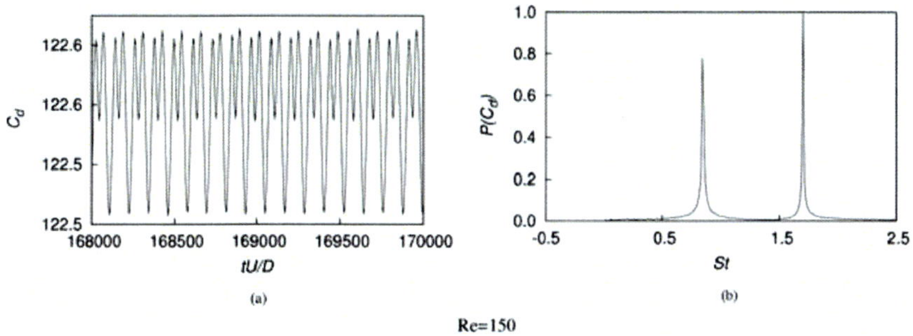

Fig. 8 **a** Time series of C_d and **b** power spectrum of C_d signal at Re = 150

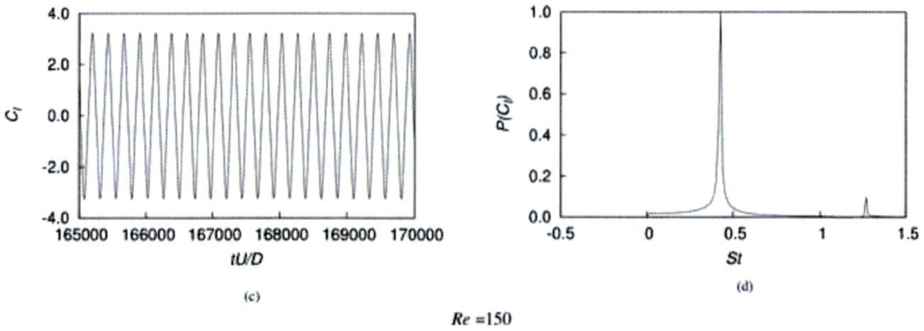

Fig. 9 **c** Time series of C_l **d** power spectrum of C_l signal at Re = 150

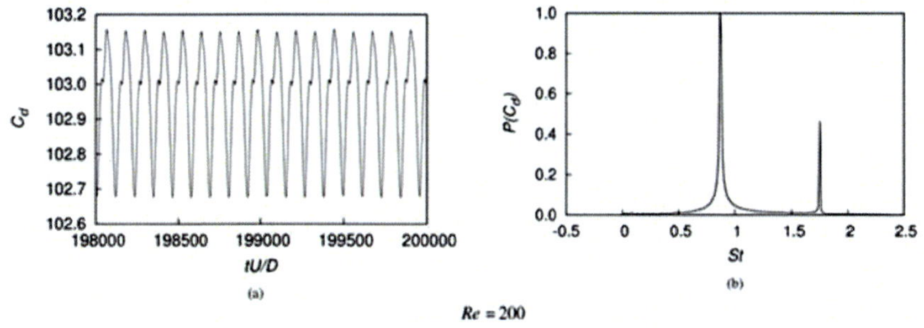

Fig. 10 **a** Time series of C_d and **b** power spectrum of C_d signal at Re = 200

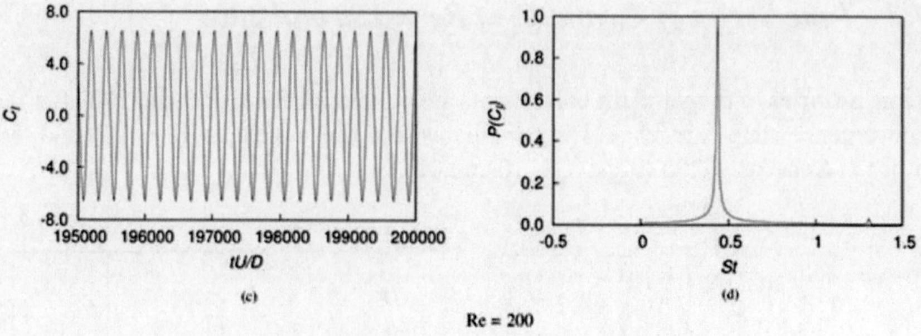

Fig. 11 **c** Time series of C_l **d** power spectrum of C_l signal at Re = 200

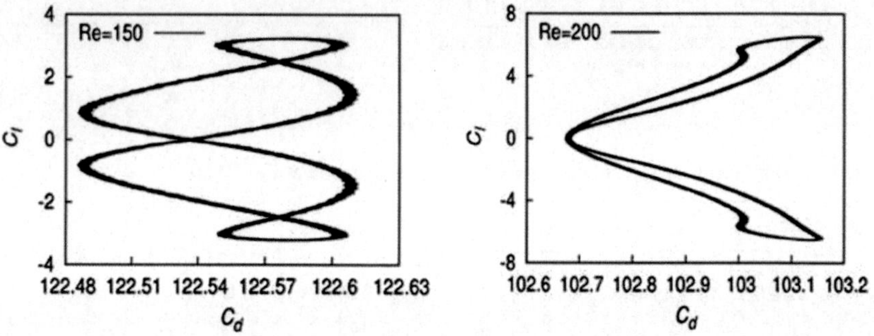

Fig. 12 Phase plots of drag and lift at Re = **a** 150 and **b** 200

7.3 Phase Plots of Drag and Lift at Re = 150 and 200

The Strouhal number is calculated from the lift coefficient data at Re = 150 and 200 and B = 0.9. Four loops of $C_d - C_l$ are generated for Re = 150 as compared to two loops of $C_d - C_l$ for Re = 200 (Fig. 12).

8 Conclusions

A Petrov–Galerkin finite element is used for discretization of the governing equation. Results have been presented for steady and unsteady flow regimes in highly confined flow around a circular cylinder from $40 \leq Re \leq 200$.

- For high blockage, there is no pressure recovery and this causes a large value of drag.
- Very strong vortices are shed from the cylinder and the wall effects resulting in an increase in the amplitude of drag and lift coefficients
- Blockage increases the drag and the Strouhal number.
- Surface pressure is minimum near the $\theta = 100°$.

References

1. Dennis SCR, Chang GZ (1970) Numerical solutions for steady flow past a circular cylinder at Reynolds numbers up to 100. J Fluid Mech 42:471–489
2. Badr HM, Dennis SCR, Young PJS (1989) Steady and unsteady flow past a rotating circular cylinder at low Reynolds numbers. Comput Fluids 17:579–609
3. Ingham DB, Tang T (1990) A numerical investigation into the steady flow past a rotating circular cylinder at low and intermediate Reynolds numbers. J Comput Phys 87:91–107
4. Tang T, Ingham DB (1991) On steady flow past a rotating circular cylinder at Reynolds numbers 60 and 100. Comput Fluids 19:217–230
5. Kang S, Choi H, Lee S (1999) Laminar flow past a rotating circular cylinder. Phys Fluids 11:3312–3322
6. Mittal S, Kumar B (2003) Flow past a rotating cylinder. J Fluid Mech 476:303–334
7. Sahin M, Owens RGR (2004) A numerical investigation of wall effects up to high blockage ratios on two dimensional past a confined circular cylinder. Phys Fluids 16:5–20
8. Sen S, Mittal S, Biswas G (2009) Steady separated flow past a circular cylinder at low Reynolds numbers. J Fluid Mech 620:89–119

Comparative Assessment of Unmodified Diesel Engine Characteristics with Commercial Diesels

Vipul Saxena, Naveen Kumar, and Raghvendra Gautam

Abstract As a result of the current ecological situation, it has become necessary to change the nature of fuel to limit the harmful effects of sulfur, which directly leads to SO_2 emissions and reduces the engine's life due to corrosion. In light of the recent mandate, the amount of sulfur in diesel was reduced from 50 to 10 ppm or even less to comply with green climate conditions. However, diesel desulfurization promotes a loss in lubricity due to the low level of aromatic and polar compounds in diesel. Diesel's low lubricity affects the chemical lubrication process of injectors, engine performance, and combustion. The purpose of this research was to compare engine characteristics between commercial Bharat Stage IV and Bharat Stage VI diesel. According to the results of the tests, Bharat Stage IV diesel has 3.52% better brake thermal efficiency and 4.23% less brake-specific fuel consumption at peak load than Bharat Stage VI diesel. As a part of the combustion process, in-chamber pressure, the pace of pressure rises, and the pace of dissipation of heat were measured, and it was discovered that Bharat Stage IV diesel has better combustion characteristics than Bharat Stage VI diesel on an unmodified diesel engine due to its better lubricity.

Keywords Performance · Combustion · Bharat stage · Desulfurization

1 Introduction

In modern times, the development of any nation is dependent on the availability and utilization of energy. Crude oil is a significant source of energy, accounting for more than one-third of global primary energy demand. Crude oil consumption is increasing day by day as the economy's prime mover, primarily in developing countries such

V. Saxena (✉) · N. Kumar
Centre for Advanced Studies and Research in Automotive Engineering, Delhi Technological University, Delhi, India
e-mail: vipul1611@gmail.com

R. Gautam
Department of Mechanical Engineering, Delhi Technological University, Delhi, India

V. Saxena
Department of Mechanical Engineering, JSS Academy of Technical Education, NOIDA, India

D. P. Mishra et al. (eds.), *Recent Trends in Thermal and Fluid Sciences*, Lecture Notes in Mechanical Engineering, https://doi.org/10.1007/978-981-19-3498-8_19

as India and China. Diesel engines play critical roles in the Indian economy because they are used in heavy trucks, city buses, trains, electric generators, and underground mining equipment [1]. Because of their greater proficiency and roughness, diesel engines have proven useful in transportation and military applications [2]. Furthermore, such engines are major contributors to unsafe outflows, and outflows from vehicles harm health and air quality. Lead, CO, NOx, particulate matter, carbon dioxide, and hydrocarbons are the primary components of an inner burning motor's fume gases [3, 4].

In the current global situation, the automobile industry is attempting to incorporate the most recent outflow guidelines to control the ever-increasing contamination. The implementation of BS-VI outflow standards, which was managed in April 2020 for significant urban areas in India, was a quick and significant step toward a cleaner and greener climate. These standards can be achieved only by a reduction in sulfur content in diesel from 50 to 10 ppm or even less. The most visible technology for that is diesel desulfurization. Along with sulfur, various polar and aromatic compounds are removed from the diesel as a result of this process. These compounds are associated with diesel lubricity. When two fluids with similar viscosities come into contact, wear, or scrape, the one with less contact, wear, or scraping is considered to have better lubricity [5, 6].

As a result, the purpose of this work is to compare the potential of Bharat Stage IV commercial diesel to Bharat Stage VI commercial diesel on an unmodified diesel engine. The comparison is based on engine characteristics such as performance and combustion. A more specific investigation would be a significant addition to the existing literature. This study evaluates the effect of Bharat Stage VI diesel on engine performance and combustion characteristics on the unmodified engine with the help of various physicochemical properties.

2 Experimental Setup

In this study, a comparison of Bharat Stage IV and Bharat Stage VI commercial diesel was performed. BS-IV diesel entered the market in 2017, with BS-VI following in April 2020. Table 1 summarizes the various physicochemical properties of two samples. A constant-speed, Kirloskar manufacture, single-chamber, air-cooled, direct infusion engine was chosen for the current experiment. The complete particulars of the motor are listed in Table 2, which include a rated power output of 5.8 kW @ 1500 rpm.

Both fuel experiments were carried out in an attempt to separate the burning and performance parameters. By observing burette particles and using a sophisticated watch, the speed was kept at around 1500 rpm and the time was kept for 2 cubic centimeters (cc). For varying the load, electrical loading arrangements were the part of the test rig. To make inferences, the illustrated charts were plotted after the performance. A Kistler piezoelectric pressing factor transducer and a crank angle encoder were utilized to complete the burning experiment. Figure 1 shows an engine test rig diagram.

Table 1 Comparison of various physicochemical properties of commercial diesels

Property (unit)	Sample 1	Sample 2	Measuring instrument	Testing method
Density (Kg/m^3)	834.62	832.37	DMA 4500 Density Meter	ASTM D4052
Kinematic Viscosity (mm^2/s^2)	2.565	2.476	Visco Bath, petro test	ASTM D445
L.H.V (MJ/kg)	46.27	43.89	Parr 6100 Calorimeter	ASTM D240
Sulfur (ppm)	35.33	3.45	XOS Sindie-7039	ASTM D7039
Cetane Index	46	49		ASTM D4737
Oxygen (wt %)	0.04	0.03	CHNSO Elemental Analyzer	ASTM D5373

Table 2 Motor particulars

Make	Kirloskar (Shakti)
Type	CAF-8 (Single-Cylinder, 4-Stroke, DI Engine)
Model	KSA-2057-P-1170
Compression ratio	17.5:1
Fuel injection pressure	200 bar
Fuel injection timing	23°bTDC
Bore diameter	95 mm
Stroke length	110 mm
Orifice diameter	13.4 mm

Fig. 1 Engine test rig

3 Outcomes and Discussion

The burning and performance character of BS-IV and BS-VI commercial diesel were thoroughly investigated in an unaltered CI engine.

3.1 Performance Parameters

With variable loading conditions, BTE, BSFC, and EGT were measured in a constant-speed engine.

Brake Thermal Efficiency. It computes the rate at which the fuel's warmth content is converted into useful power production. Despite this, the BTE of the several tests conducted under low stacking settings was nearly identical and insignificantly BS-VI diesel less than BS-IV diesel, and this could be owing to the reduced heating value of BS-VI diesel, as seen in Fig. 2 (Table 3).

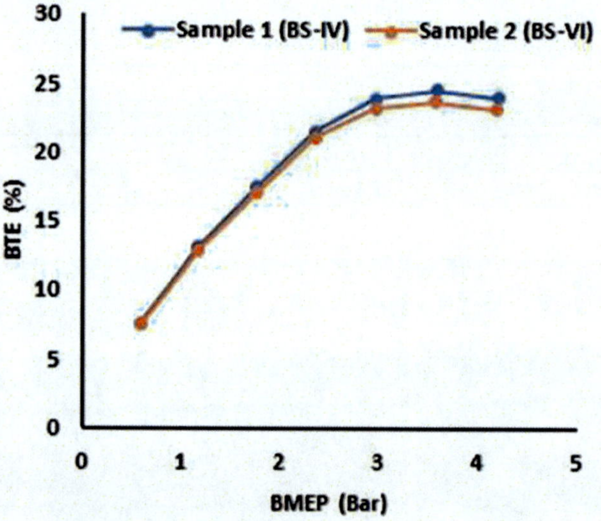

Fig. 2 BTE versus BMEP

Table 3 Performance output parameter (brake thermal efficiency)

	Sample 1 (BS-IV)		Sample 2 (BS-VI)	
Value	BMEP (bar)	BTE (%)	BMEP (bar)	BTE (%)
Minimum	0.582768	7.6067422	0.582388	7.497924
Maximum	3.583325	24.562862	3.571357	23.72761

Furthermore, at a 60% loading scenario, while the load is gradually increased, the BTE for BS-VI diesel is found to be almost equal to that of BS-IV diesel as shown in Fig. 2. In general, the BS-IV diesel has a higher warm efficacy when the motor is operated at medium to high loading conditions, implying enhanced and complete combustion [7, 8]. This could be because there is enough oxygen available to completely combust the fuel, increasing combustion efficiency.

Brake-Specific Fuel Consumption. The BSFC calculates environmental friendliness by calculating how effectively fuel is used to deliver unit power [9]. It's the ratio of fuel consumption to braking power in numbers. Figure 3 shows that fuel increments are used at lower loads; however, Fig. 3 shows the least BSFC is observed to be of BS-IV during transitional to high loading scenarios. It's possible that a higher warm discharge rate and an increased in-chamber pressure boost the motor's power output under medium to high load conditions.

In diesel engines, the fuel pump supplies fuel on a volumetric basis; however, because the density of BS-IV diesel is higher, the amount of fuel supply should be increased at the start to ensure proper combustion. BS-IV diesel has lower BSFC at the highest load than BS-VI diesel, possibly due to more consistent and productive burning at high loads, as well as a decrease in heat release rate [10]. It is reasonable to assume that BS-IV diesel consumes 4.23% less fuel than BS-VI diesel at peak load as per Table 4.

Fig. 3 BSFC versus BMEP

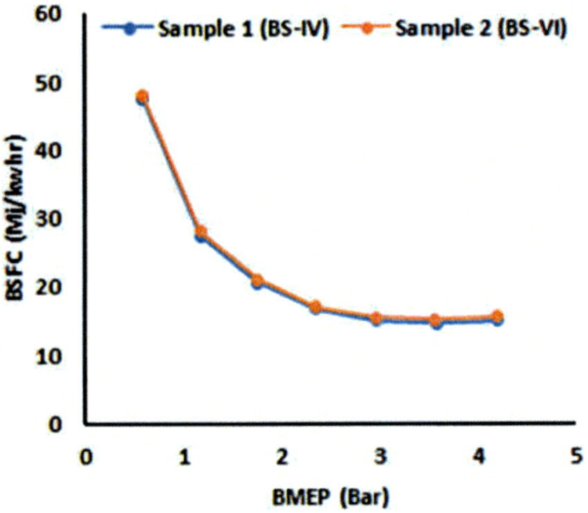

Table 4 Performance output parameter (brake-specific fuel consumption)

	Sample 1 (BS-IV)		Sample 2 (BS-VI)	
Value	BMEP (bar)	BSFC (Mj/kwhr)	BMEP (bar)	BSFC (Mj/kwhr)
Minimum	3.583325	15.182189	3.571356	14.566272
Maximum	0.582768	48.013289	0.582387	47.326436

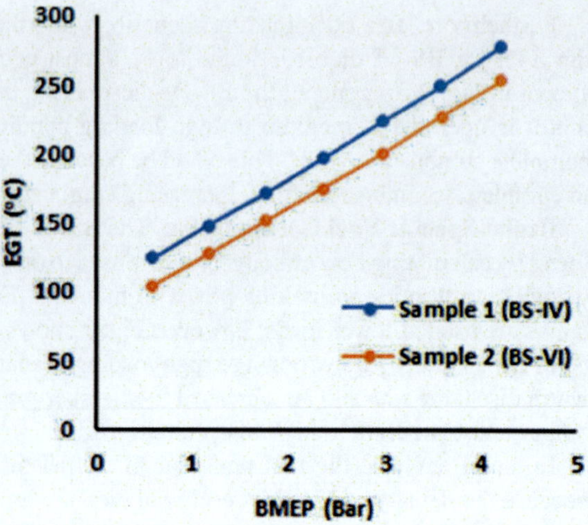

Fig. 4 EGT versus BMEP

Table 5 Performance output parameter (exhaust gas temperature)

	Sample 1 (BS-IV)		Sample 2 (BS-VI)	
Value	BMEP (bar)	EGT (°C)	BMEP (bar)	EGT (°C)
Minimum	0.582768	125	0.582387	104
Maximum	4.203082	278	4.194602	254

Exhaust Gas Temperature. While optimizing the motor's performance, EGT may be one of the most significant characteristics to consider [11]. Figure 4 shows that the nature of EGT for BS-IV diesel ranges from 104 to 254 °C and 125 to 278 °C for BS-VI diesel from low to high loading as given in Table 5. When the load is continually expanding, when EGT rises, it indicates that more fuel should be fed into the chamber [12].

3.2 *Combustion Analysis*

During the combustion examination of Bharat Stage IV and Bharat Stage VI commercial diesel, the variation in in-chamber pressure, heat discharge rate, and rate of pressure climb with crank angle were all taken into account.

In-cylinder Pressure. For BS-IV and BS-VI commercial types of diesel, Fig. 5 displays fluctuations in in-chamber pressure as a function of crank angle. Furthermore, it was discovered that the pressing factor patterns for tested energize under top-loading conditions are similar. The pinnacle chamber pressures for BS-IV and BS-VI diesel were 70.8 bar and 69.19 bar, respectively. Because of the adequate oxygen supply, BS-IV diesel-burning begins in advance, resulting in a superior pressure increase toward the top dead center (TDC) [12].

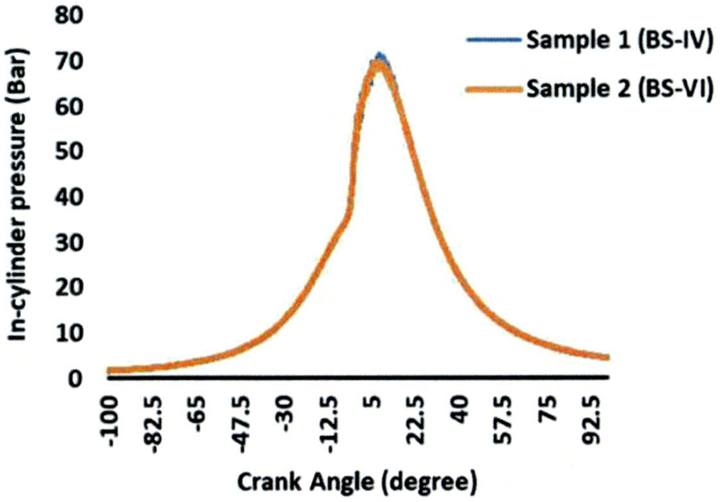

Fig. 5 Deviation in In-cylinder pressure corresponding to crank angle

Heat Release Rate. The heat release rate (HRR) decreases as the burning process advances, but accelerates after ignition is initiated, as seen in Fig. 6. The most extreme net HRRs for BS-IV and BS-VI diesel were 68.7 J/°CA and 64.8 J/°CA, respectively. The cetane index determines the quality of the gasoline start; a higher cetane index indicates a shorter ignition delay. Following the start delay measure, premixed burning is followed by a post-dissemination measure, in which the regulated fuel–air

Fig. 6 Deviation in heat release rate corresponding to crank angle

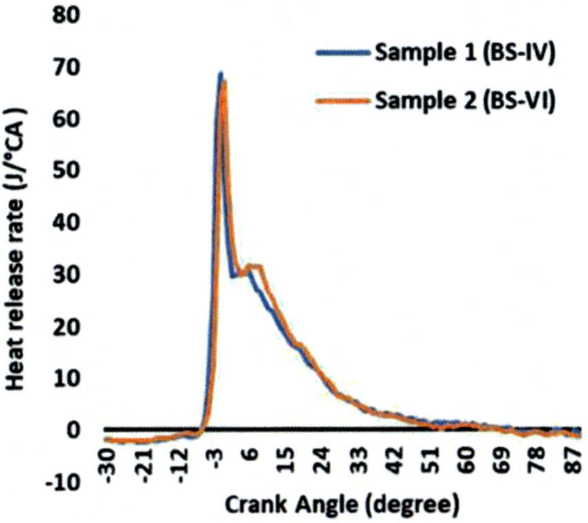

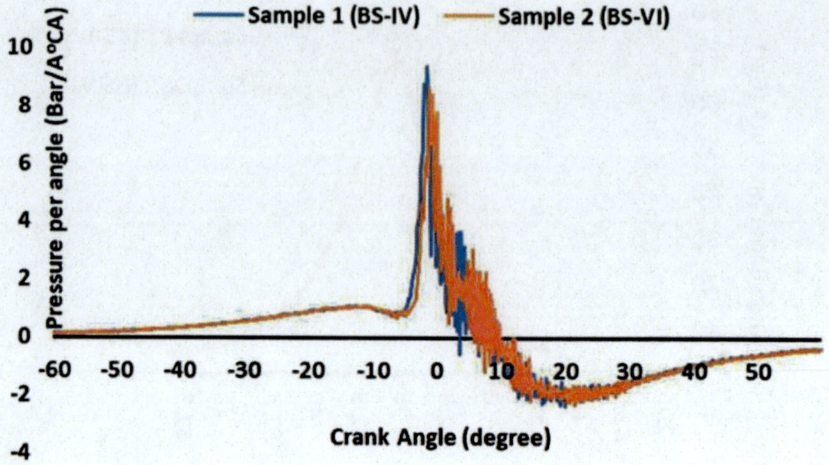

Fig. 7 Deviation in the rate of pressure rise corresponding to crank angle

combination obstructs HRR, as evidenced by reduced thickness and higher dissipation rate [12]. Performance qualities improved at top loads for tried fills due to effective and complete combustion, which can be found in the performance area.

Rate of Pressure Rise. With the motor running at a constant 1500 rpm, Fig. 7 displays the rate of pressure rise for a variety of BS-IV and BS-VI commercial diesel. The apex pressure for BS-IV diesel fuel is 9.35 bar/A°CA at maximum engine load and for BS-VI diesel fuel is 8.65 bar/A°CA. Because the start delay causes the ignition to occur before TDC, the pinnacle pressure for BS-IV diesel fuel is higher during burning cycle, and thus, heat is delivered quickly during premixed consumption [12].

4 Conclusions

This work offers a comparative assessment based on the trial examination. Two commercial diesel energizes were evaluated based on their performance and combustion characteristics on an unmodified diesel engine. The following conclusions can be reached based on the obtained results.

- BS-IV diesel fuel had a 3.52% higher brake thermal efficiency than BS-VI diesel during the tests. Low heating value and higher friction losses in BS-VI diesel due to low lubricating characteristics produce this effect.
- The brake-specific fuel consumption in BS-IV diesel is lower at moderate to peak load. At peak load, it uses 4.23% less specific fuel than BS-VI diesel.
- The generation of NOx is influenced by the temperature of the exhaust gas; hence, the chances of NOx formation are higher for BS-VI diesel on unmodified diesel engines.
- BS-IV diesel has somewhat more in-cylinder pressure at TDC than BS-VI diesel due to its higher calorific value and cetane index.

References

1. Panda K, Ramesh A (2021) Diesel injection strategies for reducing emissions and enhancing the performance of a methanol-based dual fuel stationary engine. Fuel 289:119809. https://doi.org/10.1016/j.fuel.2020.119809
2. Chauhan BS, Kumar N, Cho HM (2010) Performance and emission studies on an agriculture engine on neat Jatropha oil. J Mech Sci Technol 24(2):529e35
3. Walsh MP (2001) Global trends in diesel emissions regulation—a 2001 update. SAE Technical Paper 2001–01–0183
4. Kevin CS (2000) Simultaneous reduction of NOx and particulate emissions by using multiple injections in a small diesel engine. SAE Paper 01–3084
5. Luo H, Fan W, Li Y, Nan G (2013) Biodiesel production using alkaline ionic liquid and adopted as a lubricity additive for low-sulfur diesel fuel. Bioresour Technol 140:337–341
6. Fajar R, Sugiarto B (2012) Predicting fuel properties of partially hydrogenated jatropha methyl esters used for biodiesel formulation to meet the fuel specification of automobile and engine manufacturers. Kasetsart J Nat Sci 46(4):629–637
7. Raheman H, Ghadge VS (2007) Performance of compression ignition engine with mahua (Madhuca indica) biodiesel. Fuel. https://doi.org/10.1016/j.fuel.2007.02.019
8. Vibhanshu V, Karnwal A, Deep A, Kumar N (2014) Performance, emission and combustion, analysis of diesel engine fueled with blends of mahua oil methyl ester and diesel. SAE Tech Pap. https://doi.org/10.4271/2014-01-2651
9. Azad AK, Rasul MG, Khan MMK, Sharma SC, Bhuiya MMK (2016) Recent development of biodiesel combustion strategies and modeling for compression ignition engines. Renew Sustain Energy Rev. https://doi.org/10.1016/j.rser.2015.12.024
10. Rajasekar E, Selvi S (2014) Review of combustion characteristics of CI engines fueled with biodiesel. Renew Sustain Energy Rev. https://doi.org/10.1016/j.rser.2014.04.006
11. Ashraful AM, Masjuki HH, Kalam MA, Fattah IMR, Imtenan S, Shahir SA (2014) Production and comparison of fuel properties, engine performance, and emission characteristics of biodiesel from various non-edible vegetable oils: a review. Energy Convers Manag. https://doi.org/10.1016/j.enconman.2014.01.037
12. Gautam R, Kumar N, Sharma P (2013) Comparative assessment of performance, emission and combustion characteristics of blends of methyl and ethyl ester of jatropha oil and diesel in compression ignition engine. SAE Technical Paper, 01–2664

Optimization of Jet Angle in Crossflow Injectors

Anil Antony Karathra, K. J. Antony Alwin, Adams Joseph D'coutho, Joseph Antony Edezhath, and Aswith R. Shenoy

Abstract A crossflow or transverse flow injector is a type of fuel injector used in rocket engines. This study is focused on how a change in the angle of liquid jet injection affects the mixing of propellants in these injectors. The results are analyzed using two fuel-mixing parameters, i.e., the spray cone angle and jet impingement distance. CFD methods have been employed to replicate an experimental test setup, and simulations have been computed for injection angles of 30°, 45°, and 60° in the direction of flow, and 90°. Water has been selected as the material to substitute for the injected liquid propellant, and nitrogen gas has been selected as the mainstream gas. The results of this work help in determining the optimal angle of injection of liquid propellants in this type of injector. Further, this helps in better atomization of propellants which results in better combustion and reduced release of pollutants.

Keywords Injector · Atomization · CFD

1 Introduction

Study of the atomization process is an unavoidable factor in the design of an injector. From the atomization study, the information regarding the mixing of the propellants, spray characteristics, droplet size, and the nature of the distribution can be derived.

The objective of this work is to determine the optimal angle for injection of a liquid jet propellant using a jet in a crossflow mechanism, so as to ensure maximum atomization and mixing. The study is done by simulating the injection process in the ANSYS Fluent CFD suite. The optimal angle is found by studying the paths followed by the liquid jets after injection. A simulation is run for each angle of injection keeping the boundary conditions unchanged, and the results from each simulation are compared with each other to determine the final result.

The atomization and mixing of fuel are one of the key factors, which decides engine performance, reliability, and efficiency. Hence, it is important that the fuel

A. A. Karathra (✉) · K. J. Antony Alwin · A. J. D'coutho · J. A. Edezhath · A. R. Shenoy
Department of Mechanical Engineering, Albertian Institute of Science and Technology, Kochi, Kerala 682022, India
e-mail: antony3anil@gmail.com

D. P. Mishra et al. (eds.), *Recent Trends in Thermal and Fluid Sciences*, Lecture Notes in Mechanical Engineering, https://doi.org/10.1007/978-981-19-3498-8_20

gets pre-mixed in the injector, in order to maintain a good supply of well-mixed fuel in the combustion chamber. Previous findings suggest that co-axial injectors are the most preferable for injection of propellants in a combustion chamber. This is mostly because of their simple design, low pressure losses, and better combustion stability, when compared to other injectors.

One of the most effective ways of improving the atomization of a liquid jet in transverse flow is by changing its angle of injection. The atomization characteristics of liquid jet injection at different angles were studied by Fuller et al. [1] and showed that the liquid jet breakup process happened in two regimes. The aerodynamic forces related to the gaseous crossflow accelerate the liquid jet in the flow direction and form unstable waves that lead to the breakup of the liquid jet. When it comes to a non-aerodynamic breakup, the liquid jet disintegration is mainly due to the liquid turbulence and liquid inertial forces. For an injection angle of 90°, a liquid jet breakup occurs due to aerodynamic regimes. But in the case of smaller angles of injection, say at 30°, the liquid disintegration is due to the non-aerodynamic regimes, leading to the formation of large kinks.

H. Almeida et al. [2], in their study, also suggested that when the angle of liquid injection decrease from 90 to 30° the breakup process becomes less effective. The relative velocity between the liquid jet and crossflowing gas decreases as the injection angle decreases. The aerodynamic forces acting on the liquid jet were also observed to be following the same pattern, causing an increase in drop size (SMD) hence decreasing the quality of atomization.

The spray cone angle or spray cone half-angle has previously been used as a key parameter in studies related to combustion and sprays by various researchers. Chung et al. [3] suggested that the burning conditions were greatly affected by the fluctuations in cone angle. Additionally, in his work R. K. Mishra [4] showed that an increase in spray cone angle up to 100° helps in improving the combustion quality. As the angle goes beyond 100°, a drop in combustion temperature was observed.

JICF has been studied previously using the CFD method by various scholars worldwide. Many of these simulations make use of the Direct Numerical Simulation (DNS) approach, Reynolds averaged Navier–Stokes (RANS) approach, or Large Eddy Simulation (LES). Su and Mungal [5] mentioned that crossflowing jet simulations are being explored using LES because of their inherent ability to handle temporal unsteadiness and spatial complexity.

A. M. Ruiz et al. [6] performed a highly resolved Large Eddy Simulation of crossflowing jets based on a previously conducted experiment by Su and Mungal [5]. This simulation helps in understanding how the jet trajectory can be mapped based on the momentum flux ratio. It also gives an idea of how the mesh grid affects the results of the simulation. Further, E. Liu et al. (2020) studied the relationship between the fuel–air mixing and flow characteristics at different velocity ratios for a methane jet in crossflow. They concluded that jet and crossflow interact intensely in the near field, and that the penetration of the jet depends on the velocity ratio (R). At lower values of 'R', the fuel bends at the outlet of the injector, while at higher velocity ratios the fuel penetrated deeper providing more region for the air–fuel mixture.

The main disadvantage of LES is that it requires a significant computational ability to compute the solutions of a simulation. Due to the same reason, the SST k-ω model was employed for all the simulations. Discrete Phase Modeling (DPM) was enabled instead of the standard Volume of Fluid (VOF) approach to reduce computation time. The liquid jet was introduced into the crossflowing gas as inert particles in the simulation, and the results were obtained by measuring the distance between the gas inlet and the point of impingement of the jets.

2 Methodology and Procedure

There are various parameters based on which mixing of two fluids can be analyzed, some of them being mean drop size (SMD), spray distribution, spray pattern, spray cone angle, etc. For the purpose of this study, only two parameters have been taken into consideration: the distance from the outlet of the injector to the point of impingement of the jets and the spray cone angle.

When two liquid jets are injected transversely through an orifice into another fluid flowing across the orifice, the jets meet at a point downstream in the direction of flow. The point of impingement of these jets depends mainly on the angle of injection and the flow velocities. Also, this parameter is a key factor that decides the extent to which the two fluids get mixed. The closer the point of impingement is to the outlet of the injector, the better will be the mixing.

Another important parameter that decides the quality of mixing of two fluids is the spray cone angle. The wider the cone angle is, the better will be the mixing. It is to be noted a cone angle that is too wide isn't ideal since this may cause the fine droplets to hit the injector walls and form larger droplets or lumps.

2.1 Design of Injectors

Four different configurations of gas-centered co-axial injectors have been considered for the purpose of this study. These injectors make use of the JICF mechanism to mix two propellants of which one is gaseous and the other is a liquid. The gas is injected through the central part, while the liquid propellant is injected through small orifices.

The orifice diameter is set as 1 mm, and the number of orifices will be six. These liquid injection orifices are arranged in two rows, each having three orifices. The radial angular distance between the orifices is 120° and the second row is provided with a 60-degree radial offset in order to prevent collision of the jets.

The distance between the two rows of liquid injection orifices is set as 8.06 mm. This is because for a liquid jet in crossflow, the axial breakup length for a single jet is equal to 8.06 times of Liquid jet diameter [7]. Further, the ratio of the length of the liquid injection orifice to the diameter of the orifice is taken as 2 to prevent hydraulic flip and cavitation [8].

Table 1 Design data of injector

Angle of injection	Max range of MFR (q)	Transverse breakup length (Y/d_j) (at max MFR)	Center post diameter	Axial breakup length (X/d_j) (at max MFR)	Distance from last row of injection to exit
30	11	3.26	12	3.73	15
45	11	6.52	12	14.89	20
60	11	9.78	12	9.931	14
90	11	10.874	12	7.44	12

Table 2 Fixed design parameters

Sr. No	Parameter	Value
1	Liquid injection orifice diameter	1 mm
2	Number of liquid injection orifices	6 Nos
3	Number of row of orifices	2 rows of three orifices each
4	Radial angular distance between the two injectors in single row	120 Degrees
5	Axial distance between the two rows	8.06 mm
6	Ratio of length of liquid orifice to diameter of orifice	2

The distance between the last row of injection and the exit of the injector is decided based on the axial breakup length, and the diameter of the injector is based on the transverse breakup length (Tables 1 and 2).

For an angle of injection of 90°, the transverse breakup length is

$$Y/d_j = 3.44 \times q^{0.5}[3] \tag{1}$$

And, for all other forward injection angles, it is

$$Y/d_j = 6.33 \times q^{0.314} \times \sin^2\theta[19] \tag{2}$$

The axial breakup length for all angles of injection is calculated as

$$X/d_j = 6.34 \times q^{0.07}/\sin^2\theta\ [19] \tag{3}$$

3 Grid Independence and Simulation Setup

A grid independence study is performed to find the effect of the number of grids on the results of the simulation. This process helps in finding the grid, which yields a

Table 3 Mesh statistics

METRIC	30°	45°	60°	90°
Element quality	0.76617	0.64544	0.7665	0.76682
Avg. aspect ratio	2.2061	2.2138	2.2058	2.205
Avg. skewness	1.1046e−002	1.0838e−002	1.0622e−002	1.038e−002
Avg. orthogonal quality	0.99942	0.87852	0.9974	0.9998
Nodes	44,341	44,341	44,341	44,341
Elements	43,744	43,744	43,744	43,744

good result with minimum computational time. In this work, the independent study was carried out on grids of sizes of approximately twenty-one thousand, thirty-five thousand, forty-four thousand, and fifty-three thousand.

The results of the study showed that an increase in grid size beyond forty-four thousand does not yield any significant variation in the results, while a grid size of nearly twenty-one thousand could not capture the required physics of the simulation. Although a grid size of thirty-five thousand yielded good results, it had slight deviations from the results obtained with the finer grid of size forty-four thousand. Hence, the grid size was finalized to be nearly forty-four thousand.

The detailed mesh statistics for all the geometries is shown in the Table 3.

3.1 Solver Setup

The simulations were run in the SST k-omega viscous model with two Eulerian phases, one being gaseous nitrogen and the other being water. Inlet velocities of the two fluids were used as the boundary conditions, with nitrogen having a velocity of 158.47 m/s and water having a velocity of 7.0416 m/s. The liquid jet was injected into the transverse flow as inert particles of diameter 0.0002 m by enabling the Discrete Phase Model (DPM). Also, the operating density was set at 1.225 kg/m^3 for all the simulations.

4 Results

The results obtained from the simulation of each angle were compared side by side to find the best angle of injection. As mentioned earlier, the spray cone angle and the distance of impingement of the jets are two key parameters on which the results of this study are based upon.

The results obtained are presented in the Table 4.

The mesh used for the simulation is shown in Figs. 1 and 2, while Fig. 3 shows the velocity contour obtained from the simulation. Figures 4, 5, 6, and 7 show the

Table 4 Results

Angle of injection (degrees)	30	45	60	90
Impingement distance (mm)	76	29	13	04
Spray cone angle (degrees)	2.97	6.83	10.46	17.17

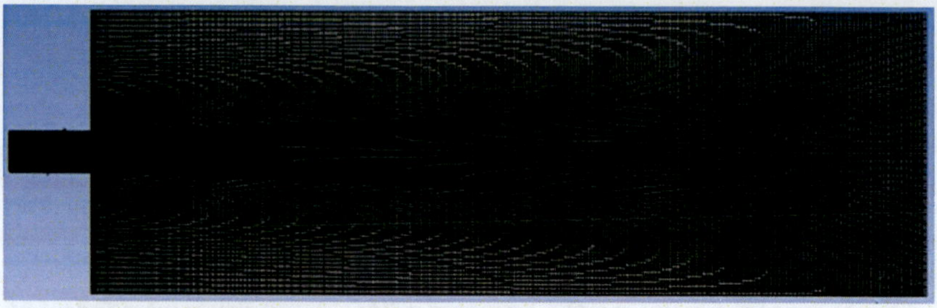

Fig. 1 Mesh used for simulations

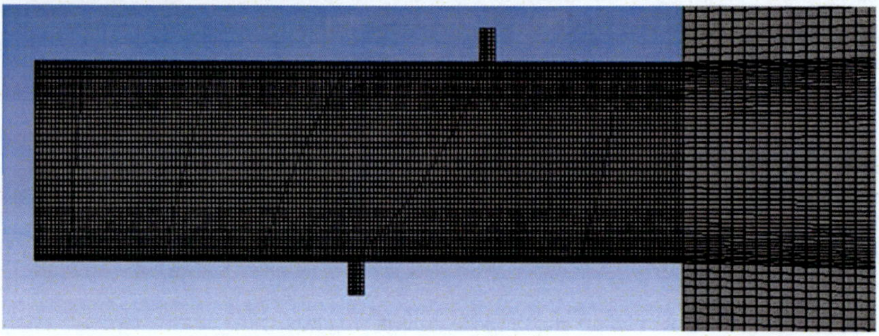

Fig. 2 Mesh (Injector Part)

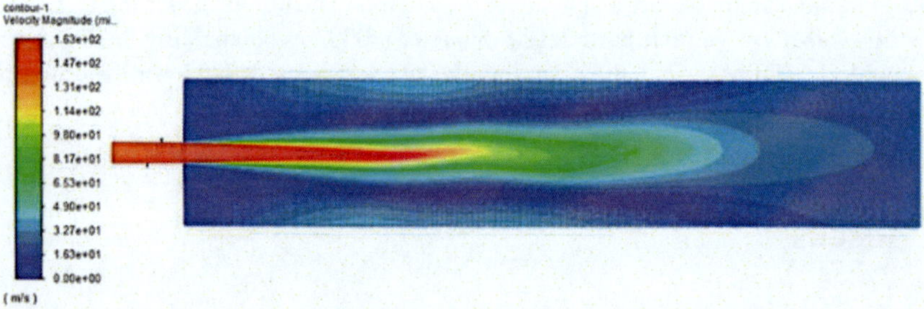

Fig. 3 Velocity contour of 90-degree injection

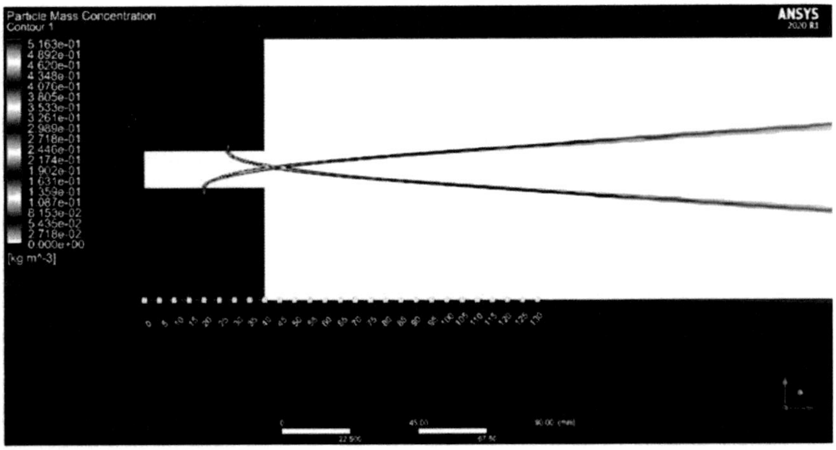

Fig. 4 Path followed by the water particles in 90-degree injection

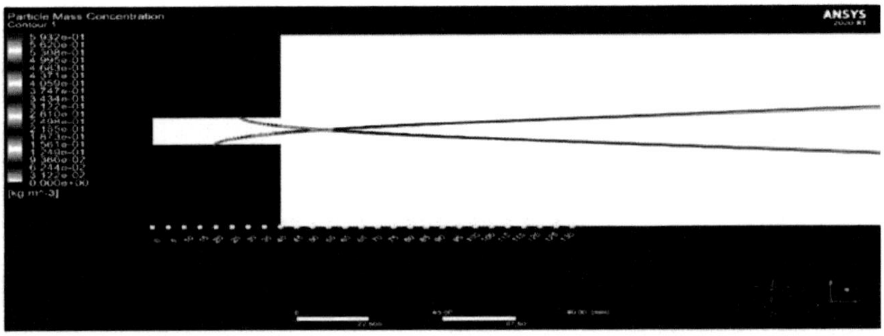

Fig. 5 Path followed by the water particles in 60-degree injection

path followed by the injected water droplets for each angle. The presented data was obtained by processing the simulation results.

The analysis of the data shows that as the injection angle increases from 30 to 90°, there is a decrease in impingement distance and an increase in spray cone angles. This shows that as the injection angle increases, the quality of mixing of the propellants increases, which thereby increases the combustion quality.

5 Conclusion

In this study, the effect of injection angle on the mixing of propellants has been modeled and analyzed using CFD simulations. The following conclusions can be drawn from the results:

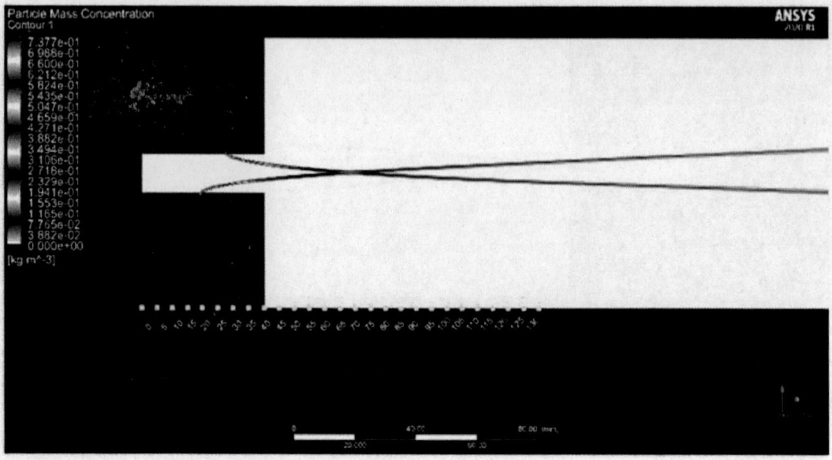

Fig. 6 Path followed by the water particles in 45-degree injection

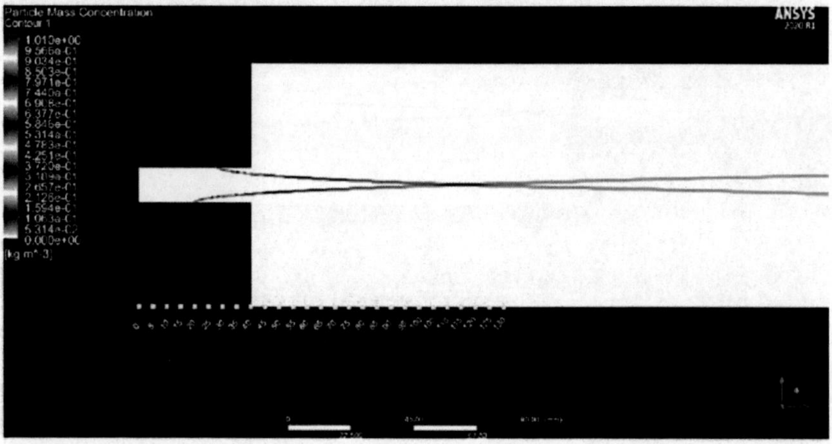

Fig. 7 Path followed by the water particles in 30-degree injection

- The angle of injection of a liquid jet in crossflow injectors directly affects the combustion process.
- An increase in injection angle from 30 to 90° caused the spray cone angle to increase significantly.
- The optimal angle for injection was found to be 90°, since this angle makes the highest value of spray cone angle, and has the lowest impingement distance (Figs. 8 and 9).

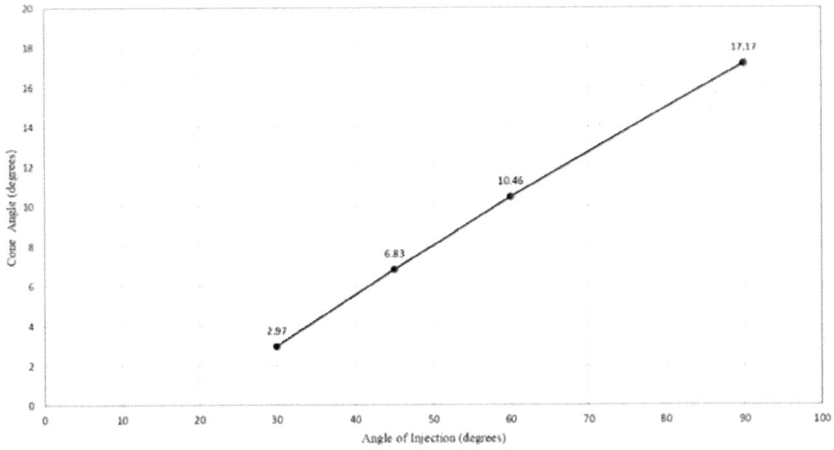

Fig. 8 The plot shows an increase in spray cone angle as the angle of injection increases

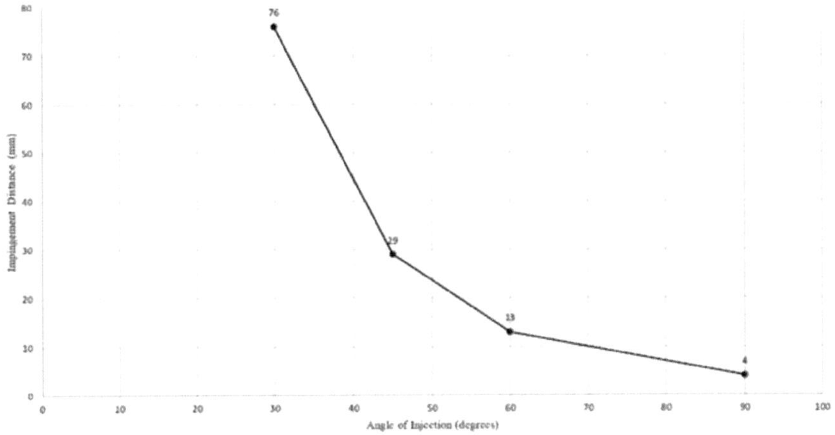

Fig. 9 The plot shows a decrease in impingement distance as injection angle increases

References

1. Fuller RP et al (2000) Effect of angle on atomization of liquid jets in transverse airflow. AIAA J 38(1)
2. Almeida H et al (2010) Effect of the liquid injection angle on the atomization of liquid jet in subsonic crossflow. In: ILLAS-Europe 2010, 23rd annual conference on liquid atomization and sprays system, Brno Czech Republic
3. Chung Y, Kim H, Jeong S, Yoon Y (2016) Dynamic characteristics of open-type swirl injector with varying geometry. J Propul Power 32(3):583–591
4. Mishra RK, Kishore Kumar S, Chandel S (2016) Effect of spray cone angle on flame stability in an annular gas turbine combustor. Int J Turbo Jet Engines 33(1):35–44

5. Su LK, Mungal MG (2004) Simultaneous measurements of scalar and velocity field evolution in turbulent crossflowing jets. J Fluid Mech 513
6. Ruiz AM et al (2015) Flow topologies and turbulence scales in a jet-in-cross-flow. Phys Fluids 27(4)
7. Wu P-K et al (1997) Breakup process of liquid jet in subsonic cross flows. J Propuls Power 13(1)
8. Song J et al (2011) Effect of orifice internal flow on liquid jet in subsonic cross flow. J Propuls Power 27(3)

Influence of Truncation on the Performance of Aerospike Nozzles Using Numerical Technique

Prahlad V. Deshpande, G. N. Nagaharish, Abdulrajak Buradi, and Acharya Madhusudhan

Abstract Aerospike Nozzles (ANs) or altitude compensating nozzles have been under research and development for the Single Stage to Orbit (SSTO) launching vehicles since the 1950s. Plug nozzles are variants of the Aerospike Nozzle (AN) which are truncated along the length. Computational Fluid Dynamics analysis was employed to study the effect of truncation on an AN, with different percentages of truncations such as 0, 20, 40, 60 and 80% of the length of nozzle. For designing the aforesaid AN geometries with different percentages of truncations, Autodesk Fusion 360 software was used. The behaviour of the flow and study of other parameters like velocity, Mach number and temperature were analysed using Ansys Fluent 19.1 software. The analysis was done using the K-Epsilon (k-ε) turbulence model. It is one of the most widely used turbulent analysis models. It has been observed that due to the change in the truncation of the nozzle, we see that the above-stated parameters differ in each case. It is found that the nozzle profile with 40% truncation exhibits superlative results compared to the other nozzle profiles.

Keywords Aerospike nozzle · Plug nozzles · Altitude compensating nozzle · Single stage to orbit (SSTO) · Computational fluid analysis · K-epsilon

Nomenclature

t	Time
υ	Total velocity
ρ	Density
θ	Temperature
θ_0	Absolute temperature
ϕ	Eddy viscosity
ϵ	Rate of deformation

P. V. Deshpande · G. N. Nagaharish · A. Buradi (✉) · A. Madhusudhan
Centre for Computational Fluid Dynamics, Mechanical Engineering Department, Nitte Meenakshi Institute of Technology, Yelahanka, Bangalore 560064, Karnataka, India
e-mail: arbnitk@gmail.com; abdulrajak.buradi@nmit.ac.in

D. P. Mishra et al. (eds.), *Recent Trends in Thermal and Fluid Sciences*, Lecture Notes in Mechanical Engineering, https://doi.org/10.1007/978-981-19-3498-8_21

v_x	Velocity component in x-direction
τ	Shear stress
α	Bulk viscosity coefficient
v_y	Velocity component in y-direction
μ_0	Dynamic viscosity at 273 K
μ	Dynamic viscosity at a specified temperature
A_e/A_t	Area ratio at the Exit
P_i	Nozzle inlet pressure
P_o	Nozzle exit pressure
A_t	Area at the throat
A_e	Area at the area
f_x	Body force in x-direction
E_i	Internal energy
k	Thermal conductivity
f_p	Pressure force
f_y	Body force in y-direction
f_z	Body force in z-direction
S_c	Sutherland's constant
v_z	Velocity component in z-direction
$\dot{Q}$	Volumetric heat addition per unit mass

1 Introduction

The combustion system in rockets is different from the conventional turbojet engines which are air-breathing. The rocket engines (REs) are solely dependent on the propellants for producing thrust. Nozzles play integral roles in amplifying the thrust produced by the REs. Nozzles are used largely in REs to hasten exhaust gases to supersonic speeds in the axial directions. Nozzles are broadly classified into Bell nozzles, Convergent–Divergent nozzles or De Laval nozzles and Aerospike nozzles (ANs). An Aerospike nozzle (AN) is a sort of nozzle with the capability of constant elevation compensation. ANs are named after their design which consists of "spikes". An AN profile can be interpreted as an inside-out bell nozzle. The conical design facilitates transforming and amplifying the effect of oxidization of rocket fuels that develops at high pressures. This results in high-speed effluvium discharge from the nozzle outlet. In the case of bell nozzles, the exit pressure of the exhaust gases is limited by the nozzle geometry. In ANs, there is no geometrical restriction to the exhaust gases. Due to this, the expansion of exhaust gases varies to match the ambient pressure at all altitudes. This makes them altitude compensating nozzles. In well-developed conditions, we will have a single stage to orbit launch vehicles.

The presence of truncation of spikes in ANs is to adjust the atmospheric pressure and theoretically assimilate the carrier rockets. This affects the flow properties on the surface of the spike with deviation in the plug length [1]. At different altitudes, the aerospike models constantly maintain aerodynamic efficiency, thus reducing the fuel usage by 25–30% at lower altitudes. The study conducted on these AN models involved the varying of the angle of attack and the spike length to interpret the variation of fluid properties with respect to the spike lengths [2]. At higher altitudes, the AN is able to increase the exhaust of an engine to a higher active nozzle–area ratio. Improved performances can be attained relative to the bell nozzles, particularly for prolonged circumstances. Similarly, in over-enlargement circumstances, the losses may not be identical as in the case of the bell nozzle [3]. In under-prolonged circumstances, for the design settings, this kind of nozzle performs as a perfect nozzle. ANs are frequently mentioned in an altitude recompensing nozzle owing to their ability of preserving efficiency with the alteration of elevation [4]. As compared to a conventional bell-shaped nozzle, the performance of AN is superior, i.e., overall 90% [3, 5]. A space vehicle employed with AN saves about 25–30% of fuel at fewer altitudes [3, 4]. Due to atmospheric pressure, at fewer altitudes the efficiency of AN is much higher that limits the increase of the exhaust gas.

In [6, 7], the aerospike is considered as the main issue The study involved heat transfer analysis of multiple design profiles with variation in the distances. In [8], the effect of truncation in aerospike is performed numerically for different pressures. In the study conducted by Karthikeyan et al. [9] on ANs, they have attempted to improve the flow straightness and the thrust. They have done flight research of their design using high power solid fuel, and it was concluded in their study that, due to the greater actual aerospike throat area compared to the designed throat area, the aerospike chamber pressure and thrust were subsided. Sanoob et al. [10] have conducted a study on the ANs which emphasizes the spike length optimization of the ANs. They have also considered multiple pressure ratios for their optimization. Their outcome illustrates that the full spike model delivers the maximum performance. But when truncated, up to 60% truncation, the behaviour is analogous to that of a full spike. Muwaaz and Shaik [11] carried out their study on the flow analysis of expansion deflection nozzles. They have conducted numerical analysis for the truncated models of under-expanded, ideal and over-expanded nozzles. They have studied the variation of thrust with respect to truncation for the aforementioned cases. From their study, they have concluded that in the case of under-expansion, the base pressure compensates for the thrust loss; the effect of truncation in this case is trivial. In the case of over-expanded nozzles, the thrust intensification is directly proportional to the truncation of the spike.

Reddy et al. [12] have conducted a study on the effect of the spike geometry on the performance of the spiked nozzle. They have done a computational study on the effect of truncation and angle on the performance of a spiked nozzle. They have considered angular models of 35° and 45° and truncated models of 20% and full spike lengths (100% truncation). The authors found that the performance in the case of angular spikes, the velocity output is substantial than the truncated models. The study on ANs under different working conditions is performed by Dumitrescu et al. [13]. They used different truncated models to study the performance of ANs under different working conditions, i.e., for the pressure and temperature. The study conducted by Rommel et al. [14] was on the flow-field analysis of a plug nozzle, for multiple pressure ratios. From their simulations, they have found that the specific impulse is superior in the case of truncated plug nozzles. In this paper, we have presented a detailed study on the effect of truncation on the profile performance of different truncated ANs using the CFD technique.

2 Methodology

2.1 CAD Model Design

There are two different methods, namely the approximate method and the method of characteristics established on rocket assumptions have been employed to design the AN for different percentages of truncations. We used Autodesk Fusion 360 software for our design purposes. We have designed a 2-D model of an Aerospike Nozzle section. The Nozzle designs have been truncated for every 20% up to 80% of length. We have considered this design as we had the scope to incorporate some changes as required. This design has been considered for a pressure ratio (Pi/Pa) of 20:1. Figure 1 and Table 1 represent the geometry and design parameters of the axisymmetric AN, respectively.

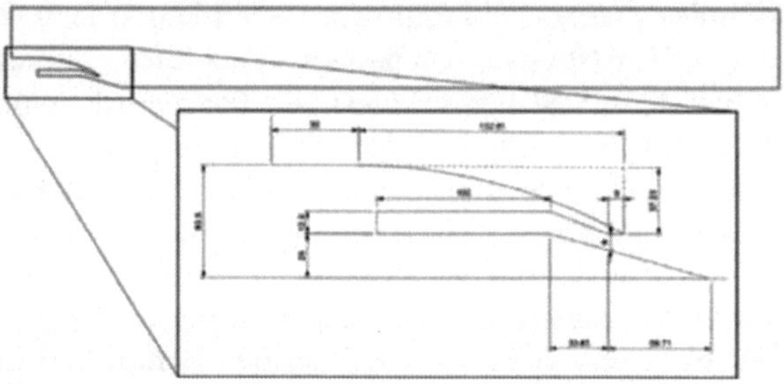

Fig. 1 Geometry of the nozzle

Table 1 Design parameters

Design parameters	Values
Throat area, A_t	92.916 mm^2
Outlet area, A_e	922.702 mm^2
Throat to outlet area ratio, A_e/A_t	9.930
Nozzle inlet pressure, P_i	20 bar
Nozzle exit pressure, P_o	1.01325 bar

2.2 Governing Equations

Nozzles are elements which are always bound by the continuity and momentum equation. This implies that the mass, momentum and energy are always conserved in a nozzle.

The continuity equation is given as

$$\frac{\partial \rho}{\partial t}+\rho\left(\frac{\partial v_x}{\partial x}+\frac{\partial v_y}{\partial y}+\frac{\partial v_z}{\partial z}\right)=0 \tag{1}$$

The Energy equation in (3) dimensions is given as

$$\begin{aligned}\frac{\partial}{\partial t}\left[\rho\left(E_i+\frac{v^2}{2}\right)\right]+\nabla\cdot\left[\rho\left(E_i+\frac{v^2}{2}\vec{v}\right)\right]=\rho\dot{Q}&+\frac{\partial}{\partial x}\left(k\frac{\partial\theta}{\partial x}\right)+\frac{\partial}{\partial y}\left(k\frac{\partial\theta}{\partial y}\right)\\&+\frac{\partial y}{\partial z}\left(k\frac{\partial\theta}{\partial z}\right)-\frac{\partial(v_x f_p)}{\partial x}-\frac{\partial(v_y f_p)}{\partial y}\\&-\frac{\partial(v_z f_p)}{\partial z}+\frac{\partial v_x\tau_{xx}}{\partial x}+\frac{\partial v_x\tau_{yx}}{\partial y}+\frac{\partial v_x\tau_{zx}}{\partial z}\\&+\frac{\partial v_y\tau_{xy}}{\partial x}+\frac{\partial v_y\tau_{yy}}{\partial y}+\frac{\partial v_y\tau_{zy}}{\partial z}\\&+\frac{\partial v_z\tau_{xz}}{\partial x}+\frac{\partial v_z\tau_{yz}}{\partial y}+\frac{\partial v_z\tau_{zz}}{\partial z}+\rho\vec{f}\cdot\vec{v}\end{aligned} \tag{2}$$

The Momentum equations in Cartesian coordinates (x, y and z directions) are given as

$$\rho\frac{\partial v_x}{\partial t}=-\frac{\partial p}{\partial x}+\frac{\partial\tau_{xx}}{\partial x}+\frac{\partial\tau_{yx}}{\partial y}+\frac{\partial\tau_{zx}}{\partial z}+\rho f_x \tag{3}$$

$$\rho\frac{\partial v_y}{\partial t}=-\frac{\partial p}{\partial y}+\frac{\partial\tau_{xy}}{\partial x}+\frac{\partial\tau_{yy}}{\partial y}+\frac{\partial\tau_{zy}}{\partial z}+\rho f_y \tag{4}$$

$$\rho\frac{\partial v_z}{\partial t}=-\frac{\partial p}{\partial y}+\frac{\partial\tau_{xz}}{\partial x}+\frac{\partial\tau_{yz}}{\partial y}+\frac{\partial\tau_{zz}}{\partial z}+\rho f_z \tag{5}$$

We have also used Sutherland's law of viscosity to determine the viscosity of the working fluid in the nozzle. Sutherland's law gives the temperature–viscosity relationship of a fluid. A constant is also introduced called Sutherland's constant. It is unique for every fluid.

Mathematically, Sutherland's law can be expressed as

$$\frac{\mu}{\mu_o}=\left(\frac{\theta}{\theta_o}\right)^{3/2}\left(\frac{\theta_o+S_c}{\theta+S_c}\right) \tag{6}$$

2.3 Computational Modelling

Computational Mesh. The 2D Nozzle profile design has been meshed using Ansys 19 (Academic). A structured mesh of quadrilateral cells was generated. The grid was generated for a fluent model with a medium mesh grid for the smooth and perfect discretization of the fluid domain. The intensity of the grid is equal along the whole domain. The 2-D model has been used as the model is axis-symmetric. The mesh with around 1.1 lakh elements gave better convergence and higher resolution. The grid obtained is shown in Fig. 2.

Mesh Independent Study. The Ansys 19 (Academic) version has been used for the generation of the grid and the CFD study. The mesh, element size and the number of elements have been selected after many iterations from coarse to fine meshing depending on the time for computation and the accuracy of the results. The selected mesh for our study is around 110,000–120,000 elements. This mesh is taken as a 2-D fluid domain as the axisymmetric solver has been adopted. We can see from Fig. 3 with the graph of maximum velocity versus Number of Elements that with a

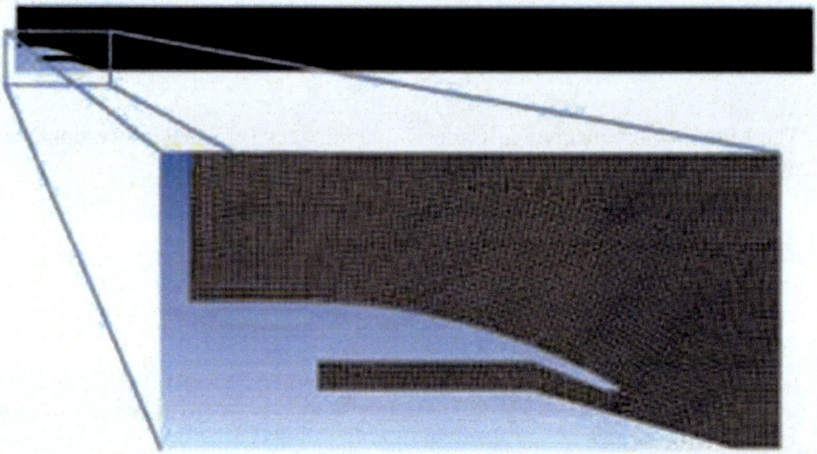

Fig. 2 Meshed view of the 0% truncated model with enlarged view of the mesh near the nozzle outlet

Fig. 3 Maximum velocity versus number of elements

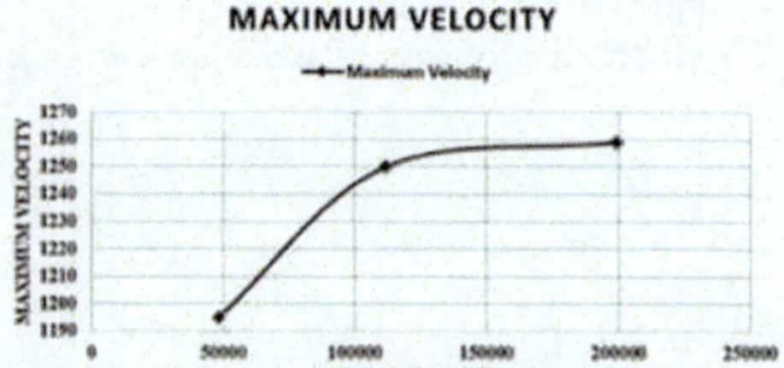

Table 2 Mesh independent study data

Mesh type	Number of elements	Maximum velocity in m/s
Coarse mesh	48,189	1195
Medium mesh	111,629	1250
Fine mesh	199,365	1259

coarse mesh the maximum velocity is low and as we move further and reduce the element size and generate medium mesh with around 110,000–120,000 elements, we get a value very close to the fine mesh that was generated with nearly 2 lakh elements. Since the difference between the maximum velocity in medium mesh and the fine mesh was not much when compared to the maximum velocity of the coarse mesh with the fine mesh and also fine mesh took a comparatively longer time to compute the same number of iterations when compared to the medium and the coarse mesh, the medium mesh has been considered for faster and good results (Table 2).

2.4 *Boundary Conditions*

The meshed nozzle profile was imported into Ansys Fluent, one of the premier CFD solvers.

The Inlet, as shown in Fig. 4, has been set to the "pressure inlet" condition. The detailed list of boundary conditions used in this study is given in Table 3.

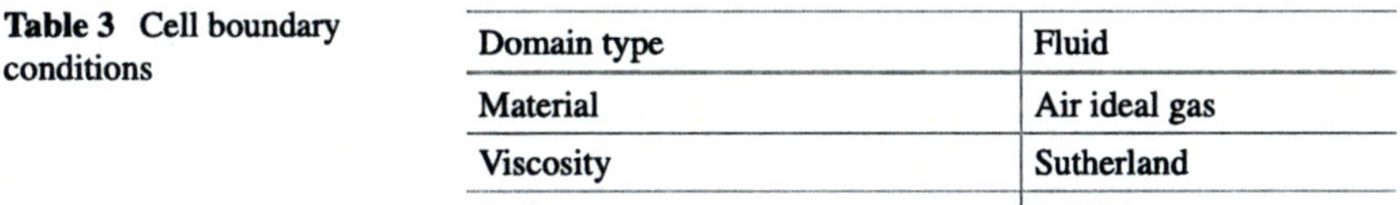

Fig. 4 Named selections of the fluent solver

Table 3 Cell boundary conditions

Domain type	Fluid
Material	Air ideal gas
Viscosity	Sutherland
Reference pressure	1.013 bar
Turbulence model	k-epsilon
Turbulence wall function	Realizable

2.5 Fluent Setup and Turbulence Model

The study is conducted for a transient density-based flow with a pressure ratio of 20. The density-based flow has been used since the flow is supersonic and incompressible. Transient flow has been used to study the turbulent fluctuations in the flow. The parameters like velocity, pressure and Mach number values are not only dependent on the position in the coordinate system but also the time. Only one-half section of the nozzle profile has been designed, so to study the result of the whole nozzle profile the axisymmetric model has been used.

The turbulence model used is k-ε (K-epsilon model). The K-epsilon turbulence model has been used as it is one of the most widely used CFD models. K-epsilon is considered to be one of the simplest models as it requires only two transport equations (PDEs). Our model has a confining flow at the nozzle, i.e., we are forcing the exhaust to enter the nozzle and exit only in a particular direction. We have a turbulent flow characteristic which leads to recirculation in the outlet. For the evaluation, we found K-epsilon to be most suitable one. The exact k-ε contains several unknown and unmeasurable terms. For a practical approach, the standard k-ε turbulent model is used which is ideal to understand the relevant processes. This will minimize the number of unknowns and will present a set of equations that can be applied to a large number of turbulent applications.

For Turbulent Kinetic energy, k,

$$\frac{\partial}{\partial t}\rho k + \frac{\partial y}{\partial x_i}\rho k \upsilon_i = \frac{\partial y}{\partial x_j}\left[\frac{\phi}{\sigma_k}\frac{\partial \varepsilon}{\partial x_j}\right] + 2\phi \epsilon_{ij}\epsilon_{ij} - \rho\varepsilon \tag{7}$$

For dissipation, ε,

$$\frac{\partial}{\partial t}\rho\varepsilon + \frac{\partial y}{\partial x_i}\rho\varepsilon\upsilon_i = \frac{\partial y}{\partial x_j}\left[\frac{\phi}{\sigma_\varepsilon}\frac{\partial \varepsilon}{\partial x_j}\right] + \lambda_{1\varepsilon}\frac{\varepsilon}{k}2\phi\epsilon_{ij}\epsilon_{ij} - \lambda_{2\varepsilon}\rho\frac{\varepsilon^2}{k} \tag{8}$$

The above equations contain some adjustable constants σ_k, σ_ε, $\lambda_{1\varepsilon}$ and $\lambda_{2\varepsilon}$. The values for these constants have been calculated through several iterations of data fitting for a wide range of turbulent flows as

$$\sigma_k = 1.00, \sigma_\varepsilon = 1.30, \lambda_{1\varepsilon} = 1.44, \lambda_{2\varepsilon} = 1.92$$

With reference to the above values, Eqs. (7) and (8) can be re written as:

For Turbulent Kinetic energy, k,

$$\frac{\partial}{\partial t}\rho k + \frac{\partial y}{\partial x_i}\rho k \upsilon_i = \frac{\partial y}{\partial x_j}\left[\phi\frac{\partial \varepsilon}{\partial x_j}\right] + 2\phi\varepsilon_{ij}\epsilon_{ij} - \rho\epsilon \tag{9}$$

For dissipation, ε,

$$\frac{\partial}{\partial t}\rho\varepsilon + \frac{\partial y}{\partial x_i}\rho\varepsilon\upsilon_i = \frac{\partial y}{\partial x_j}\left[\phi\frac{\partial\varepsilon}{\partial x_j}\right] + 2.88\frac{\varepsilon}{k}\phi\epsilon_{ij}\epsilon_{ij} - 1.92\rho\frac{\varepsilon^2}{k} \tag{10}$$

The fluid domain is set with air as the working fluid. The Air is set to ideal gas conditions. A pressure inlet has been set with 20 bar with a temperature of 1000 K, and the pressure outlet has been set to sea-level conditions with 1.01325 bar with a temperature of 287 K. The operating conditions have been set to zero. The study is based completely on the sea-level conditions.

3 Results and Discussions

In this section, we describe the results of computational analysis. The detailed computational analysis is performed using Ansys Fluent 19 (Academic). The analysis has been performed by keeping the spike length, pressure inlet and percentage of truncation as the key parameters to improve the geometry. The AN performance is studied for various parameters. The flow characteristics such as velocity, Mach number and temperature contours are studied and reported.

3.1 *Velocity Distribution Through AN for Different Percentages of Truncations*

Figure 5 illustrates the velocity distribution contour plots of AN for different percentages of truncations (viz., 0, 20, 40, 60 and 80%). The velocity range has been set

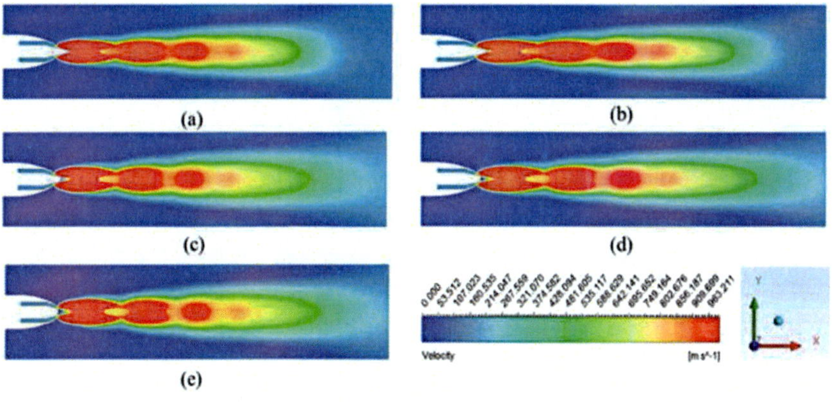

Fig. 5 Velocity contours obtained for different percentages of truncations. **a** 0%, **b** 20%, **c** 40%, **d**, 60% and **e** 80%

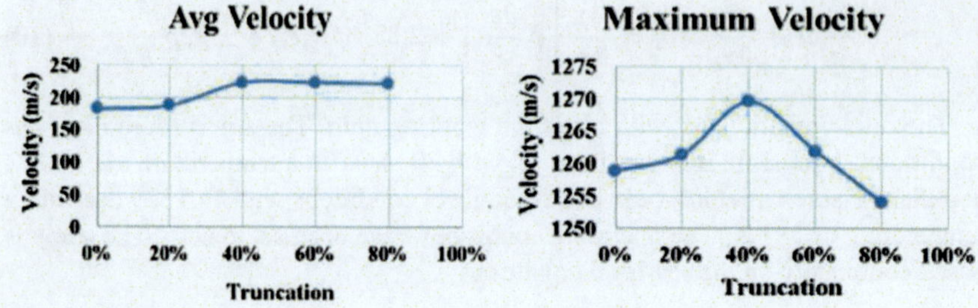

Fig. 6 Average velocity and maximum velocity distributions for different percentages of truncations

from 0 m/s to 1000 m/s for better visualization of the exhaust plumes. The shock diamonds are also visible in all percentages of truncations. From the above velocity contour, Fig. 5b, c for 20 and 40% truncation, we can see that the exhaust gases continue to flow almost the same way as that in the AN (0% truncated nozzle) as shown in Fig. 5a. For 60% (Fig. 5d) and 80% (Fig. 5e) truncation, there is a blue region at the base of the plug nozzle. This region shows the presence of recirculation in the base region. Recirculation adds to the overall thrust.

Figure 6a, b shows the Average velocity versus Truncation and Maximum Velocity versus Truncation graphs. The values of average velocity and maximum velocity are taken with respect to the entire domain. There is an increase in the maximum velocity up to 40% truncation which is the peak. On further increasing the truncation, there will be a decrease in the maximum velocity. The increase in the maximum velocity is because, as the truncation of the nozzle increases from 0 to 40% there is enough surface on the nozzle plug to gain the velocity in the direction along the spike or the plug of the nozzle. But beyond 40% truncation, i.e., for 60% and 80% truncated models there is recirculation present at the plug base region. This recirculation reduces the velocity drastically as the exhaust gases do not flow away but are enclosed in the base region.

3.2 *Velocity Streamline Distribution Through AN for Different Percentages of Truncations*

From the velocity streamline contours for 0% truncated nozzle in Fig. 7a and 20% truncated nozzle in Fig. 7b, we can see that most of the gases coming from the inlet do not flow over the complete surface of the nozzle, i.e., most of the gases get separated from the surface before the gases reach the tip of the nozzle. In the 40% (Fig. 7c), 60% (Fig. 7d) and 80% (Fig. 7e) truncated models, the exhaust gases flow over the complete surface. When compared with the results of other models, the 40% truncation model of the nozzle shows very good flow characteristics as the flow is completely over the surface which is best for the complete use of the whole surface

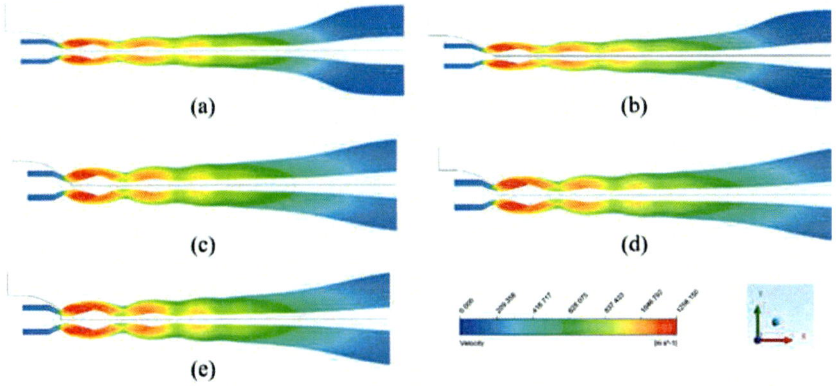

Fig. 7 Velocity streamlines obtained for different percentages of truncations. **a** 0%, **b** 20%, **c** 40%, **d** 60% and **e** 80%

for producing high thrust. In the full spike nozzle (0% truncated model) and the 20% truncated model, we can see that the exhaust gases' flow separates from the surface before the end of the surface.

3.3 Temperature Distribution Through AN for Different Percentages of Truncations

Figure 8 illustrates the temperature distribution contour plots of AN for different percentages of truncations (viz., 0, 20, 40, 60 and 80%). The temperature contours

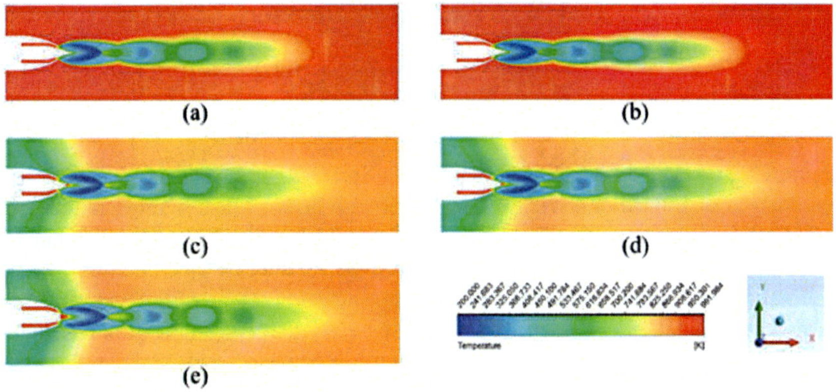

Fig. 8 Temperature distribution contours obtained for different percentages of truncations. **a** 0%, **b** 20%, **c** 40%, **d** 60% and **e** 80%

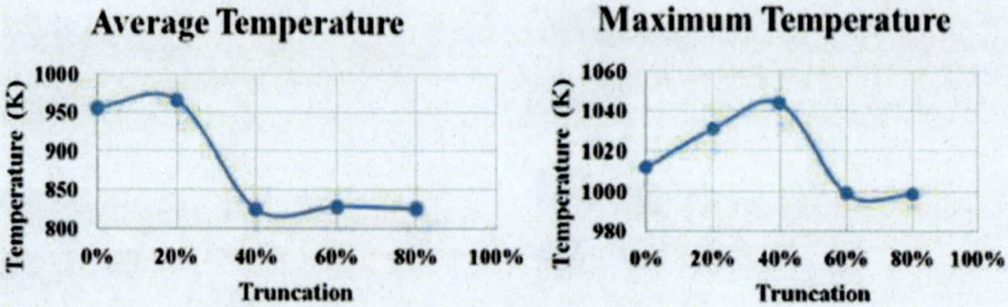

Fig. 9 Average temperature and maximum temperature distributions for different percentages of truncations

are in the range from 200 to 1000 K. The first case, as shown in Fig. 8a, has high temperature at the tip and nozzle surface. In the truncated models, Fig. 8b, c, i.e., for 20% and the 40% truncated model, it is seen that the base temperature is high when compared to the other parts of the nozzle wall. But in the case of 60 and 80% truncation, we can see in Fig. 8d, e, the temperature is higher compared to the other truncated models (20 and 40%). This is mainly due to the presence of recirculation. The hot gases near the surface tend to recirculate in the base region which is of high temperature. So, the cooling has to be done at the base region.

The above two plots (Fig. 9a, b) show the variation of average temperature versus truncation of the nozzle and maximum temperature versus the percentage of truncation of the length of the nozzle. From Fig. 9b, we can see that the maximum temperature is highest at 40% truncation of the nozzle, also from the average velocity versus truncation plot (Fig. 9a), we can see that the average temperature is less for 40% truncated model. This is advantageous as in a 40% truncated model maximum temperature is very high since there is enough wall for the exhaust gases to flow over the surface.

3.4 *Mach Number Distribution Through AN for Different Percentages of Truncations*

Figure 10 illustrates the Mach Number distribution contour plots of AN for different percentages of truncations (viz., 0, 20, 40, 60 and 80%). Mach number is defined as the ratio of the local flow velocity to the sonic velocity, i.e., Mach number is directly proportional to the outlet velocity. In Fig. 10, the low Mach number zones are found near the base region of the spike for all percentages of truncations. The effect of recirculation can be also seen for 0% truncation (Fig. 10a), it is seen that near the nozzle tip, a Mach number of around 1 is obtained. The Mach number at the base keeps reducing its size as truncation reduces due to the stagnation created. This results in recirculation. For our study, we have selected the range from 0 to 2.5

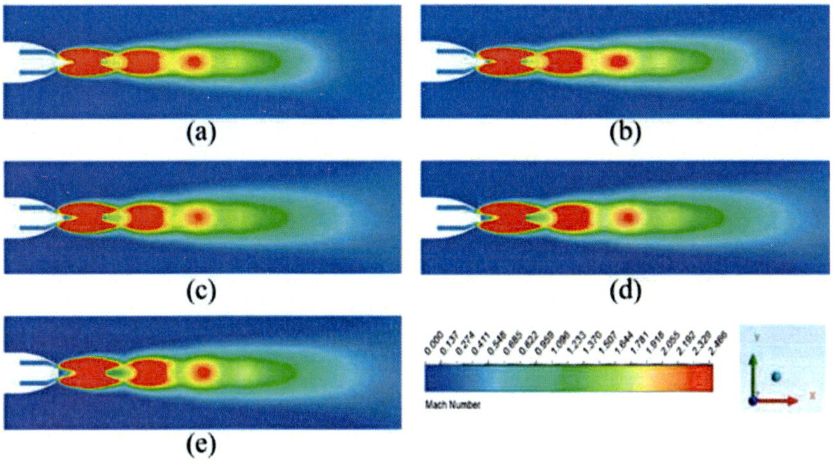

Fig. 10 Mach Number distribution contours obtained for different percentages of truncations. **a** 0%, **b** 20%, **c** 40%, **d** 60% and **e** 80%

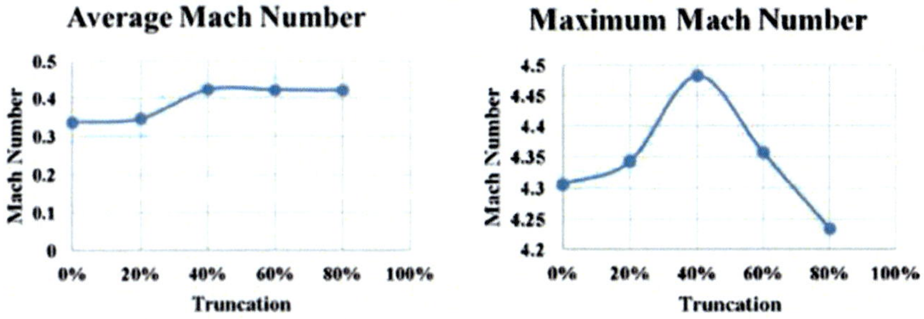

Fig. 11 Average Mach Number and maximum Mach Number distributions for different percentages of truncations

Mach. The Average Mach Number versus Truncation and Maximum Mach Number versus Truncation plots are in Fig. 11a, b.

3.5 *Effect of Recirculation Due to Truncation*

Recirculation is the phenomenon that occurs due to the sudden discontinuity of the surface on which the fluid flows. Due to high velocity, the fluid will tend to stick to the surface. Now as the truncation increases, the discontinuity between the flow and the surface will also increase. This will lead to more recirculation. Initially, the vector plot was put for 20, 40, 60 and 80% truncated models to find recirculation in the truncated region. From the above images, we can see that there is no recirculation

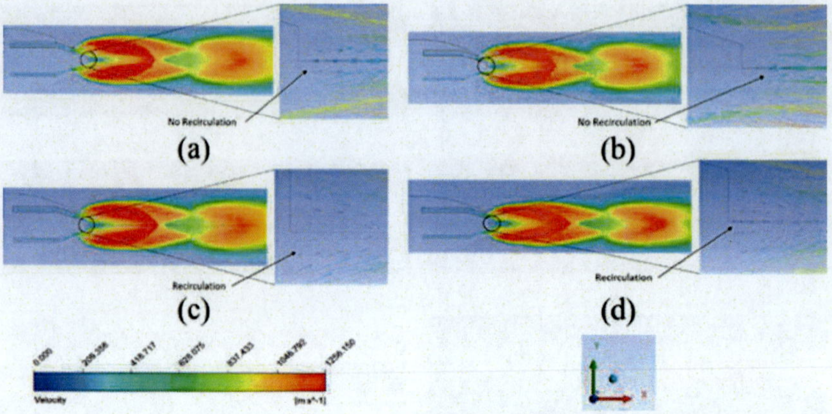

Fig. 12 Recirculation vectors obtained for **a** 40%, **b** 50%, **c** 60% and **d** 80% truncation

till the 40% truncated model; in the 60% truncated model (Fig. 12), we can see that there is a circulation of very small magnitude. For the 80% truncated model (Fig. 12), we can see there is a clear recirculation. Since there is slight recirculation that was visible from 60%, analysis was done to a 50% truncated model to check the presence of recirculation. It is seen from Fig. 12 that there is no recirculation. The recirculation adds to the thrust produced by the Aerospike nozzle. When the exhaust gases are released at a high velocity and directed towards the spike part in the 0% truncation model, there are enough surfaces that the fluid can follow to hold the profile.

As there is truncation, the surface for the fluid flow decreases. Up to 50% of the surface of the plug, part of the nozzle is enough for the exhaust gases to maintain the direction as that of the aerospike nozzle. From 60% truncation, the velocity and the pressure of the gases in the direction normal to the surface are more but the surface discontinuity is faster so the fluid layers that are very close to the surface tend to pull into the truncated region and start recirculating; this recirculation tends to provide a virtual surface to compensate for the truncated part of the plug part of the nozzle. This recirculated part holds the shape and is also bound by the surrounding high-pressure gas flow so the pressure acting on the plug will increase and thus helps in increasing the thrust.

4 Conclusion

In this present study, the effect of truncation on AN performance is studied using the numerical technique. The k-ε turbulence model was employed to solve the equations governing the laws of fluid flow. Five different AN models are studied with 0, 20, 40, 60 and 80% of truncations. From the above-obtained contours and graphs, we have understood and observed that the effect of truncation on AN has a maximum impact on the flow behaviour. The flow field is found to be uniform. This can be inferred due to the inconsequential recirculation zones for the 0, 20 and 40% truncated model. From the above-obtained contours and graphs, it is inferred that the outlet conditions for the 40% truncated condition are the greatest compared to the other four conditions. In the exhaust plume, there are the expansion waves, compression waves and the wake region. In the wake region, there is a lot of recirculation of the exhaust plumes.

At 0% truncation, there is very less space for the recirculation to take place. So whatever recirculation does take place is constructive in nature, i.e., increases the velocity. At 20% truncation also the case is similar, but here there is a little more place for the wake region. This again causes constructive recirculation to increase the velocity outlet. When comparing 20% truncation with 0% truncation, the contours display that the area covered by the dead zone (no recirculation) is more for the 0% truncation model. At 40% truncation, the balance of the recirculation to the other waves is perfect leading to the maximum outlet velocity. Now, as the truncation is increased, the wake region area is high due to which the activity of recirculation will increase leading to the destructive effect of recirculation. In this case, the destructive effect is the reduction in outlet velocity which happens suddenly after crossing 40% truncation.

References

1. Immich H, Caporicci M (1997) FESTIP technology developments in liquid rocket propulsion for reusable launch vehicles. In: 32nd joint propulsion conference and exhibit, p 3113
2. Nguyen T, Pieper J (1996) Nozzle flow separation. In: Proceedings of the 5th international symposium of propulsion in space transportation, Paris, France
3. Kumar KN, Gopalsamy M, Antony D, Krishnaraj R, Viswanadh CB (1996) Design and optimization of aerospike nozzle using CFD. In: IOP conference series: materials science and engineering, vol 247, no 1. IOP Publishing, p 012008
4. Gopala D, Kowsik V, Joseph C, Arun Justin MP, Sainath R, Murugan DT, Ilakkiya S (2013) Thrust force analysis of spike bell nozzle. Int J Appl Res Mech Eng 3(1):54–59
5. Swathi G, Sandeep CS, Snigdha M, Sravanthi G, Govardhan D (2017) Three-dimensional computational flow simulation of truncated aerospike nozzle considering different plug lengths. Indian J Sci Technol 10(13):1–4
6. Gopala Krishnan G, Akhil J, Nagaraja SR (2015) Heat transfer analysis on hypersonic reentry vehicles with spikes. In: Proceedings of the 23rd national heat and mass transfer conference and 1st international ISHMT-ASTFE heat and mass transfer conference
7. Sreekanth N, Akhil J, Nagaraja SR (2016) Design and analysis of secondary spike on blunt head. Indian J Sci Technol 9(45)
8. Immich H, Caporicci M, Immich H, Caporicci M (1997) Status of the FESTIP rocket propulsion technology programme. In: 33rd joint propulsion conference and exhibit, p 3311
9. Karthikeyan T, Aravindhkumar SK, Kumar JA (2016) Design and analysis of aerospike nozzle to improve thrust in hybrid rocket engine. Int J Eng Trends Technol 36(7)
10. Sanoob SN, Prince MG, Sundar B (2013) Numerical analysis of aero-spike nozzle for spike length optimization. Int J Res Eng Technol 1(6):1–14
11. Muwaaz SA, Shaik N (2016) Design and numerical flow analysis of expansion deflection nozzle. Int J Eng Dev Res 04(2):1081–1094

12. Reddy PJ, Bandaru K, Mohan KSK, Prabhudev BM (2019) Investigation on effect of spike geometry on spiked nozzle performance using computational tools. In: AIP conference proceedings, vol 2080, no 1. AIP Publishing LLC, p 030005
13. Dumitrescu O, Gherman B, Dragan V (2018) Study of full and truncated aerospike nozzles on performances at different working conditions. In: IOP conference series: materials science and engineering, vol 400, no 4. IOP Publishing, p 042017
14. Rommel T, Hagemann G, Schley CA, Krulle G, Manski D (1997) Plug nozzle flow field analysis. J Propul Power 13(5):629–634

Simulation Modelling of a Stand-Alone Triple Basin Solar Desalination System

Rahul K. Meena, Pushpendra Singh, and Anil Kumar

Abstract This paper aims to design and analyse a stand-alone triple basin solar desalination system (TBSS). Geometry for the triple basin solar desalination system equipped with parabolic concentrator is modelled on ANSYS. Simulation analysis is carried out on ANSYS Fluent v19.2 at the atmospheric conditions of Delhi (28.7041° N, 77.1025° E). All modes of heat transfer and evaporation and condensation phenomena are examined. Therefore, the dimensions and geometry of the fins are varied as cylindrical and triangular. Heat-absorbing material in the fins is tested by varying it with charcoal and river sand. Within this analysis domain, simulation results of the designated systems were calculated and a comparison is carried out. Simulation results found that peak temperature is higher at the edge of the fin in triangular geometry (73.112 °C) compared to cylindrical fins (66.09 °C). Maximum temperature of air and water-vapour mixture inside the system is 351.5 K, and the minimum temperature is 304.25 K. A parabolic concentrator causes the maximum sunlight to concentrate on the surface of the glass casing and that trapped radiation increases the temperature of the system. The peak temperature of the system raises to 78.35 °C for the current system. Maximum water production rate in a TBSS is 16.78 kg m^{-2} day^{-1}.

Keywords Desalination · Triple Basin solar still · ANSYS · Water temperature · Solar heat flux · Water productivity

Nomenclature

Nu	Nusselt Number
Tigs	Inner glass temperature of solar still

R. K. Meena (✉) · P. Singh
Department of Mechanical Engineering, Delhi Technological University, Delhi 110042, India
e-mail: rj26@outlook.com

A. Kumar
Centre for Energy and Environment, Delhi Technological University, Delhi 110042, India
e-mail: anilkumar76@dtu.ac.in

D. P. Mishra et al. (eds.), *Recent Trends in Thermal and Fluid Sciences*, Lecture Notes in Mechanical Engineering, https://doi.org/10.1007/978-981-19-3498-8_22

hew	Evaporative HT coefficient
hcw	Convective HT coefficient
hrw	Radiative HT coefficient
Pw	Partial saturated vapour pressure at water temperature, N/m^2
Pg	Partial saturated vapour pressure at glass temperature, N/m^2
Tw	Temperature of water
Tg	Temperature of the inner glass surface
k	Thermal conductivity of humid air W/mK
L	Latent heat of vapourization, J/kg
Ma	Molecular weight of dry air, kg/mole
Mw	Hourly distillate yields, kg/m^2
Pt	Total vapour pressure in the basin, N/m^2
I	Solar heat flux, W/m^2
Greek β	Expansion factor of vapour
μ	Dynamic viscosity of vapour
ρ	Density of vapour

1 Introduction

Water is a fundamental need of all living creatures, including humans and animals, to survive and live. Access to clean drinking water forms a priority. Availability of water for sanitation, agricultural activities, production, manufacturing, etc. forms a vital need for human sustainability and growth. Availability of water for various utilities can turn troubles into potential—facilitating people with timely reaching to office, and contributing to enhanced health levels for children, adults, and families worldwide. Under the current scenario, 78.5 crore people 1/9th lacks reach to safe drinkable water [1] and 200 crore people 1/3rd lacks access to a toilet [2]. Nearly 10 lakh people lose their lives each year due to sanitation and hygiene-related diseases, which could be curtailed to certain limits with access to safe water or sanitation. Every 2 min, a person dies from a water-borne disease [2].

Solar desalination is a prominent process that results from the production of potable water free of cost using solar energy. This energy is used to evaporate water in a device commonly known as "solar still". Different versions are still used for spraying seawater, in desert survival kits, and for home water purification. For those who are unhappy with the quality of drinking water supplied by the municipality and the other purification methods are available to them, solar employed distillation or groundwater, tap water is a fun, efficient energy alternative. The degrading quality of water day by day due to human civilization and industrial wastes is a major problem for the modern world. Therefore, the reusability of the water and treatment of contaminated water is of vital importance for the current generation so that our future generation can get water at least for their basic needs.

Velmurugan et al. [3] performed an experiment to analyse the effect of incorporation of the fins at the lower side of the base of the basin and found that the system's yield has significant improvement, i.e. it increased from 1.88 to 2.8 kg/m^2. Sponges increase productivity by 15.3%, wick-type SS improves it by 29.6%, and fin-type solar still enhances the productivity by 45.5% when compared with conventional solar still.

Setoodeh et al. [4] developed a 3D model of a system of solar still. In this model, a different phase method was used for the combined process of evaporation and condensation with the assistance of a computational fluid dynamics simulation tool. Using this tool, a process of the simulation of single slope solar still was accomplished to get results. In the structure of a solar still, the volume of fluid (VOF) was owned to give liquid and prepared by combining the air, water, and vapour system at the quasi-steady-state condition. The building of geometry and meshing was prepared by using ANSYS Workbench 11. A simulation was done using tetrahedral meshing, and different meshing having a different number of elements were developed to calculate better results, 32,322, 47,179, 64,694, and 85,315 cells. The energy and mass transfer concepts, continuity, and momentum at steady-state condition were used for numerical modelling. As a result, it was predicted CFD is strong equipment for designing process, variables investigation, and can be used for removing difficulties during geometry creation of solar still.

Energy storing materials (ESMs) are the materials that store the energy when the source (sunlight in our case) is available and releases it when a source is absent (sunshine-off hours). Materials that can store sensible heat like iron scraps, red brick pieces, washed stones, cement concrete pieces, quartzite rock, etc. Murugavel et al. [5] examined the effect of ESM and found that quartzite rock of ¾ inch size is effective heat-storing material.

Rajaseenivasan et al. [6] comparative analysis on single basin SS with double basin SS arrangement and establish that double basin system is 84% more effective than single basin solar still. Also performed review analysis on different systems to incorporate to improve the distillate and improve the efficiency and found certain arrangements such as double effect system is more efficient than single effect, provision for cooling water, multi-basin system, parabolic concentrators, vacuum tubes enhances the productivity and increases the efficiency of the system significantly. They also carried out a performance study on solar still with square and circular fins on the basin surface when wick material is covered over fins. They found that the production of the system comes out to be 4.55 kg/m^2/day as compared to 3.16 kg/m^2/day for conventional solar still.

Panchal et al. [7] demonstrated a transient mathematical model for the solar still with a triple basin setup. He found that the evaporation rate is maximum for the minimum mass of water at the surface of the basin without any dry spots, and it decreases with the increasing mass of water. He also found that the velocity of the air above the surface of water facilitates the rate of evaporation by a certain instant, but it is not dependent on the water mass in the lower and middle basins. Above that, there is no significant improvement in the rate of evaporation. Maximum distillate production in the triple basin setup was calculated as 12.635 kg/m^2/day, which justifies the result with experimental data. Srithar et al. [8] experimentally analysed the triple basin solar desalination performance still (TBSS) with energy-storing material stored inside the triangular fins below the top and middle basins. They also incorporate a cooling water arrangement as cover cooling (CC), parabolic dish concentrator (PDC), and PV panel. He analysed the TBSS with CC, TBSS with PDC, TBSS with CC, and PDC with charcoal and river sand filled in the triangular fins. TBSS with river sand and charcoal in fins improves the distillate production by 25.6 and 34.2% compared to conventional TBSS setup.

Khare et al. [9] created a multi-phase, 3D computational fluid dynamics model of simple solar still by using ANSYS Fluent v14.0 and also experimented on it. The simulation and exploratory system of single slope solar still were compared with each other for the climate condition of Jaipur. The assistance of ANSYS Workbench accomplished the physical structure surfaces and their meshing. The meshing of geometry has 3D hexahedral shape elements, and 1.5 million elements were created during meshing with a growth rate of 1.2. The energy and mass transfer concepts, continuity, and momentum for the steady-state condition were used for numerical modelling with some assumptions. In a simulation, the overall efficiency and other parameters were checked.

The objective of the study is to create a 3D computational fluid dynamics model of a triple basin solar desalination system equipped with a parabolic concentrator to analyse the pattern of evaporation and condensation within the solar still at the environmental conditions of Delhi (27.0248° N, 74.2169° E). The paper is validated with already published work by Srithar et al. [8]. The model is formed using ANSYS Workbench and simulation was carried out with the help of ANSYS Fluent v19.2. Multiple variables inside the still were considered and observed at each stage. It was examined for water production of the TBSS system for the entire day of 21 June 2020.

1.1 Ansys

ANSYS is a Workbench-based simulation tool which works on the multiphysics model. Hence, multiphysics Simulation technology provides us with the potential to explore more and predict whether a designed product will function normally in real-world conditions. It has the ability to design the future based on the basic principles of modelling, physics, mathematics, and computer science. Simulation enables engineers to see how their designs perform in millions of real-world scenarios, while reducing or even eliminating the need for expensive physical testing. It's like being able to see the future beyond its implication, hence allowing us to achieve innovations that are completely different from past methodologies. Present work is carried out on ANSYS Fluent v19.2 which is the latest edition of the Workbench simulation tool software [12].

2 Modelling of Triple Basin Solar Desalination Apparatus with Parabolic Concentrator

A system of 3-phase 3D models is designed for the fluid volume of the multiphase model. The still was established by means of evaporation and condensation processes under transient conditions, such that only vapourization of liquid arises at the surface and its interface was deliberately considered for modelling. Act of solar still is dependent on different parametric variables such as coefficients of internal heat and mass transfer, are checked. Mass transfer coefficient and internal heat in solar still are dependent on convection, evaporation, and radiation. Consequently, the convective, evaporative, and radiative HT coefficients are the main heat transfer coefficients under different modes of heat transfer. An RNG k-ε turbulence model was accounted for with typical wall functions for both liquid and vapour phases. Time, energy, mass transfer, and volume-average continuity equations were established numerically in this analysis for each phase.

2.1 Numerical Simulation of Triple Basin Simulation

Results were obtained by making some assumptions and equations used by computational fluid dynamics to obtain results below in governing equations.

2.2 Governing Equations

Under SS conditions, typical calculations depend on momentum, energy, continuity, and mass transfer conservation concepts.

2.2.1 Energy Equations [10]

The assumptions of energy balance equations are.

i. Inside-glass surface of still and basin water has no increase in temperature.
ii. No escaping of water from any type of hole or crack is present in the solar stills.
iii. The water is preserved at a constant level inside the basin.
iv. The heat capacity of the insulated material, absorber, and the glass surface is very low.

The equation of energy for the mixture is mentioned hereunder [10]

$$\frac{\partial}{\partial t}\sum_{k=1}^{n}(a_k\rho_k E_k) + \nabla\cdot\sum_{k=1}^{n}\{a_k\overline{v}_k\ (\rho_k E_k + p)\} = \nabla\cdot\left(k_{eff}\nabla T\right) + S_E \quad (1)$$

where k_{eff} is the effective conductivity $(a_k(k_k + k_t)$.

2.2.2 Continuity Equation [10]

$$\frac{\partial}{\mathrm{dt}}(\rho_m) + \nabla.(\rho_m\overline{v_m}) = 0 \quad (2)$$

where $\overline{v_m} = \frac{\sum_{k=1}^{n} a_k\rho_k\overline{v_k}}{\rho_m}$.

2.2.3 Momentum Equation [10]

Equation of Momentum for all phases is accomplished by conjoining each equation of momentum of all phases.

$$\frac{d}{dt}(\rho_m\overline{v}_m) + \nabla(\rho_m\overline{v}_m\overline{v}_m) = -\nabla\cdot p + \nabla[\mu_m(\nabla\overline{v}_m + \nabla v_m^T)] + \rho_m\overline{g} + \overline{F} + \nabla\cdot\left(\sum_{k=1}^{n} a_k\rho_k\overline{v}_{dr,k}\cdot\overline{v}_{dr,k}\right) \quad (3)$$

ρ_m Mixture density.
$\overline{F}$ External and Gravitational body forces.
μ_m Viscosity at molecular level of the mixture.

2.2.4 Volume Conservation Equation [10]

It is a restraint that results in the volume fractions by adding to unity:

$$r_G + r_L = 1 \quad (4)$$

r_L volume fraction of liquid, and
r_G volume fraction of gas.

2.2.5 Mass Transfer Equation [10]

The sources of enthalpy in a system for multiple phases are

$$H_p = -m_p i_q j\left(h_p^i\right) \quad (5)$$

$$Hq = m_{p^iq}j\left(h_p^i + h_p^{f^i} - h_q^{f^j}\right) \tag{6}$$

2.2.6 Pressure Constraint [10]

The liquid and gas phases share equal pressure field:

$$P_G = P_L = P \tag{7}$$

2.2.7 Heat Transfer Equation [10]

The buoyancy force is the primary cause of heat transfer. The heat transfer coefficient that is evaporative, radiative, and convective across the contact and the gas phase was prepared with the help of the null equation model and the liquid phase is demonstrated by (2017)

$$h_{total} = h_{r\omega} + h_{cw} + h_{e\omega} \tag{8}$$

$$h_{Cw} = 0.8840\left(Tw - Tg + \frac{(P_\omega - p_g)(T_w + 273)}{268 \cdot 9 \times 10^3 - P_w}\right)^{1/3}$$

where $P_w = \exp\left(25.317 - \frac{5144}{T_w+273}\right)$, $Pg = \exp\left(25317 - \frac{5144}{Tg+273}\right)$

$$h_{e\omega} = 16.273 \times 10^{-3} h_{Cw} \cdot \frac{P_\omega - P_g}{T_w - T_g} \tag{9}$$

and

$$h_{rw} = \varepsilon_{eff}\sigma\left[(TW + 273)^2 + (Tg + 273)^2(Tw + Tg + 546)\right] \tag{10}$$

where $\varepsilon_{eff} = \frac{1}{\varepsilon_g} + \frac{1}{\varepsilon_\omega} - 1$.

2.2.8 Water Production [10]

Distinct interacting mass fluxes were used for mass transfer modelling. The rate of vaporization of the water depicts the Rate of freshwater production.

Hence, the mass flux equation between the two phases is

$$\dot{m}_{ew} = \frac{\dot{q}_{ew} \cdot A_w \cdot t}{h_{fg}} \tag{11}$$

where $h_{fg} = 2 \cdot 4935 \times 10^6 \left[1 - 9.4779 \times 10^{-4} T + 1.3132 \times 10^{-7} T^2 - 4.7974 \times 10^{-9} T^3\right]$

$$\dot{q}_{ew} = h_{ew}(T_w - T_g)$$

2.2.9 Solar Irradiance [10]

$$\delta = 23 \cdot 45 \sin(360 * (n + 284)/365)$$

$$\omega = (\text{solar time} - 12:00) \times 15$$

where Solar time $=$ Standard time $\pm 4(\mathrm{L_{st}} - \mathrm{L_{Loca}}) + \mathrm{E}$

$$\mathrm{Ig} = \mathrm{I_b} + \mathrm{I_d} \tag{12}$$

where $\mathrm{I_b} = \mathrm{I_{bn}} \cos \theta_z$ and, $\mathrm{Ig} = \mathrm{I_{bn}} \cos \theta_z + \mathrm{I_d}$.

2.2.10 Convergence Criterion [11]

Newton–Raphson Method is used as a convergence criterion which is the most widely used scheme.

Let a function f(x),

$$x_{n+1} = x_n - \frac{1}{f'(x_n)} f(x_n)$$

Non-linear Equation, $f(x) = 0 \Leftrightarrow x = g(x)$.

For Convergence Criteria, use Lipschitz condition and $x_n = g(x_{n-1})$

$$|g'(x_n)| < k < 1 \Rightarrow |x_n - x^e| \le k|x_{n-1} - x^e|$$

Fast convergence, $|g'(x^e)| = 0$

$$g(x) = x + h(x) f(x), h(x) \neq 0$$

$$g'(x^e) = 1 + h(x^e) f'(x^e) + h'(x^e) f(x^e)$$

$$= 1 + h(x^e) f'(x^e)$$

$$g'(x^e) = 0 \Rightarrow h(x) = -\frac{1}{f'(x)}$$

Newton-Rapson Iteration, $x_{n+1} = g(x_n) = x_n - \frac{f(x_n)}{f'(x_n)}$.
This function shows quadratic convergence.

2.2.11 Computational Fluid Dynamics (CFD) Model of Modelling

CFD is an appreciated tool utilized to estimate and study the shape of the fluid stream of moist air and the variation of temperature of the adjacent walls. For envisioning the field flow of any situation, this computational technique can prepare the design, guide, and get the controlling step. In the present study, CF dynamics is utilized to construct geometry, and the results of the simulation were obtained using ANSYS v19.2.

The proposed system consists of a triple basin solar still (TBSS), parabolic dish concentrator (PDC), cover cooling arrangement (CC), and triangular/cylindrical fins incorporated below the top and middle basins as main components. Design details and parameters are mentioned as follows.

2.3 Geometry Creation

ANSYS is a constructive engineering simulation tool that is frequently utilized to find solutions to engineering problems and their optimization. Figure 1a, b shows the

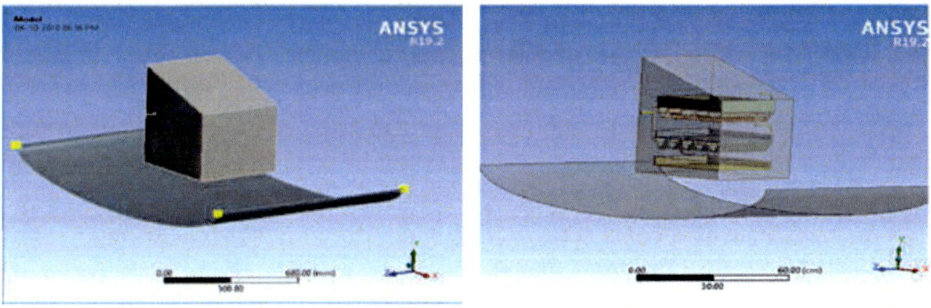

a. Triple basin solar desalination model

b. Triple basin solar desalination model with internal geometry

Fig. 1 Modelling of triple basin solar desalination system

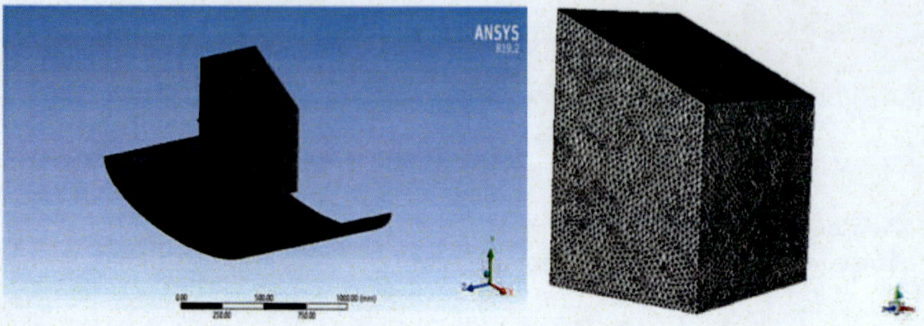

Fig. 2 The unstructured mesh of triple basin setup and concentrator

3D geometry of a triple basin solar desalination setup with parabolic concentrator; ANSYS Design modeller was used to model in ANSYS 19.2.

This model contains a concentrator, and triple basins separated into compartments at the top, middle, and lower. Upper and middle basins contain triangular fins and pipes to flow saline water through them. This system is modelled as per dimensions.

2.3.1 Meshing

The meshing of the physical model of the TBSS system was carried out by using ANSYS 19.2 Workbench Meshing. Figure 2 displays a triple basin setup with an unstructured mesh, comprising 2,878,069 elements with an element size of 10 mm and 542,951 nodes with a progression of 1.2 and using 3D tetrahedral meshing at the basins and outer cover. At the same time, the concentrator has a hexagonal meshing pattern.

2.3.2 Boundary Conditions and Preliminary Conditions

ANSYS Fluent is fed with a set of instructions to process Meshing files of a triple basin still. A multi-phased structure of solar desalination still was incorporated inside the framework of the volume of fluid (VOF) for a mixture of water in the liquid state, and air and water vapour system at the transient condition with gravitational acceleration. Solar irradiance falls on the tapered glass surface of the outer casing glass cover, and on the parabolic dish concentrator. It is an important factor for the inner side of the setup, dependent on the transmissivity and absorptivity of the glass cover and reflectivity of the reflector surface of the dish concentrator (aluminium sheets). When the rays are incident on the reflective surface of the parabolic concentrator, they get reflected in concentrated form on the outer shell of TBSS, which gets absorbed inside. In the fluent, boundary conditions and physics involved under physical phenomena are specified to solve the energy equation, momentum equation, and continuity equation at all the boundaries. RNG k-ε viscous phase model was used, which forecasts

Table 1 Boundary conditions [8]

Domain	Domain type	Boundary details	Location	Boundary type
Solid	Cell	Stationary wall with heat flux	Parabolic concentrator	Wall
Solid	Cell	Fixed wall temperature	Glass cover	Wall
Solid	Cell	Adiabatic	Sidewalls	Wall
Solid	Cell	Adiabatic	Front and back wall	Wall

the extent of flow and enhanced water productivity. In the model, the near-wall conditions were dispensed by means of a "standard wall function". Adhesion forces are engaged in making droplets inside the solar still. "No-slip" boundary conditions at walls were well-defined for both the liquid phase and gas phase. The volume fraction of the air and water were taken as 0.87 and 0.13. CFD simulation had a duration of 12 h run time for this triple basin setup (Table 1).

3 Experimental Data

Triple Basin Solar Desalination System with Parabolic Concentrator and equipped with fins underneath the basin base was proposed and prepared by Srithar et al. [8].

The experimental setup incorporates two components, one evaporative unit (inner component) and the other a condensing casing (outer glass cover—outer component). All the triple basin setup parts are modelled for 4 mm thickness, and the material is selected as a window glass sheet. This glass sheet facilitates the radiation to be captured from the top and sides. Dimensions of the inner component are selected as 0.3 m × 0.36 m and 0.33 m in height. The inner component is divided into three basins with a 0.12 m gap in between.

Top and middle basins carry 4 triangular/cylindrical hollow fins of 0.075 m width and height of 0.05 m (0.075 m diameter for cylindrical fins when 4 fins are modelled). These fins were fed with charcoal and river sand in the hollow cavity to act as heat storage media. Copper pipes feed water through the fins of the upper basin. Heat storage media helps the increased rate of evaporation during sunshine hours with better efficiency and intensity. Also, it facilitates evaporation during off sunshine hours. A constant level of feed water is maintained at 0.01 m in each basin with an overflow protection mechanism. Overflow water is carried away in successive steps (upper basin to middle basin, middle basin to lower basin) with the copper piping via fins. The outer casing is modelled with 0.4 × 0.46 m^2 with 0.4 m height at one end and 0.62 m at the other end with the angle of inclination at 28.7° (latitude of Delhi, coordinates of Delhi Technological University campus). Condensation occurs at both the surfaces, the top surface and at the side cover of the outer casing. Channels to collect condensate are attached on the side. The cover cooling arrangement is also

Fig. 3 Photographic view of triple basin solar desalination system [8]

Table 2 Geometry and properties

Bounding box			
Length	X = 1601 mm	Y = 781.21 mm	Z = 1000 mm
Properties			
Volume 9.1139e+007 mm^3		Scale factor value: 1	
Statistics			
Statistics bodies 15		Active bodies 15	Nodes 542,951
Angle of slope		28.7°	Elements 2,878,069

designed to increase the temperature difference between the vapour and glass sheet. This facilitates an increase in the rate of condensation.

Snapshots of experimental work conducted by Srithar et al. [8] has been shown in Fig. 3, and all the sizes taken for simulation work are shown in Table 2.

4 Results and Discussion

In this analysis, results for simulation were calculated for each stage under clear weather conditions of the TBSS still using ANSYS Fluent v19.2. In the current simulation analysis, the distribution of temperature inside the still is observed and projected the temperature of the still for the complete day. The system is also validated for the number of fins installed and its geometry, i.e. 4 fins of triangular nature and 4 fins for the cylindrical geometry are analysed.

4.1 Simulation Results

ANSYS Fluent v19.2 was engaged to form the results using parallel run processors with double precision running at 3.0 GHz. Fluent practices a second order up-wind resolution process to transform the prevailing equations into a numerically resolvable algebraic equation.

4.1.1 Temperature Distribution

Figure 4a, b respectively visualizes the gas-phase temperature volume rendering inside the TBSS pipes and basin. Temperature variation at each stage was calculated inside the apparatus. In the temperature volume rendering, the red colour visualizes the maximum value of the temperature, while the blue colour visualizes the minimum value of the temperature for the triple basin system.

Figure 5a, b respectively visualizes the temperature distribution rendering inside the fins of triangular and cylindrical geometries. The maximum temperature inside the triangular fins is found to be 73.112 °C and the minimum is 22 °C. The maximum temperature is distributed at the narrow edges of the triangular fins. The maximum temperature inside the cylindrical fins is found to be 66.093 °C and the minimum is 23.438 °C. The maximum temperature is distributed at the lower surface of the cylindrical fins.

Figure 6a, b shows the irradiation heat flux falling on the surface.

Comparison: peak temperature is higher in triangular fins than in cylindrical fins, as shown in Table 3, but it is visible that the highest temperature is found at the edge of the fin in triangular fins. It should be noted that the cylindrical fins contain a symmetry pattern; thus, the distribution of temperature is more symmetric and can be extracted through the pipes more symmetrically (Fig. 7).

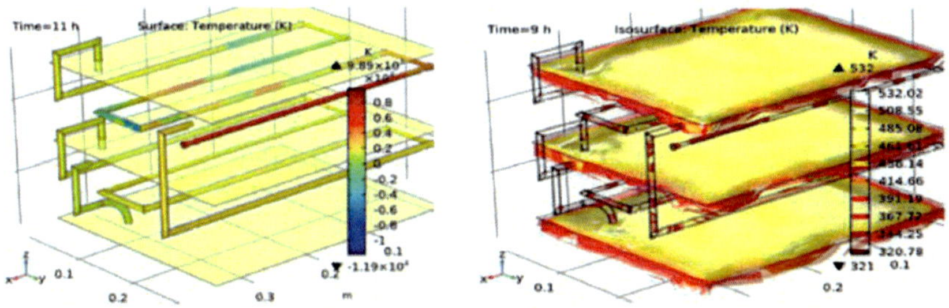

a. Temperature distribution inside the pipes and water surface

b. Iso-surface Temperature distribution inside the pipes and water surface

Fig. 4 Temperature distribution

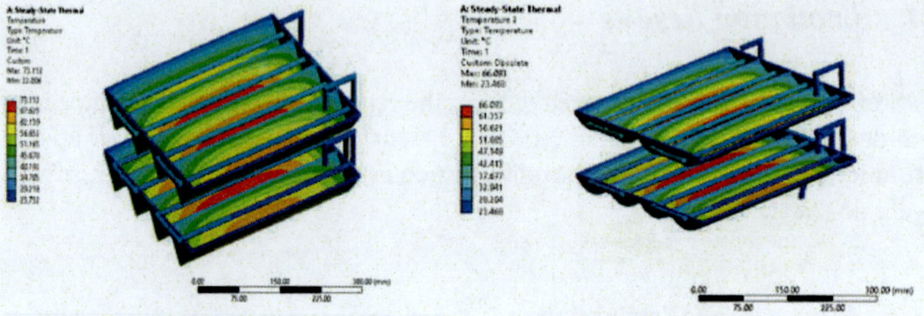

a. Temperature distribution of four triangular fins

b. Temperature distribution of four Cylindrical fins

Fig. 5 Temperature distribution of triangular and cylindrical fins

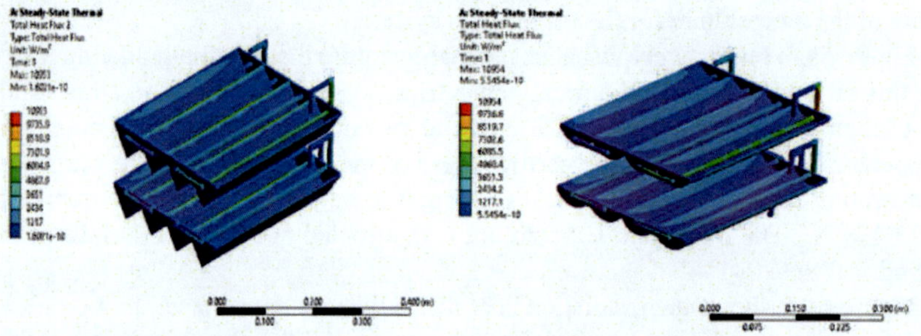

a. Irradiation heat flux distribution of four Triangular fins

b. Irradiation heat flux distribution of four cylindrical fins

Fig. 6 Irradiation heat flux falling on the surface

Table 3 Peak temperature comparison for triangular and cylindrical fins

Fins	Triangular fins-4	Cylindrical fins-4
Temperature	73.112 °C	66.09 °C

Figure 8a, b visualizes the gas-phase temperature volume rendering inside the TBSS still. Temperature variation at each stage was calculated inside the still. In the temperature volume rendering, the red colour visualizes the maximum value of the temperature, while the blue colour visualizes the minimum value of the temperature for the system. In TBSS, the temperature is increasing from bottom to top slowly. Basins are attaining a maximum temperature at the lowermost because of the mesh orientation towards solar incident rays, which were concentrated by the parabolic concentrator.

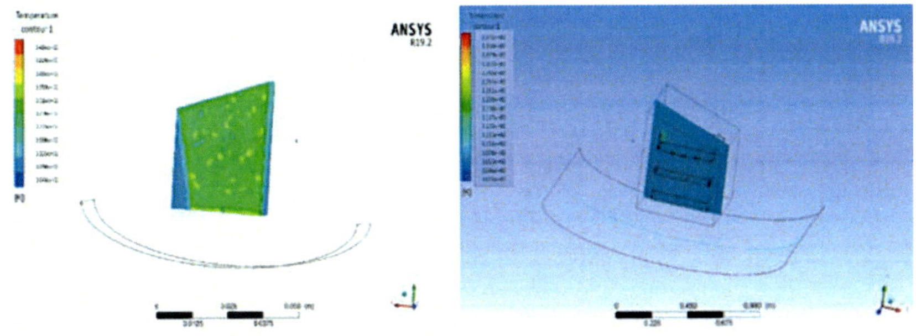

a. Temperature distribution inside the triple basin Side view

b. Temperature distribution inside pipe flow

Fig. 7 Temperature contour rendering inside the TBSS still side view and inside pipe flow

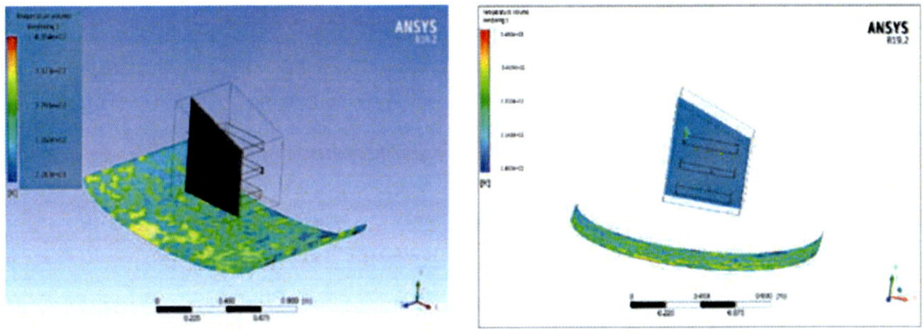

a. Temperature distribution over the surface of parabolic concentrator

b. Temperature distribution over the surface of parabolic concentrator and pipe flow

Fig. 8 Gas-phase temperature volume rendering inside the TBSS still

4.1.2 Temperature Variation with Time

Figure 9 shows the temperature variation of the Lower Basin of TBSS concerning the time. The temperature of the still is calculated for the entire day. From the simulation result, it was predicted the maximum temperature was obtained at 13:00 h., i.e. 351.5 K (78.35 °C), while the minimum temperature is obtained in the morning during sunrise and sunset, i.e. 304.25 K (31.1 °C), which is the very low temperature to convert the water into vapour. The red line in the graph shows the temperature, while the blue line shows time on the streamline. From the graph of a triple basin solar still, it is predicted that there is no time when the temperature inside the TBSS still is constant (Fig. 9). From the graph of TBSS still, it is predicted that temperature is almost constant from 12:00 to 12:30 h. Then, it increases, and at 13:00 h. It attains maximum temperature. Again, from 14:00 to 14:15 h, the temperature is almost constant, and it decreases. Maximum Temperature in the timeframe of 13:00 h is attributed to two

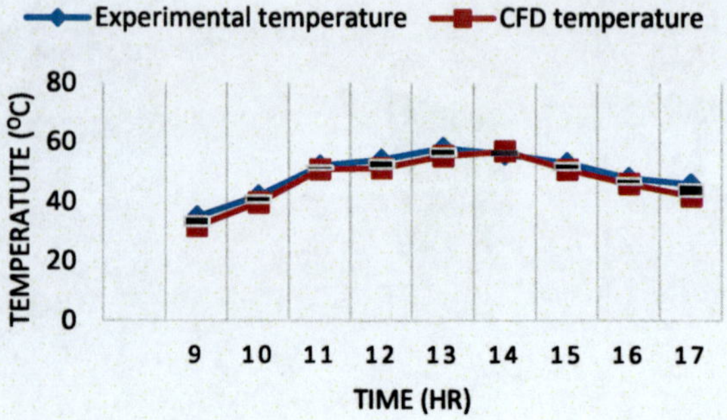

Fig. 9 Temperature variation concerning the time of TBSS

main reasons. First, maximum solar heat flux is striking the surface of the earth at this timeframe. Second, the orientation of the solar concentrator is arranged such that the solar radiations which were reflected by the parabolic concentrator cover the major portion of the surface of the basin. Both these factors were responsible for the maximum temperature at 13:00 h.

4.1.3 Water Production Rate

Figure 10 visualizes the graph of the water production rate concerning the time. The simulation of still starts running at 6.30 h. In this figure, it is noted that as the action starts at 8:00 h, time passes and water starts to warm up due to solar radiation absorbed by water. Moderately, still space heats up with water vapour and the freshwater production rate rises till 13:30 h and after that, when solar radiation decreases water moves downward slowly.

It was found that this system with charcoal filled fins and parabolic concentrator along with cooling water arrangement at the top attains water production yield at a rate of 16.78 kgm^{-2} day^{-1}.

It was observed that water production of TBSS still with single slope equipped with parabolic concentrator is more compared to conventional solar still. Some terms affect water production, such as temperature, mass flow, pressure, solar heat flux,

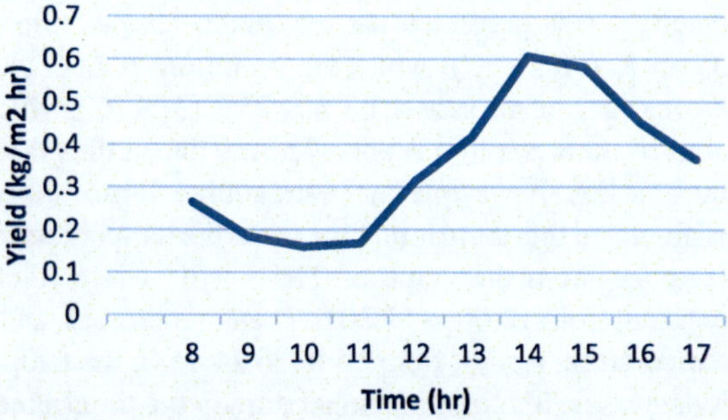

Fig. 10 Water production rate concerning time inside the TBSS still with the parabolic collector

Table 4 Percentage deviation between theoretical and experimental values

Parameters	Theoretical readings	Experimental readings	% deviation (%)
Temperature variation with time	78.35 °C	85 °C	8.48
Productivity	16.78 kgm^{-2} day^{-1}	16.94 kgm^{-2} day^{-1}	0.95

absorption coefficient, and many other factors. All these terms are checked in the simulation of the still and observed that a TBSS can produce freshwater in rural and urban areas.

4.1.4 Percentage Deviation

There is a significant deviation from the experimental and theoretical readings due to the following reasons (Table 4).

Parabolic concentrator reflector sheets material and arrangement. Weather conditions (heat flux) in real world are different as in the case of simulation. Insulation and material, cooling water flow rate and thickness variation. Impurities in saline water deviates results as simulated.

5 Conclusions

The major motive of the research is to make a computational fluid dynamics model of the TBSS system with a parabolic concentrator. A 3D model of the setup is developed using computational fluid dynamics. In the setup, evaporation and condensation processes take place. Simulations were organized on ANSYS Fluent v19.2 software for a mixture of water, air, and water vapour systems. The CFD simulation models were utilized to conduct the study for these transient conditions of the nominated system. The simulation data of the system is obtained with 200 steps over 15 min period of 12 h. In the present system, solar radiation is absorbed by water and evaporated into vapour. It condenses on the inner surface of ceiling glass.

The concluding remarks of the study are.

- The maximum temperature of air and water-vapour mix in the system is 351.5 K and the minimum temperature is 304.25 K.
- It was found that the temperature continuously varies as time passes. It reaches a peak at the noon and decreases after that. The provision of heat storage material inside the fins steadies the fluctuation over the increased interval of time.

- A parabolic concentrator causes the maximum sunlight to concentrate on the surface of the glass basin and that trapped radiation increases the temperature of the system. The peak temperature of the system raises to 78.35 °C for the current system.
- The maximum water production rate in a TBSS is 16.78 kg m^{-2} day^{-1} when the system is equipped with heat storage material, concentrator, cooling water arrangement, and proper insulation.
- The maximum temperature of the system with triangular fins is more than the maximum temperature with cylindrical fins. The maximum temperature with triangular fins is 73.112 °C and that with cylindrical fins is 66.09 °C.
- The temperature distribution is concentrated at the edge of the triangular fin but will have symmetrical distribution in the case of cylindrical fins.
- Percentage deviation is 8.48% and 0.95% for peak temperature distribution and productivity, respectively.

It was predicted that CFD results visualize that Computational Fluid Dynamics forms strong equipment for the designing process, variable investigation, and can be used for removing difficulties during geometry creation of solar still. Further work can be done in future by modifying diverse design parameters and alignment of solar still.

Acknowledgements The authors are highly grateful to the Department of Mechanical Engineering and Centre for Energy and Environment, Delhi Technology University, Delhi (India), for their facilities and assistance in compiling this manuscript.

References

1. https://www.who.int/news/item/18-06-2019-1-in-3-people-globally-do-not-have-access-to-safe-drinking-water-unicef-who
2. https://water.org/our-impact/water-crisis/
3. Velmurugan V, Deenadayalan CK, Vinod H, Srithar K (2008) Desalination of effluent using fin type solar still. Energy 33(11):1719–1727
4. Setoodeh N, Rahimi R, Ameri A (2011) Modeling and determination of heat transfer coefficient in a basin solar still using CFD. Desalination 268(1–3):103–110
5. Elango T, Murugavel KK (2015) The effect of the water depth on the productivity for single and double basin double slope glass solar stills. Desalination 359. https://doi.org/10.1016/j.desal.2014.12.036
6. Rajaseenivasan T, Elango T, Murugavel KK (2013) Comparative study of double basin and single basin solar stills. Desalination 309. https://doi.org/10.1016/j.desal.2012.09.014
7. Panchal HN, Shah PK, Sebaii (2011) Modelling and verification of single slope still using ANSYS-CFX. Int J Energy Environ 2(6):985–998
8. Srithar K et al (2016) Stand alone triple basin solar desalination system with cover cooling and parabolic dish concentrator. Renew Energy 90:157–165. https://doi.org/10.1016/j.renene.2015.12.063
9. Khare VR, Singh PA, Kumar H, Khatri R (2017) Modelling and performance enhancement of single slope solar still using computational fluid dynamics. Energy Procedia 109:447–455. https://doi.org/10.1016/j.egypro.2017.03.064

10. Kumar R, Meena RK (2020) Numerical simulation of single and double slope solar still for different performance affecting variables. IJRASET. https://doi.org/10.22214/ijraset.2020.32050
11. https://ocw.mit.edu/courses/mechanical-engineering/2-29-numerical-fluid-mechanics-spring-2015/lecture-notes-and-references/MIT2_29S15_Lecture4.pdf
12. Lee HH (2021) Finite element simulations with ANSYS workbench 2021: theory, applications, case studies. SDC Publications

Automatic Insecticide Spraying for Greenhouse Protection

V. Ankith, R. Rohith Reddy, K. Naresh, S. K. Harshad, and B. P. Dileep

Abstract In view of the diseases caused by the insects to plants in greenhouse farming, this paper designs an automatic insecticide sprayer, which can sense the motion of insects which entered the greenhouse farming field and communicate to turn on the respective mist sprayer to insecticide. Today, the Greenhouse farming rate is increasing due to the protection of crops from diseases, increasing production, minimizing production, stability and security, etc. But there are many possible ways that insects can enter greenhouse farms. So protecting the crops in a greenhouse farm will be so difficult for a farmer to spray insecticide, and the growth rate of crops will also be decreased. To decrease the burden, this automatic insecticide sprayer will be helpful to do insecticides. In this device, we use a motion sensor to detect the motion of harmful flies which entered the greenhouse farm. Since the device was automatic, there will be no need for farmers to take care of plants to protect them from insects. In this device, we used solenoid valves. It helps us to spray insecticides only in the detected region and in the other regions spraying will not be done, and we can save the wastage of insecticide.

Keywords Greenhouse farming · Insecticides · Relay · Passive infrared sensor · Solenoid valve · Water pump

1 Introduction

The assault of pests, diseases, and weeds has a negative impact on crop yield. The most common way for controlling insects, weeds, and diseases is chemical control. According to the UN Food and Agriculture Organization (FAO) and other research

V. Ankith · R. R. Reddy · K. Naresh · S. K. Harshad · B. P. Dileep (✉)
Department of Mechanical Engineering, Amrita School of Engineering, Amrita Vishwa Vidyapeetham, Bangalore 560035, India
e-mail: bp_dileep@blr.amrita.edu

V. Ankith
e-mail: ankith.vikkurthi@gmail.com

D. P. Mishra et al. (eds.), *Recent Trends in Thermal and Fluid Sciences*, Lecture Notes in Mechanical Engineering, https://doi.org/10.1007/978-981-19-3498-8_23

studies, plant pests and diseases destroy 20–40% of total world harvests. The chemicals are added to the crop by watering, sprinkling, pumping, or dusting. Spraying is one of the most powerful and safe ways to secure crops by adding a small amount of spray liquid. Manually operated low- and high-volume hydraulic sprayers, as well as power-operated hydraulic sprayers with long booms, long lances, or spray guns, are used in traditional methods to bring fluid to various targets.

This method requires more time and effort. It's tough to spray the insecticide uniformly and efficiently in the tree using standard methods. It is difficult to administer the insecticide evenly and efficiently across the tree using the standard spraying method. While this method of pest control is effective, it uses a lot of liquid per plant and takes a lot of time and work. Drip losses are also at an all-time high. As a result, a viable alternative spraying strategy should be devised to safeguard the environment from chemical pollution while also saving money. India's farm mechanization situation is complicated due to various farm holding sizes and socioeconomic inequality. The majority of farmers in India are marginal or tiny tenant farmers. A Knapsack sprayer is used for spraying, which takes longer and consumes more energy. Due to established cropping patterns, field size, and field conditions during the rainy season, farmers find it difficult to adapt to tractor-operated sprayers. To solve this dichotomy, you'll need to be more adaptable. From the perspective of the people's needs, the self-driven sprayer is the better choice because of its moderate cost and small scale, suggesting better maneuverability in small landholdings. Self-driven style sprayers will fully fill the mechanization void, allowing for faster spraying operations. This demonstrates the critical need for automated sprayers to be introduced into Indian farms.

The current row cropping concept is widely adopted by Indian farmers, and the development of automated type sprayers is urgently needed today. Besides this, finesse spraying can aid in meeting legal requirements for worker health, environmental sustainability, and food safety and quality. Keeping the foregoing in mind, the current investigation was undertaken to assess the field performance of a self-driven pesticide sprayer in field crops.

Several agricultural spraying robots for weed control and plant protection applications have been developed with the goal of reducing pesticide use while preventing crop losses due to pests. The purpose of this project is to investigate the automated methods for insecticide spraying pathways.

1.1 Literature Survey

Sumitra and Gayathri [1] Plant diseases have produced a significant post-effect situation, according to Dr. M. G. Sumithra and G. R. Gayathri in their paper, because they can effectively affect agricultural commodities in terms of both quality and quantity. The first phase requires regular and careful inspection of the plants. After that, the afflicted plants will be identified, and images of the impacted areas of the plants will be taken with detectors or cameras. After that, the photos are pre-processed, altered, and clumped together. The photos are subsequently provided as input to the processor, which compares them. If the picture is harmed, a pesticide will be applied to the afflicted area of the plant using an automatic pesticide sprayer.

Sammons et al. [2] They devised and manufactured a self-contained mobile robot that may be used to control insects and prevent illness in commercial greenhouses in this research. The platforms' capacity to maneuver themselves efficiently down greenhouse rows illustrates the success of this technology, while the pesticide spraying system consistently covers the plants with spray in the prescribed dosages. The results showed that the robot met the National Greenhouse Horticulture Center's physical requirements, allowing it to work in their greenhouses. The robot also met the deadlines and budget limits imposed on it. The robot could drive up and down the greenhouse's rails. The Induction Proximity Sensors properly sense and operate the rails. When traveling along the rails, another thesis student's spraying system was able to selectively spray designated plant groups in the greenhouse. The spray protection confirmed that the crops received an adequate and consistent dosage.

Gengaje and Deshmukh [3] In their paper, they implemented the predicated agricultural robot, where the robot scans the plant indefinitely. The cropped video is captured by a wireless camera mounted on a robot and sent to the central station detectors or cameras. The operation of the robot is decided by the person sitting in the central station. When the user discovers a flaw in the crop, the robot is given a command, the kinetic will of that robot is used, and pesticides are sprayed over the crop.

The RF transceiver will handle this. In agriculture, robots are used for industrial purposes. The primary application of robotics in agriculture is for planting, fruit picking, and driverless tractor or sprayer replacement. The main goal is to eliminate manual pesticide spraying on real farms. It will be accomplished by transmitting crop video to a central station in order to replace humans with robots. Instead, a central station uses a real-time processor to monitor robot movements and pesticide spraying. This will reduce the plant's reliance on pesticides. Such devices are most commonly used to apply agriculture to rising manpower. The device has the advantages of high speed, high quality, dependability, and low storage costs. The future context for this system will be to design the system using robot intelligence. The pesticide tank capacity is increased in proportion to the farm area.

Lee et al. [4] In this paper, they proposed various spraying methods and also the flow rate required. Improper sprayer calibration is one of the most common problems encountered by farmers and researchers, particularly when spray application does not work properly and consistently during the spraying process. As a result, when spraying in the intended field without proper calibration, the results will be abysmal with high variability. Numerous variables must be considered during sprayer calibration, and each one is critical to obtaining reliable results. Depending on the type of sprayer, there are many other ways to calibrate it. Calibration of a self-propelled or pull-behind sprayer, for example, entails deciding sprayer discharge, forward speed, and application rate. Backpack sprayers, on the other hand, are calibrated slightly differently, but with the same goal in mind. To ensure that the recommended spray application rate is accomplished, the sprayer must be calibrated according to the sprayer type.

The configuration of the backpack sprayers with one nozzle mounted in orchard sprays, which uses an airflow to assist the penetration of the spray inside the treetops, was accomplished by the following as calibration of the boom sprayers. It is necessary to replace one or more of the following factors in order to change the application rate to the acceptance of the spraying process: nozzle type, nozzle size, spraying forward speed, and operating pressure. Changing one or more of the preceding factors, on the other hand, will result in a change in the application rate amount to be applied.

Das et al. [5] In their paper, they include information on various spraying techniques of Sprayer for backpacks (knapsacks), Light-Tractor, and Multi-Purpose Farming Machine Powered by a Motorcycle.

1.2 Advantages

1. Workload of farmers will be decreased.
2. Crops will be protected from insects.
3. Insecticides will not be wasted as we have used solenoid valves.
4. They are comparatively less in cost compared to robot sprayers.

2 Methodology

2.1 Existing System

They used non-automatic and semi-automatic means for applying insecticides to crops in the existing system, such as hand sprayers and knapsack sprayers. Farmers use this technology to walk into the field and identify any insects moving in any place, then spray pesticides on the insects to kill the insecticides. After a few problems, such as farmer fatigue and knee injuries, new strategies for applying insecticides were adopted, such as remote-controlled robot sprayers, which utilize a remote to drive the robot and spray insecticides on insects when the farmer detects any insects on the farm. After that, the farmer utilized numerous automated models for controlled air application, such as an automatic thermal vapor meter, a cold fogger, and an automatic air-assisted sprayer, all of which are quite expensive.

2.1.1 Limitations

They are time-consuming, have poor implementation precision, and require stringent safety precautions because they are exceedingly dangerous to the user, who may be exposed to deadly chemicals as a result of their use in a confined space. The pesticide's

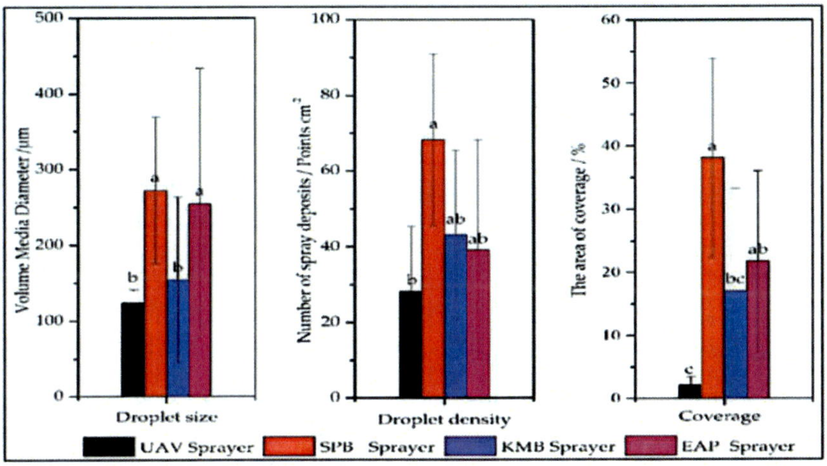

Fig. 1 Comparison of different sprayers

performance is measured by the operator's ability; manual application frequently results in uneven pesticide distribution. The high toxicity of some pesticides, as well as the poor application capability of most of the techniques used, necessitated the development of new systems with improved performance and safety features, preferably fully automated, that could be operated without the presence of the grower, and that could leave minimal residue deposits if necessary (Fig. 1).

2.2 *Proposed System*

The key to maintaining quality crop growth is to protect the farm from insects immediately. This will not be effective when the farmer detects harmful flies; since the farm will be big, the farmer alone can't look after the entire farm and the burden will also be increased, so we need to use a motion sensor which has more sensitivity, and no harmful radiations should not be emitted from the sensor [6].

We need to use a water pump which should be of smaller size for easy setup and the pump should have more suction head capacity so that it can transfer the water from the lower path (tank) to the higher path since we keep the spray nozzles on top to cover the entire path. To avoid wastage of insecticides, we need to use solenoid valves; by using these valves, the insecticides will sprinkle only on the detected region; all these instructions will be given by using Microcontroller and Relay.

So, when any fly is detected in the region, the sensor immediately detects, the water pump turns on, and the respective solenoid valves will also turn on, due to which insecticides will be sprinkled immediately on the fly, it dies, and we can protect the Greenhouse farm.

2.3 Block Diagram

The following block diagram explains the implementation of the Automatic Insecticide Spraying for Greenhouse protection using the Arduino Uno processor. In this device, we used Arduino Uno MCU since it is simple to program, and it also has a built-in analog to digital converter (ADC) and Digital to Analog Converter (DAC). Two PIR Sensors were given as input to the MCU to detect the motion of harmful flies in their respective areas. When a harmful fly passes through the sensor, it intercepts the first half of the sensor; at that time, there will be some potential difference between the two halves. By that, the PIR sensor senses the motion and gives information to Arduino. When the sensor senses a harmful fly at a distance less than 80 cm from the sensor, then Uno gives a 100 ms pulse from the digital output pin. We can't connect a pump directly to the Arduino, so here we used a relay as a switch. The relay takes the pulse from Uno and allows the water pump to start. The water pump we used is a 6–12 v non-submersible pump, which starts to spray the insecticides on the detected region with the help of a sprinkler nozzle (Fig. 2).

Solenoid valves are connected to the relay as an output, when only PIR sensor 1 detects the motion then solenoid valve 1 will turn on allowing the insecticides to pass from the solenoid valve to the sprinkler nozzle, and solenoid valve 2 will not turn on since no motion is detected, by this the insecticides will go from water pump to solenoid valve but solenoid valve 2 will not allow to spray it out and vice versa will happen when only PIR sensor 2 detects and when both sensors Detects, both solenoid valves turns on and allows to spray the insecticides. The power supply of DC water pump and DC Solenoid valves was given by using DC adapters of 12 and 24 v.

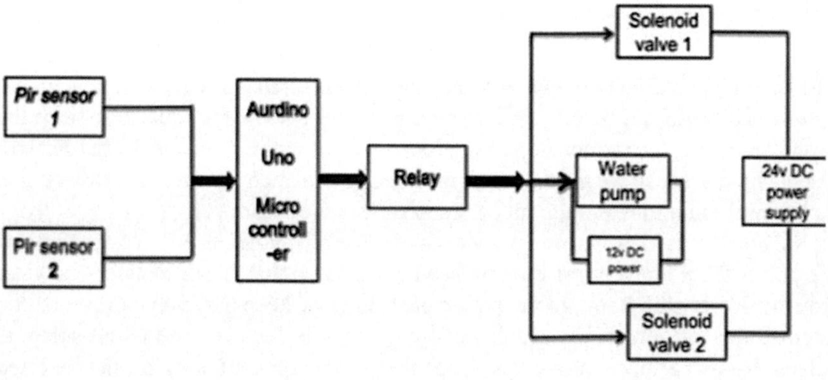

Fig. 2 Block diagram

3 Hardware Implementation

The Arduino is a single board microcontroller based mostly on ATmega328. It includes 14 digital IO pins in which 6 can be used for PWM, a 16 MHz ceramic resonator, a USB connection, an ICSP header, a power jack, 6 analog inputs, and a reset button. This consists of all the required help wanted for the microcontroller. To start, they're connected to a PC with a USB cable.

Arduino Uno Board isn't like any other boards, and it'll now no longer use FTDI USB-to-serial motive force chips in them. It is featured via the Atmega16U2 programmed as a USB-to-serial converter. The Arduino is a single board microcontroller based mostly on ATmega328. It includes 14 digital IO pins in which 6 can be used for PWM, a 16 MHz ceramic resonator, a USB connection, an ICSP header, a power jack, 6 analog inputs, and a reset button. This consists of all the required help wanted for the microcontroller. To start, they're connected to a PC with a USB cable. Arduino Uno Board isn't like any different boards and it'll now no longer use FTDI USB-to-serial motive force chips in them. It is featured via the Atmega16U2 programmed as a USB-to-serial converter.

4 Design and Analysis

For this device, we used PIR sensors over other sensors like microwave motion sensor, Ultrasonic sensor, and Vibration sensor. These all sensors can detect the object which was interrupted in the particular region but Microwave motion sensor was not environmentally friendly; the plants get affected by the microwave radiation emitted by the sensor, and it affects plant growth and by using PIR sensor, no radiations will be emitted and it was just done by the potential difference between the two slots of the sensor when it detects.

For the ultrasonic sensor, the area it can detect was less and it can detect only in the 90° direction which was exactly straight to the sensor, whereas in PIR sensor, the area it detects was more and it detects 110° of the area, 55° from the right, and other 55° from the left. The vibration motion sensor was expensive, the area it covers for detecting was so less, and it is so hard to use for our device. So, we use a PIR sensor for motion detecting and give it as an input to the MCU [7].

This PIR sensor consists of 3 pins: VCC 5–12 v DC pin, Output pin, and Ground Pin; since the output pin was 3.3 v TTL, it can be used with any MCU hardware like Arduino, Raspberry, etc.

We used Arduino hardware since it was less expensive compared to raspberry. So, we connected the pins as shown in the above figure. There are two rotary buttons in the PIR sensor, for adjusting delay time and sensitivity. We used 2 sensors to detect in two different places, so we used a breadboard for maintaining the same electrical connections between sensors 1 and 2, since the Arduino will not have more connections to connect. The output of our device is to spray the insecticides in the area it detected, so we use the relay module as a switch since Arduino or any MCU can't supply a lot of power required to the water pump, if we connect directly the Arduino will burn instantly since it can provide a maximum current of 20 ma, so for water pump which was with 1 amp, under high loads it exceeds the limit and Arduino board will be burned, so relay module will be used as a switch to drive our 12 v DC water pump 24 v DC solenoid valves. The relay which we used was an 8-switch relay. To this relay, we gave a connection of 1 DC water pump and 2 solenoid valves [8].

We gave code to Arduino in such a way the relay switch of the DC motor will turn on when at least any of the sensors start sensing; it will turn on when sensor 1 or sensor 2 or even both sensors start sensing, and the respective solenoid valves will turn on when its PIR sensor starts sensing. We also gave pause time for 5 s to the Arduino such that after the sensor stops to detect the motion, the spraying of insecticides will be continued for the next 5 s; in that 5 s, if anything motion does not get detected, then spraying will be turned off; this pause time was given because, to start the sensor immediately it almost takes 1.5 s after detecting; to maintain good efficiency, we gave pause time of 5 s.

A 6-12 V DC diaphragm pump is used, since it can maintain the ability to create a good vacuum on the suction line. The diaphragm delivers a good seal and consistent vacuum on the suction line; It maintains good head capacity so that it easily lifts the fluids from ground level to the upper level in which it is located. It is so easy to design a prototype with this water pump, since it is less costly, simple in structure, and efficient in working. A Solenoid valve is an electromechanically controlled valve. In this solenoid valve, it has a solenoid, which is an electric coil and can move the ferromagnetic core at its center. When power passes through the electric coil, it creates a magnetic field and this exerts upward force on the plunger and it opens the orifice. This principle was used in the solenoid valve to open and close the valves.

We used this Solenoid valve because it helps us to shut off or to allow fluids; when we need to pass the insecticides to more than one place, then with one water pump we can't make it turn on and off in every particular place; so if we place the solenoid valves to every pipeline where insecticides release, then with the help of MCU, we can set when the sensor detects in only one region, and then the respective solenoid valve will be turned on.

A nozzle sprinkles at the end of the pipe because it modifies the flow of the insecticides; we used this nozzle to increase the rate of flow, speed, and pressure [9]. Since the diameter of the nozzle will be so less, the insecticides will be sprayed in the form of mist and covered the entire region. The nozzle which we used has a diameter of 0.5 mm.

5 Result and Discussion

This device is very helpful in agriculture purposes where it is needed to spray the insecticides on different crops from saving them from diseases caused by insects

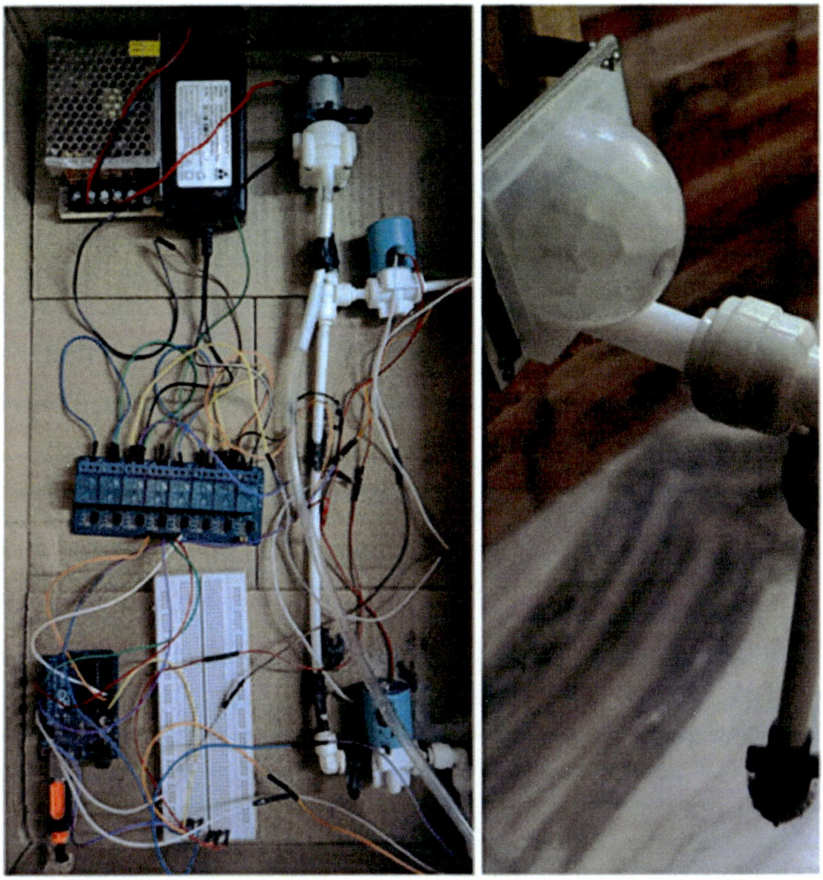

Fig. 3 Physical prototype

[10]. Currently, the systems which are being used for insecticide purposes are need more human effort and not comfortable to do this tasks to a farmer and it takes lot of time to do entire insecticides on the field.

This insecticide sprayer device has many sensors and many solenoid valves that allow us to spray the insecticides only in a particular region without wasting the insecticides automatically (Fig. 3).

6 Expenses

For making this device as a prototype, it costs around 7000 Rs. The cost of this setup increases or decreases based on the area of the farm (Table 1).

Table 1 Expenses report

S. No	Name of the component	Cost	Quantity	Total cost
1	PIR sensor	350	4	1400
2	Arduino board	2000	1	2000
3	DC diaphragm pump	700	1	700
4	Solenoid valves	250	4	1000
5	DC adapters (Water pump, solenoid valves)	500	2	1000
6	Eight switch relay module	800	1	800
7	Nozzles	40	4	160

7 Conclusion and Future Scope

The automatic insecticide sprayer device proposed in this report is ultimately expected to contribute to the automatic spraying of insecticides to decrease the burden on farmers and to maintain the effective growth of plants. It is also cost-effective and eco-sustainable because it reduces waste emissions. This model can be developed by adding image processing and the help detect the leaf disease in the early stage of crops. We can expand the detection range to 0.01 cm and use it in larger fields for smaller flies and pests.

References

1. Sumitra MG, Gayathri GR proposed in their paper—leaf disease diagnosis and pesticide spraying using agricultural robot (AGROBOT)
2. Sammons PJ, Tomonari F, Andrew B proposed in their paper Autonomous Pesticide Spraying Robot for use in a Greenhouse
3. Gengaje SR, Deshmukh SM ARM-based pesticide spraying robot
4. Lee IN, Lee KH, Lee JH, You KH Autonomous greenhouse sprayer navigation using automatic tracking algorithm
5. Das N et al Agricultural fertilizers and pesticides sprayers spraying methods
6. Automated farming using microcontrollers and sensors. Int J Sci Res Manage Stud (IJSRMS), Abdullah Tanveer, Abhish Choudhary, Divya Pal
7. Varman SAM, Baskaran AR, Aravindh S, Prabhu E (2017) Deep learning and IoT for smart agriculture using WSN. In: 2017 ieee international conference on computational intelligence and computing research (ICCIC), pp 1–6. IEEE
8. Ergonomic evaluation of knapsack sprayer used in agricultural application. Int J Sci Eng Res
9. A comprehensive review on automation in agriculture using artificial intelligence
10. Choudhary R et al (2020) Agriculture automation system. J Phys Conf Ser 1706(1). IOP Publishing (2020)
11. Prasad S (2018) Adoption and innovation of technology in agriculture. Int J Pure Appl Math 118(9):529–534
12. Durga P, Narayanan G, Gayathri B, Ramesh MV, Divya P (2017) Modelling a smart agriculture system for multiple cropping using wireless sensor networks. In: 2017 IEEE global humanitarian technology conference (GHTC), pp 1–7. IEEE

13. Radhakrishnan S, Ramanathan R (2018) A support vector machine with Gabor features for animal intrusion detection in agriculture fields. Procedia Comput Sci 143:493–501
14. Priyadharshini G, Rajya S, Bhat SA, Vigneswar M, Ranjith R (2018) Convolution neural network based farm protection system. J Adv Res Dyn Control Syst 10(03)

Numerical Investigation to Predict the Effect of Plasma Actuator on Flow Control to Enhance Aerodynamic Characteristics and Acoustics of Symmetric Airfoil—A RANS Approach

D. Vinoth Kumar and S. Vijayaraghavan

Abstract Flow control methodologies are an important area of study for the last few decades in the history of modern aerodynamics. Control of flow means a slight shift in flow structure by means of active or passive methods that serve a preferably significant aerodynamic advantage. The aim of this study is to predict flow behavior over an airfoil under the effect of a plasma actuator; an extensive numerical investigation on a symmetric airfoil with a plasma actuator's effect over it has been conducted using Ansys Fluent tool. Experimental data from NASA turbulent flow for NACA0012 is used as a benchmark and validated using the Spalart–Allmaras RANS model. User Defined Functions (UDF) with plasma source term are defined in C language which numerically simulates the Lorentz force over the airfoil which in turn is due to the effect of plasma actuators. Aerodynamic coefficients like Drag, Lift coefficients, and flow field are compared and the effect on the Lift/Drag ratio is evaluated in the presence of the plasma actuator's effect. This is indicated by the change in drag coefficient which is reduced by 13% and the lift coefficient increased by 1.2%. To evaluate the acoustics of the airfoil, Ffowcs Williams–Hawkings (Fw-H) integral method is used, Sound Pressure Level (SPL) is compared in Fast Fourier Transforms (FFT) with frequency domain, and the SPL is used to compare the effect of plasma actuator on acoustics. The average reduction in SPL is around 3% due to plasma actuators for Angle of Attack of 5°–15°, Maximum reduction of 5% for the angle of attack of 15° for plasma actuators located at 0.2 of the chord.

Keywords Symmetric airfoil · Flow control · Turbulent flow · Plasma actuator · User defined functions · Lorentz force · Acoustics · Sound pressure level

D. Vinoth Kumar (✉)
Aerospace Engineering Department, Indian Institute of Technology, Mumbai 400076, India
e-mail: vinovirtues@gmail.com

S. Vijayaraghavan
Mechanical Engineering Department, Anna University, Chennai 600025, India

D. P. Mishra et al. (eds.), *Recent Trends in Thermal and Fluid Sciences*, Lecture Notes in Mechanical Engineering, https://doi.org/10.1007/978-981-19-3498-8_24

Abbreviations

CFD	Computational Fluid Dynamics
UDF	User Defined Function
RANS	Reynolds Average Navier–Stokes
FFT	Fast Fourier Transform
Fw-H	Ffowcs Williams–Hawkings
C_L	Lift Coefficient
C_D	Drag Coefficient
SPL	Sound Pressure Level
C_P	Pressure Coefficient
X/C	Non-Dimensional Chord Length
AoA	Angle of Attack
T	Temperature
Re	Reynolds Number
ρ	Density
ν	Kinematic Viscosity
$U_{Freestream}$	Free Stream Velocity
M	Mach Number
R	Gas Constant
γ	Gamma (Specific Heat Capacity Ratio)
C	Speed of Sound
$\vec{J}$	Current Density
$\vec{B}$	Magnetic Field
$\vec{E}$	Electric Field
$\varnothing$	Electric Potential
A	Magnetic Vector Potential
$F_{Lorentz}$	Lorentz Force
б	Electrical Conductivity
∇	Gradient Operator

1 Introduction

The aerospace industry is one of the most technologically sophisticated sectors in the world, but there are also many obstacles confronting it. Among several factors, economic and climatic challenges are key factors influencing the industry. The ability to meet the operating necessities in all these fields relies heavily on one of the main aspects of every airplane, which is its aerodynamic efficiency. The event of flow separation is one of the main restricting factors for aerodynamic efficiency. Flow separation is a situation under which the flow over an aerodynamic surface splits from the surface due to a high adverse pressure gradient, forming a wake region

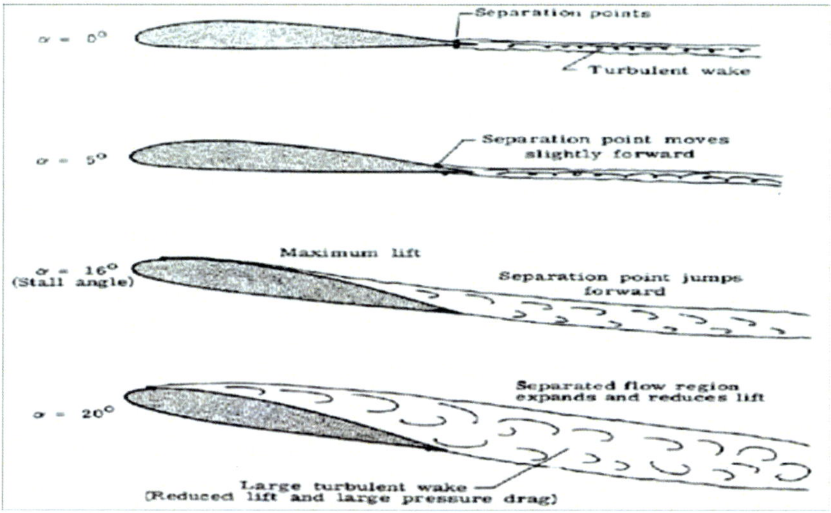

Fig. 1 Flow separation phenomenon over an airfoil [2]

behind the object [1]. This phenomenon happens over the wing of the airplane which typically has an airfoil section. At higher angles of attack, over and above the critical angle of attack, the separation happens on the suction side of airfoils, as shown in Fig. 1.

1.1 Problem Statement

As a result of flow separation, the lift force created by the wing is drastically diminished and the drag force increases at the same time due to wake formation. This situation is called separated flow, and may lead to several unsafe effects, such as high rates of descent or lack of balance [3]. High dynamic loads may also be induced during stalls on the wings. Modern larger aircraft use high lift devices to increase the aerodynamic efficiency. This involves trailing edge flaps and leading edge slats and holes, which essentially adjust the form of the wing to provide higher lift at take-off and landing [4]. Such devices may be effective, but are heavy with several moving parts, i.e. increasing the complexity and often relatively difficult to control, and comparatively weigh more. Using additional flow regulation and control mechanisms, the efficiency of high lift devices can be enhanced. One such means for flow control is a plasma actuator which controls the flow over an airfoil by active means, with typical power levels of approximately 2–40 W per linear foot of actuator's span. Such plasma actuators can be operated either in a "Steady/Continuous" or Unsteady/Pulsating way. In the first type, the input driving frequency is maintained above the fluid response frequency and therefore a constant body force is acted on the fluid. In unsteady or pulsating type, a higher driving frequency is switched ON and OFF at ultra-lower frequencies of the order of Hertz. This can be used to excite fluid instabilities that act to further amplify the actuator effect. In the pulsating type, very short duty cycles can be maintained, which can reduce the actuator power to a significant extent. For example, in applications of separation control in which pulsating type forcing is used, a 10% duty cycle is sufficient to suppress the flow separation, thereby lowering the power requirement by 90% compared to steady or continuous operation. In the present work, the Steady/Continuous mode actuator's numerical model is used as the objective is to understand the effect of constant body force on the flow domain.

2 Literature Review

The separation of flow from an airfoil surface disrupts the flow structure due to adverse pressure gradient and leads to increased pressure drag component. The intrinsic fluid viscosity causes skin-friction drag, where the fluid sticks to the wing area and the corresponding frictional shear pressure exerts a friction drag [5]. A drag force is generated as a consequence of the difference in pressure when the flow is separated and this is called pressure drag. Methods for flow control are generally classified into active and passive methods [6]. The techniques that require external energy expenses are categorized as active while those that can adjust the flow without expending energy in the form of a surface or extra affinity are referred to as passive methods. The use of surface modifications is based on a number of passive methods including roughness, helical dimples, longitudinal grooves, transceiver plates, and secondary cylinders [7]. Various approaches, including vortex engines, suction, blowing, air jets, vibrating strings, and intermittent excitation are used as mechanisms to suppress the flow separation thereby improving the aerodynamic efficiency of the airfoil. Most of these have proved to be successful and have seen functional use. One such example is a vortex generator. However, current flow control systems have their disadvantages, such as parasitic drag, mechanical difficulty, and high usage of resources [8]. Thus, research is underway in the area of separation management. The dielectric barrier discharge plasma actuators have recently come to the notice of another applicant for separation control. A plasma actuator is a tape-shaped unit that can be fastened to an aerodynamic board. It produces plasma by adding a high-voltage signal, which influences the flux in the vicinity of the actuator. Experiments have shown that plasma actuators will delay the separation of airfoils [9]. There are two distinct modes of operation of plasma actuators with dielectric barrier discharge, i.e. pulsed and continuous mode. The two forms of actuators have separate operational concepts and therefore different flow control capacities. In this work, an attempt is made to study the effect of the continuous mode plasma actuator's effect on turbulent flow over the airfoil. In high voltage direct current plasma, the electrons are driven in the electric field from cathode to anode, and they ionize the gas by neutral molecule collisions. An avalanche occurs when electrons are propagated from negative to positive terminals. Thereafter, a current flow is generated [10]. Depending on its polarity, the process of maintaining constant ionization near the higher voltage terminal is dependent. It is known as negative corona if the large electrical field is centered at the cathode. It is known as positive corona if the powerful electric field is near the anode [11]. Then ions are repelled by Coulombian forces toward the cathode, and constitute the drift zone, where it can be concluded that there is no recombination because the electrical field is small [12]. There have been past studies on the electrical properties of the surface discharge with respect to various atmospheric parameters like air pressure, humidity of air, atmospheric temperature, humidity, and dimension-based parameters like the shape, position, and the distance between the electrodes [13]. The stability of the discharge depends on the diameter of the anode with respect to the cathode. As the anode diameter gets smaller, the discharge gets more stable [14]. The discharge is not affected by the temperature until it gets hot at around 60 °C. Above this temperature, the pressure also affects the discharge [15]. The humidity in the air affects the discharge greatly. With the increase in humidity, the discharge becomes more unstable. The humidity must be as low as possible to maintain the stability of the discharge [16]. When there is a presence of free air, it affects the discharge and reduces the glow. The air must be minimal to maintain an arc [17]. In a plasma actuator setup, the method of ionizing the air is classically called a single dielectric barrier discharge. The ionized air (plasma) creates a body force in the ambient air in the presence of an electric field gradient that induces a virtual aerodynamic structure as seen in Fig. 2. This changes the pressure distribution over the surface of the actuator; there is minimal ionization in the air adjacent to the electrodes and little or no heating of the air.

The body force F_B for plasma is represented by

$$F = \frac{-\epsilon_0 \varnothing}{(\lambda_D)2}\left(\vec{E}\right) \tag{1}$$

where ϵ_0 is the permittivity of free space, λ_D is the Debye length, $\varnothing$ is the electric potential, and $\vec{E}$ is the electric field vector [18]. Computational Fluid Dynamics (CFD) is a numerical technique to solve fluid flow problems. The governing equations for fluid flow are in partial differential equations (PDEs) from which the conservation of mass, momentum, and energy is shown below. CFD converts all PDEs into

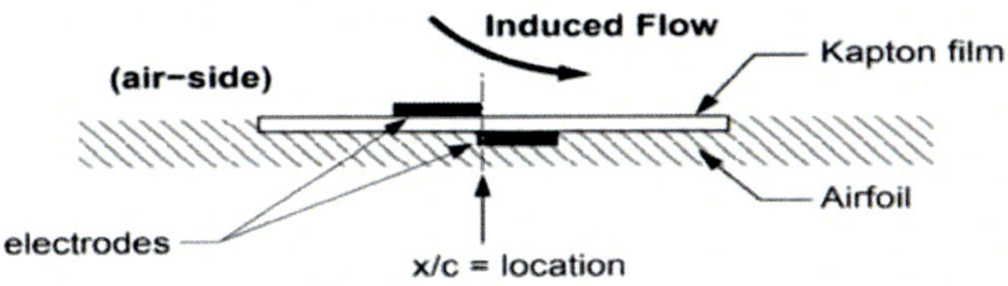

Fig. 2 Arrangement of actuator electrode [18]

algebraic equations that can be solved by using solution algorithms like SIMPLE and PISO.

$$\frac{\partial \rho}{\partial t} + \frac{\partial}{\partial x_i}(\rho v_i) = 0 \quad (2)$$

$$\frac{\partial}{\partial x_j}(\rho v_i v_j) = \frac{-\partial P}{\partial x_i} + \frac{\partial}{\partial x_j}\left[\mu\left(\frac{\partial v_i}{\partial x_j} + \frac{\partial v_j}{\partial x_i}\right) - \frac{2}{3}\mu\frac{\partial v_k}{\partial x_k}\delta_{ij}\right] \quad (3)$$

$$\frac{\partial(\rho e)}{\partial t} + \frac{\partial}{\partial x_j}\left(\rho v_j C_p T - k\frac{\partial T}{\partial x_j}\right) = u_j\frac{\partial P}{\partial x_j} + \left[\mu\left(\left(\frac{\partial v_i}{\partial x_j} + \frac{\partial v_j}{\partial x_i}\right) - \frac{2}{3}\frac{\partial v_k}{\partial x_k}\delta_{ij}\right)\right] \quad (4)$$

For acoustic computations, in the present study Fw-H methodology is used [19] which is a generic form for acoustic field computations. This method used is similar to the conservation of mass and momentum by incorporating the effects of pressure fluctuations using the Heaviside function denoted as H(f) as shown below

$$\frac{1}{a_o^2}\frac{\partial^2 p'}{\partial t^2} - \nabla^2 p' = \frac{\partial}{\partial t}\{[\rho_o u_n + \rho(u_n - v_n)]\delta(f)\} - \frac{\partial}{\partial x_i}\{[p_{ij} n_j + \rho u_i(u_n - v_n)]\delta(f)\} + \frac{\partial^2}{\partial x_i \partial x_j}(T_{ij} H(f)) \quad (5)$$

This equation has a lot of additional terms in analogy with the pressure transport equation for fluid flow, which is well described in The most significant to note that p' in above equation is the pressure at the far field. The right-hand side terms represent the monopole and dipole noise source terms and their advection and temporal variation. P_{ij} is the total stress tensor due to normal and shear stresses respectively, and T_{ij} is called Lighthill's stress tensor, respectively.

2.1 Objectives of This Study

The primary objective of this work is to perform a numerical simulation to predict the effect of a plasma actuator on the flow control and acoustics of an airfoil. The expected outcomes of this work are

i. To perform numerical analysis of turbulent flow over the airfoil and validate from the experimental study through literature data.
ii. To perform numerical analysis on the plasma actuators with momentum source over the airfoil surface and compare plasma actuator's effect on aerodynamic coefficients with different Angles of Attack (AoA).

iii. To study the effect of plasma actuators in continuous discharge mode on the acoustics of the airfoil and compare the SPL with and without plasma actuators.

3 Numerical Methodology

3.1 Airfoil Selection and Literature Data

In the present study, in order to estimate the effect of plasma actuators on the aerodynamic coefficients and understand the significance of plasma actuators on the same the turbulent flow and in order to validate the numerical data with that of the experimental data by Ladson [20] experiments and overflow numerical code results using Spalart–Allmaras (SA) turbulent model, the literature data of NACA 4 series airfoil with symmetric geometry, i.e. NACA0012 airfoil. The selected airfoil NACA0012 is modeled in Ansys fluent using .txt file import and the geometry is as shown in Fig. 3.

This NACA airfoil four series is controlled by 4 digits, e.g. NACA 0012, which designate the camber, position of maximum camber, and thickness. If an airfoil number is

NACA MPXX, e.g. NACA 0012

- M is the maximum camber divided by 100. In the example M = 0, so the camber is 0.00 or 0% of the chord.
- P is the position of the maximum camber divided by 10. In the example, P = 0 so the maximum camber is at 0.0 or 0% of the chord or at the leading edge.
- XX is the thickness divided by 100. In the example, XX = 12 so the thickness is 0.12 or 12% of the chord.

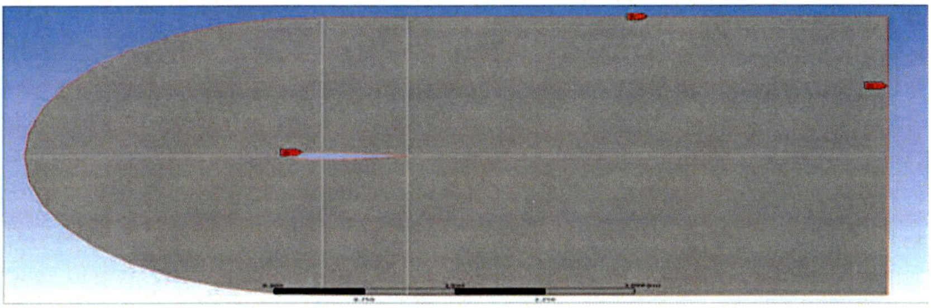

Fig. 3 Computational domain generated over NACA0012 airfoil

3.2 Computation of Turbulent Flow over Airfoil Without Plasma Actuator

The various steps involved in numerical modeling to predict the turbulent flow over an airfoil for a Reynolds number of 6 Million are as mentioned below.

Step-1: Y+ Calculation and Grid Generation

To determine the wall distance, a y^+ desired value is selected, for the S-A RANS model with a y^+ of 30 and above and to estimate the required wall distance with the computational mesh as shown in Fig. 4.

To find U from Re

$$Re = \frac{\rho U L}{\mu}$$
$$\mathrm{U} = 88.65\,\mathrm{m/s} \tag{6}$$

To estimate skin friction C_f from Re

$$\mathrm{C}_f = (2\log_{10}(Re_x) - 0.65)^{-2.3}$$
$$\mathrm{C}_f = 0.0027871 \tag{7}$$

To find wall shear stress τ_W

$$\tau_W = \mathrm{C}_f * 0.5\rho \mathrm{U}^2$$
$$\tau_W = 13.41\,\mathrm{Pa} \tag{8}$$

To calculate the friction velocity

$$\mathrm{u}_\tau = \sqrt{\frac{\tau_W}{\rho}}$$

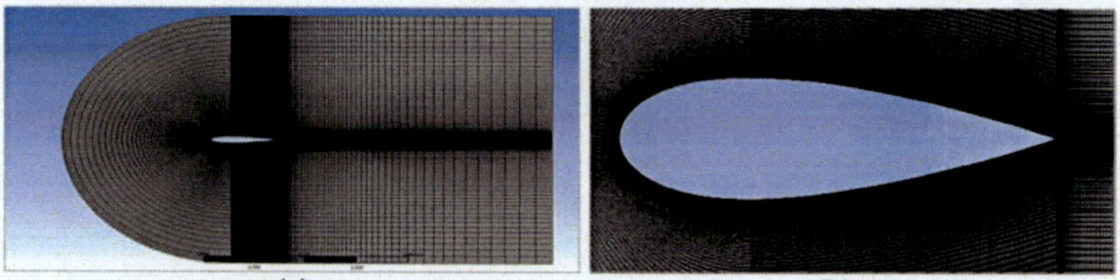

(a) (b)

Fig. 4 (a, b) Computational grid generation around NACA0012 airfoil

Table 1 Initial and boundary conditions

Parameter	Value	Unit
Temperature (T)	300	K
Reynolds number (Re)	60,000,00	–
Density (ρ)	1.225	kg/m^3
Kinematic viscosity (ν)	1.58E-05	m^2/s
Non-dimensional chord length (X/C)	0–1	–
Free stream velocity (U_{Free})	88.65	m/s
Mach number (M)	0.255337	–
Gas constant (R)	287	J/kg-K
Specific heat ratio (γ)	1.4	–
Speed of sound (C)	347.2	m/s

$$u_r = 3.309\,\text{m/s} \tag{9}$$

To find y_{wall} *with* $y^{\pm} = 30$

$$y_{wall} = \frac{y^{+}H}{\rho u_r} \tag{10}$$

$$y_{wall} = 0.000121\,\text{m} = 0.121\,\text{mm} \tag{11}$$

Step-2: Initial and Boundary Conditions

Turbulent computations in present study have been performed using the S-A model [21] in Ansys Fluent; the S-A turbulence model is a widely used and robust 1-equation eddy-viscosity turbulence model for external flow simulations. The initial and boundary conditions are as tabulated in Table 1.

3.3 Numerical Implementation of Plasma Actuator's Influence on Airfoil

Simplified equations for electromagnetic phenomenon within Plasma Actuator assuming Electro neutrality, Quasi-steady electromagnetic, and magnetohydrodynamic condition as valid [3] can be simplified as derived below.

Conservation of Electric Charge:

$$\nabla.\vec{J} = 0 \tag{12}$$

Gauss Law (Magnetic):

$$\nabla.\overrightarrow{B} = 0 \tag{13}$$

Gauss Law (Electric):

$$\nabla.\overrightarrow{E} = 0 \tag{14}$$

Ampere's law:

$$\nabla \times \overrightarrow{B} = \mu o.\overrightarrow{J} \tag{15}$$

Faraday's Law:

$$\nabla \times \overrightarrow{E} = 0 \tag{16}$$

Ohm's Law:

$$\overrightarrow{J} = \sigma.\overrightarrow{E} \tag{17}$$

where σ and μo are electrical conductivity and permeability of free space. Electric potential $\emptyset$ is defined as

$$\overrightarrow{E} = -\nabla\varnothing \tag{18}$$

Substituting it in Ohm's law,

$$\overrightarrow{J} = -(\sigma\nabla\varnothing) \tag{19}$$

Combining it with Charge Conservation equation,

$$\nabla.\sigma\nabla\varnothing = 0 \tag{20}$$

Gauss's law (magnetic) can be satisfied by defining a vector potential A such that

$$\overrightarrow{B} = \nabla \times A \tag{21}$$

Putting this and J in Ampere's law yields

$$\nabla \times \nabla \times A = -\mu o.(\sigma\nabla\varnothing) \tag{22}$$

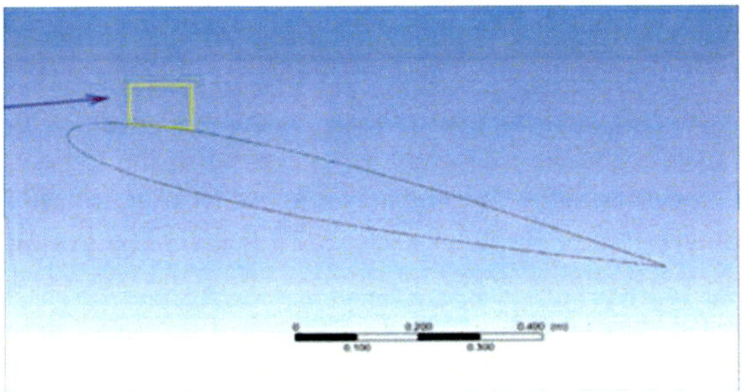

Fig. 5 Plasma actuator definition over airfoil at 0.2 C

Now, the Lorentz electromagnetic force can be computed for J and B as shown below

$$\vec{F} = q\vec{v} \times \vec{B} \tag{23}$$

When it is simplified into Vector Cross-Product Form

$$F_{LORENTZ} = \vec{J} \times \vec{B} \tag{24}$$

So, the final form of implementation of the plasma actuator's effect is in a force vector which is computed as an interaction between the electric and magnetic fields. The Lorentz force is similar to the form of net charge density and magnetic flux interaction in a space field with an arbitrary magnitude of 2100 N/m^3 in the present study at 0.2 X/C as mentioned in Fig. 5. The user-defined functions (UDFs) are developed in C computer language [22] for the dielectric barrier discharge plasma simulation. The UDF script is then dynamically linked with the Ansys Fluent tool with the available numerical model for turbulent flow over an airfoil after which computations with the effect of Plasma Actuator on the airfoil are performed. The variables are stored in the center of each cell, a procedure of interpolation, based on a second-order scheme, and are then used to assign values to the faces of the control volumes (CV). The variable gradients on each CV face are computed using a multi-dimensional Taylor series expansion with least-squares cell-based approaches. The semi-implicit method for pressure-linked equation (SIMPLE) algorithm was used for the pressure–velocity coupling. In turbulent viscosity and momentum equations, the second-order upwind scheme is used for convective terms.

3.4 *Airfoil Computational Aeroacoustic Analysis*

The Acoustic model Fw-H is used for Acoustic computations and the source of noise is the airfoil surface. This method used is similar to the transport of mass and momentum by considering the transport physics of pressure fluctuations using Heaviside function H(f). The numerical Algorithm used is PISO with Transient second-order formulation and QUICK Scheme for advection. The Other Model Parameters are as given below in Fig. 6 as an Input to the Fw-H Model.

For implementing the acoustic model to compute SPL at desired locations in the present study for AoA = 15°, the receivers are defined at suitable locations in the near zone locations in space close to airfoil walls and in a transverse plane in the wake region of the airfoil as seen in Fig. 7. As discussed in the previous chapter for conditions with plasma actuator's effect included, the UDF file is compiled as the source term for the Y-Momentum equation. Numerically recorded SPL values through the receivers are compared for the turbulent flow over the airfoil without plasma actuator's effect and with plasma actuator's effect, i.e. Lorentz force term is included.

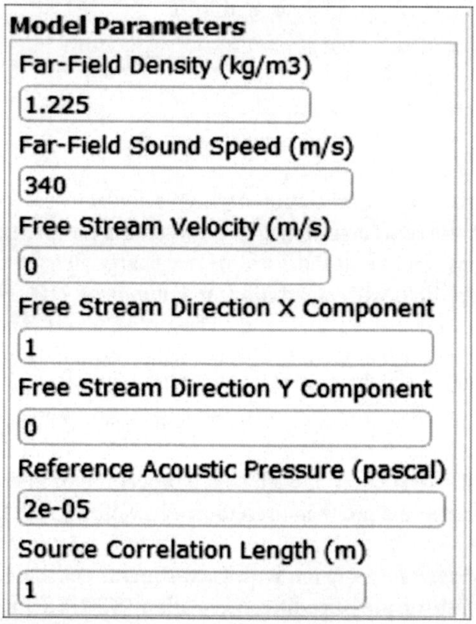

Fig. 6 Flow and fluid parameters used for Fw-H integral method

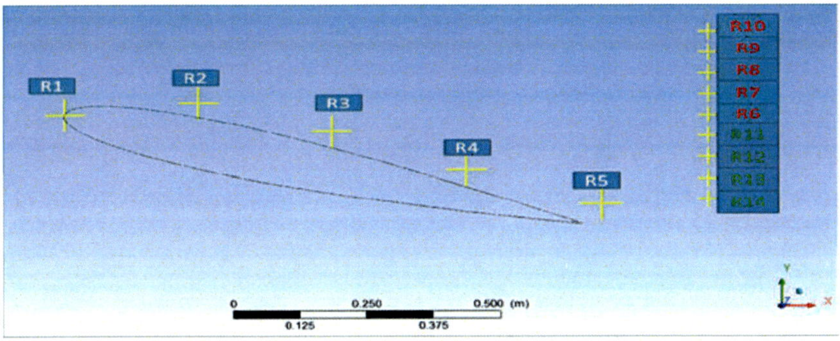

Fig. 7 Numerical acoustic signal receivers' definition over the computational domain

4 Results and Discussions

4.1 Numerical and Experimental Data Comparison for Turbulent Flow over Airfoil Without Plasma Actuator for Various AoA

The generated initial grid is with ~153,000 elements and with an average orthogonal quality of 0.96. The grid independent study is performed with this mesh, and it is found that at ~176,000 elements with Y^+ of 30 the results tend to converge with grid, i.e. there is no significant difference in aerodynamic coefficients between the previous grid and the present grid. Thus, the grid is considered as optimal for this numerical study and the values are then compared with experimental data from the literature. The experimental results obtained in accordance from the Ladson [20] and the numerical data from Thomas et al. [23] are compared with the obtained numerical results for turbulent flow over the airfoil, and the aerodynamic coefficients are compared for different AoAs −20°, −15°, −10°, −5°, 0°, 5°, 10°, 15°, and 20°. A few key flow parameters are represented in this section, Figs. 8 and 9.

The flow separation at high AoA of 20° leading to stalling conditions with a drastic reduction in the lift is represented by turbulent viscosity and the detached flow in the wake region as shown in Fig. 10. Sample C_L convergence with iterations in the present study is shown in Fig. 11 for AoA = 15°.

From Fig. 12 and Table 2 comparison, the numerical data is having a strong agreement with Ladson experimental data, and the maximum deviation is 6.8% while the agreement between the numerical data from the present study and the overflow code is also on the same extent with maximum deviation is 6.5%.

The experimentally measured pressure variations from which the calculated C_P distribution over airfoil surface from the literature is plotted against CFD results obtained in the present study are represented in Fig. 13. A strong agreement between both results is evident in Table 2.

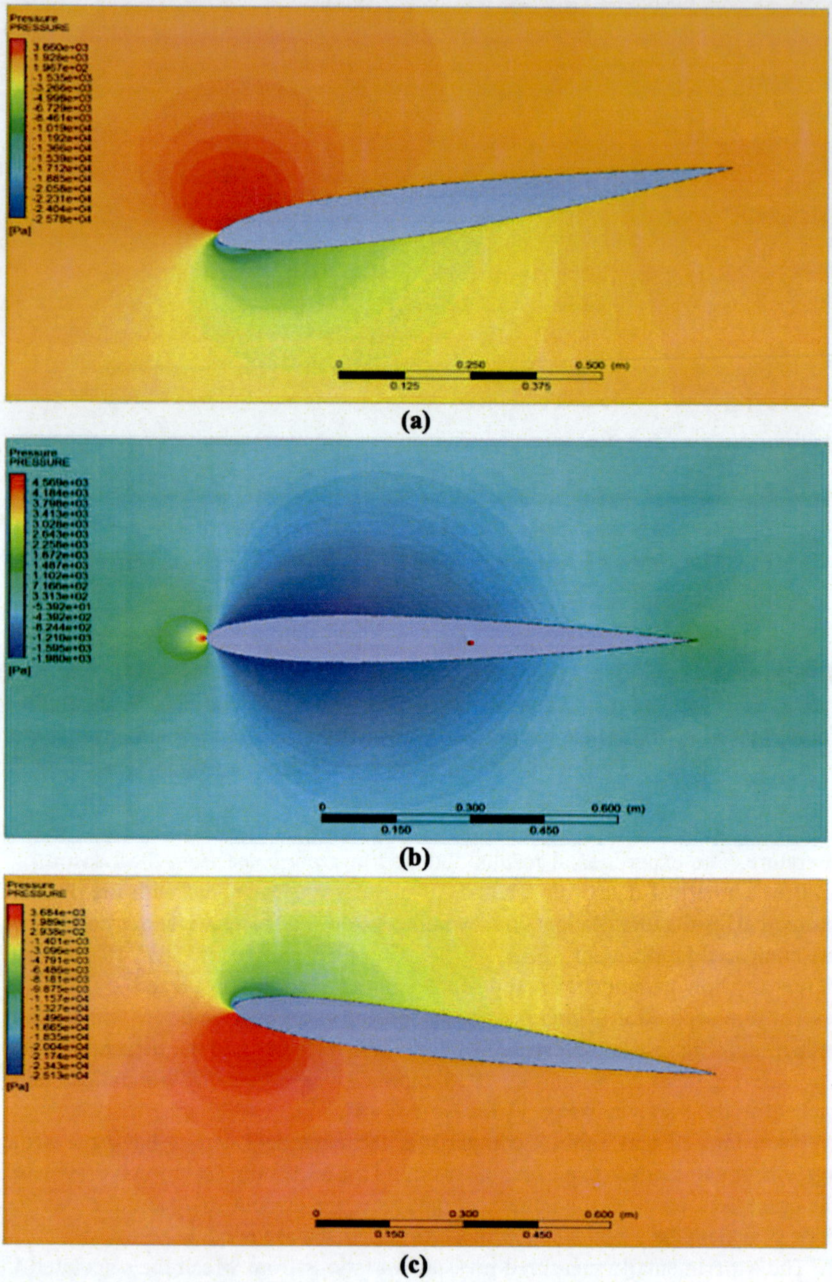

Fig. 8 **(a, b, c)** Pressure distribution over the airfoil for AoA = −10°, 0°, 10°

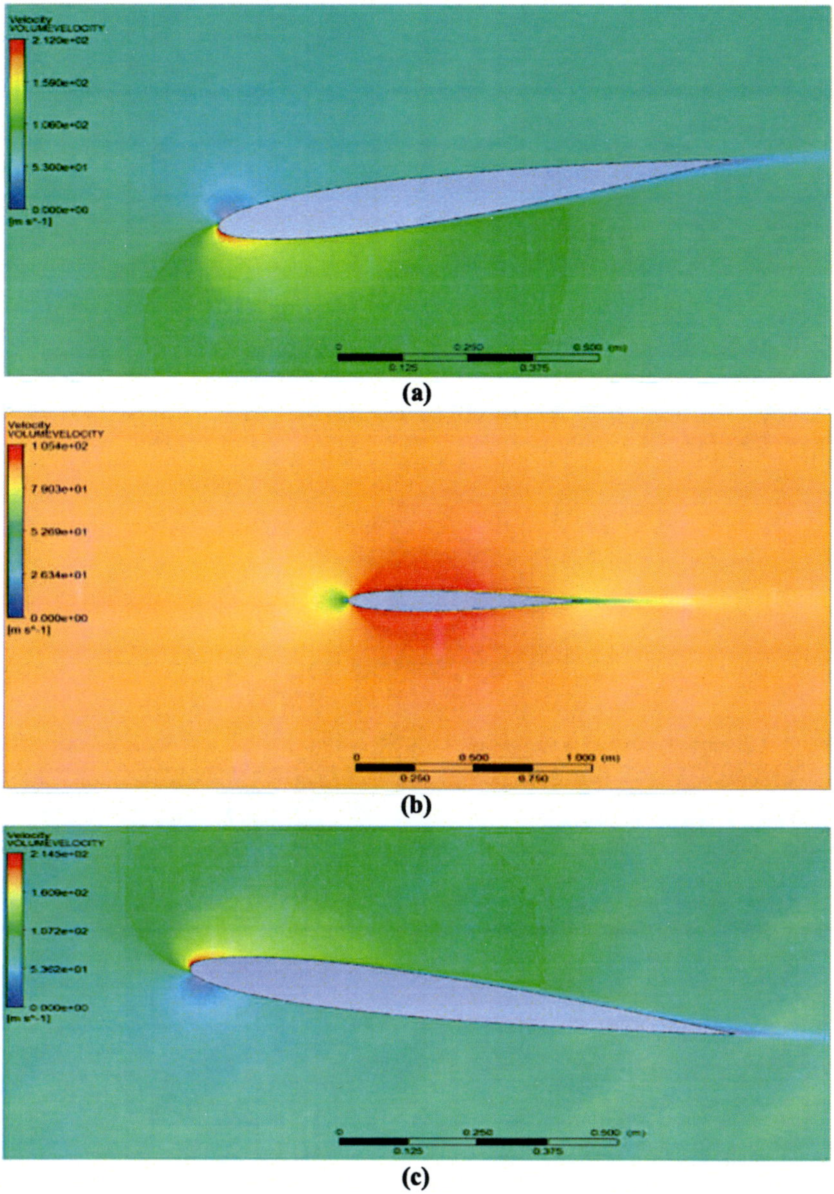

Fig. 9 (**a**, **b**, **c**) Velocity contours over the airfoil for AoA = −10°, 0°, 10°

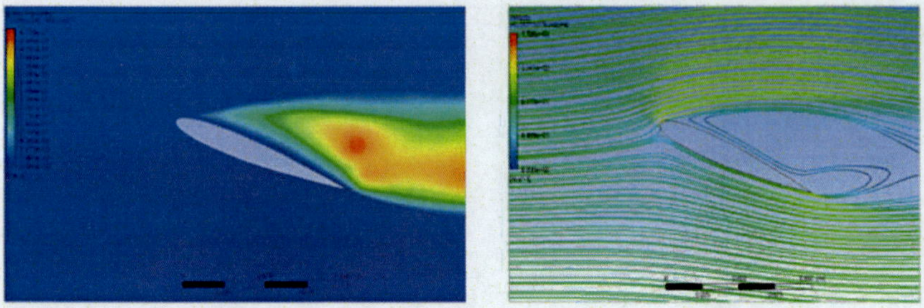

(a) (b)

Fig. 10 (a, b) Flow separation visualized at AoA = 20°

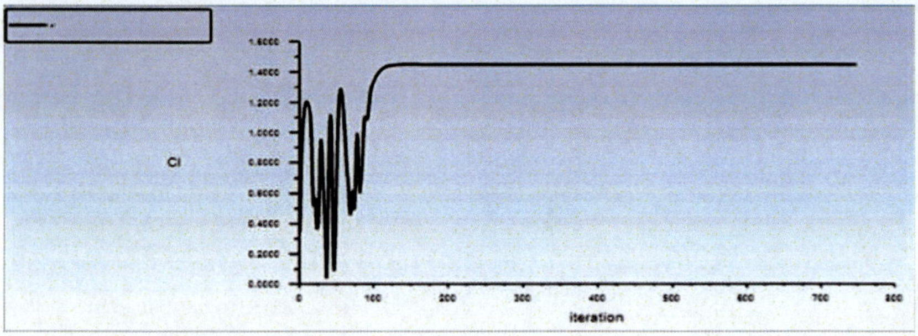

Fig. 11 CL versus iterations for AoA = 15°

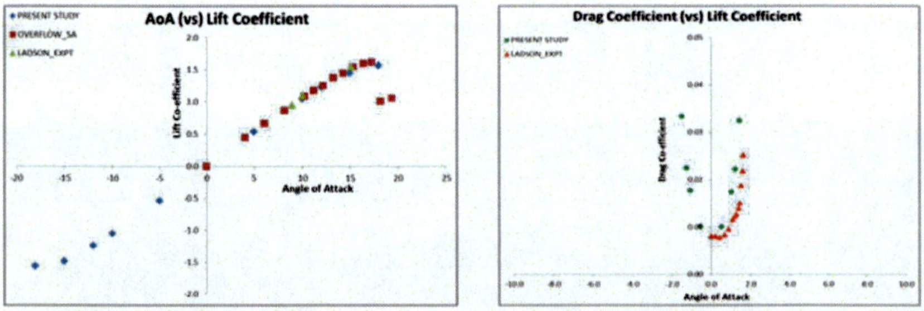

(a) (b)

Fig. 12 (a, b) C_L, C_D versus AoA comparison between numerical and experimental data from literature

Table 2 CL versus AoA comparison between numerical and experimental, numerical data from literature

Name	AoA (degrees)	C_L (present study)	C_L (Overflow)	% Difference (present study and overflow)	C_L (Ladson experiment)	% Difference (present study and Ladson experiment)
DP 1	0	−0.0041	0	4.1	0	4.1
DP 2	−5	−0.5402	(NA)			
DP 3	−10	−1.0554				
DP 4	−12	−1.2431				
DP 5	−15	−1.4856				
DP 6	−18	−1.5568				
DP 7	5	0.5355	0.541	1.02	(NA)	
DP 8	10	1.0437	1.09032	4.47	1.09839	4.98
DP 9	12	1.2261	1.25565	2.41	(NA)	
DP 10	15	1.4478	1.54194	6.50	1.55403	6.84

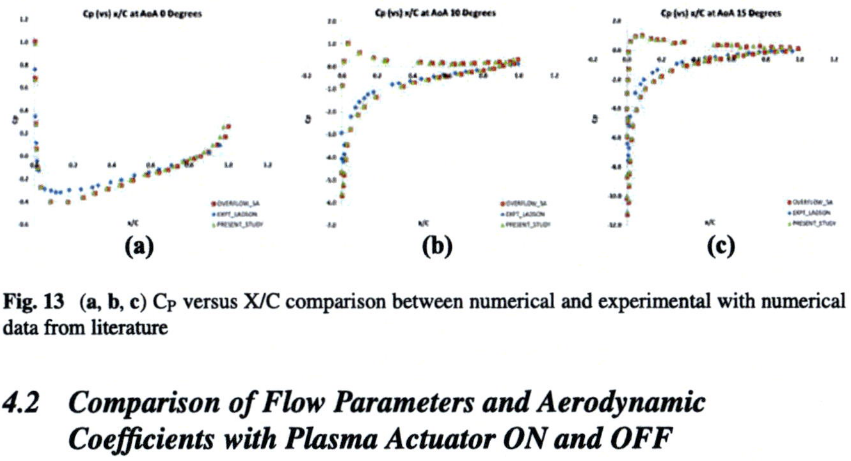

(a) (b) (c)

Fig. 13 **(a, b, c)** C_P versus X/C comparison between numerical and experimental with numerical data from literature

4.2 Comparison of Flow Parameters and Aerodynamic Coefficients with Plasma Actuator ON and OFF Conditions

It can be observed from Fig. 14 that without a plasma actuator, the maximum pressure difference between top and bottom sides for AoA = 10° is 27,814 Pa while with plasma actuator's effect, i.e. with Lorentz source term, the value is 29,415 Pa by an increase of 3–6% for 5°–15°. However, this seems to be very insignificant as mentioned in Table 3 with C_L increase of 0.6–2.7%. Table 3 shows the variation in aerodynamic coefficients for various AoA for the plasma actuator's effect in ON and OFF conditions and the comparison plot in shown in Fig. 15.

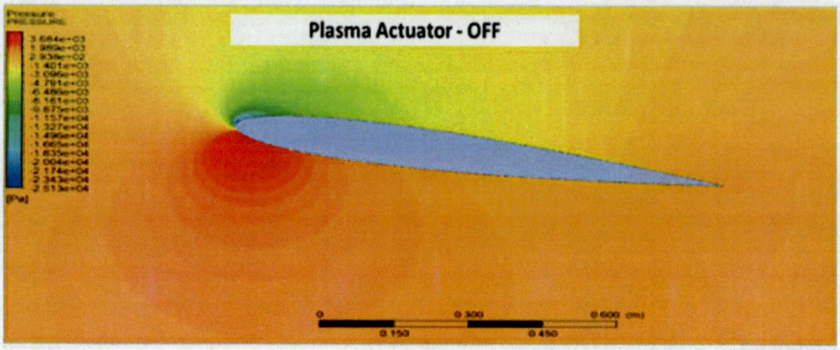

(a)

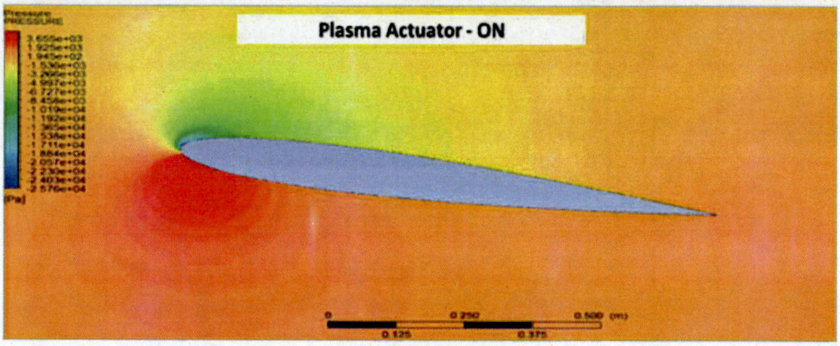

(b)

Fig. 14 **(a, b)** Pressure distribution comparison between plasma actuators OFF and ON conditions for AoA = 10°

Table 3 Flow and aerodynamic coefficients comparison between plasma actuators OFF and ON conditions

AoA	C_L		% Increase	C_D		% Decrease
	PA-OFF	PA-ON		PA-OFF	PA-ON	
0	0	0.004399	0	0.0079	0.007775	1.12
5	0.5355	0.549835	2.67	0.0101	0.008436	16.47
10	1.0437	1.055820	1.16	0.0174	0.014644	15.99
12	1.2261	1.236829	0.88	0.0222	0.019154	13.88
15	1.4478	1.456914	0.63	0.0324	0.028545	12.00

From Table 3 and Fig. 15, it can be observed that a reasonable decrease in C_D is accompanied by a slight increase in C_L, thus there is a net increase in aerodynamic efficiency with 1.1%–22.5% for AoA in the range of 0°-15°, respectively.

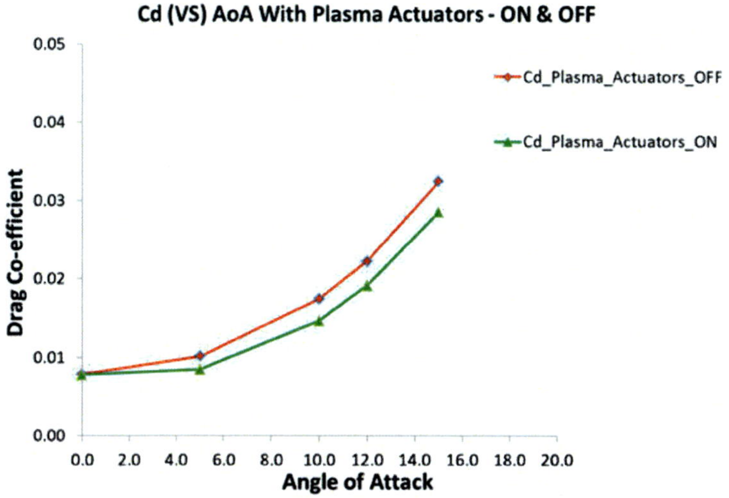

Fig. 15 CD versus AoA comparison between plasma actuators OFF and ON conditions

4.3 Numerical Acoustic Data Comparison with Plasma Actuator ON and OFF Conditions

The numerical model and methodology as discussed in 3.4 are adopted to record SPL in the receiver matrix, and the FFT data of SPL is stored as overall SPL for that receiver in the given AoA 15°. From the recorded data, the comparison between conditions without and with a plasma actuator from Table 4 and Fig. 16 depicts a maximum reduction of 5.5% in far field and 4.63% in the near-wall zone, and the average reduction in the SPL level is around 3%.

5 Conclusions

Results obtained for change in aerodynamic coefficients due to the effect of plasma actuators are studied in detail, and the effect on SPL is also observed through numerical means in the present study. The preliminary objective is to implement the desired turbulence model to closely predict the flow over the symmetric airfoil and compare the same with experimental data and literature data; the same is accomplished with the SA turbulence model. A close agreement between the numerical results and experimental data is observed. Then the further approach to implementing the effect of Plasma Actuators to perform the flow control, i.e. Active method for flow control and the plasma actuator's effect is observed with a significant reduction of C_D and a slight increase in C_L, thus the aerodynamic efficiency is improved with 1.1%–22.5% for AoA in the range of 0°–15°, respectively. An attempt was made to study

Table 4 SPL comparison between plasma actuators OFF and ON conditions

Receiver	Location		SPL without PA (dB)	SPL with PA (dB)	% Decrease
	X	Y			
1	0	0.05	147.16	145.67	1.01
2	0.25	0.03	138.11	136.58	1.11
3	0.5	−0.04	137.04	134.71	1.70
4	0.75	−0.13	135.59	132.15	2.54
5	1	−0.21	124.71	118.94	4.63
6	1.2	0	120.13	115.96	3.47
7	1.2	0.05	120.68	116.78	3.23
8	1.2	0.1	120.99	117.31	3.04
9	1.2	0.15	121.13	117.64	2.88
10	1.2	0.2	121.15	117.83	2.74
11	1.2	−0.05	119.21	114.70	3.78
12	1.2	−0.1	117.71	112.74	4.22
13	1.2	−0.15	115.24	109.60	4.89
14	1.2	−0.2	111.18	105.05	5.52

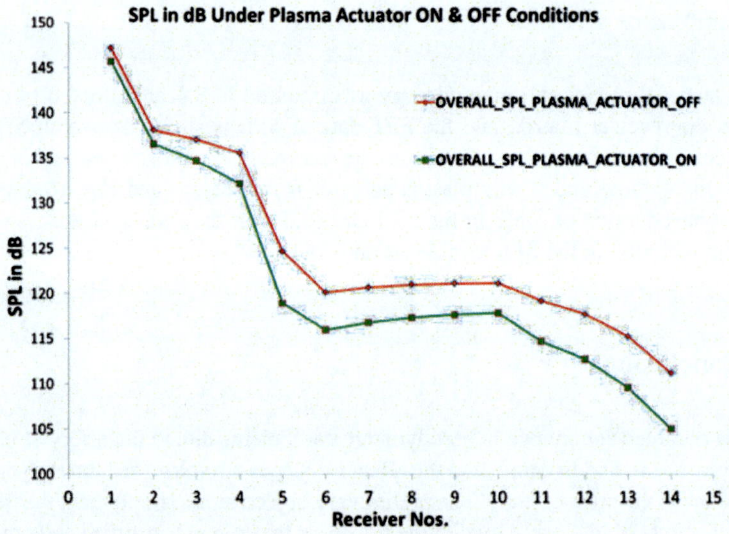

Fig. 16 SPL comparison between plasma actuators OFF and ON conditions

acoustics using the Fw-H numerical model, i.e. to predict the Sound pressure levels due to the pressure fluctuations along the flow over the airfoil body. The average reduction over the SPL is around 3% due to plasma actuators while the maximum reduction is around 5.5% for the plasma actuator at 0.2 X/C. This in turn concludes the plasma actuator's location could be a critical factor to decide its effectiveness in Noise suppression. Hence, certain scope for future research based on the present work like optimization toward location of the plasma actuators for different AoA of the airfoils and performing own experiments for different flow configurations discussed in this work.

References

1. Fatahian H, Salarian H, Nimvari ME, Khaleghinia J (2020) Computational fluid dynamics simulation of aerodynamic performance and flow separation by single element and slatted airfoils under rainfall conditions. Appl Math Model 83:683–702. https://doi.org/10.1016/j.apm.2020.01.060
2. Alrayah A, Taha OY, Seory AMA (2015) Numerical study on the effect of different backward step configurations on the aerodynamics performance of NACA 23012
3. Bangga G, Lutz T, Arnold M (2020) An improved second-order dynamic stall model for wind turbine airfoils. Wind Energy Sci 5(3):1037–1058. https://doi.org/10.5194/wes-5-1037-2020
4. Lacy DS, Sclafani AJ (2016) Development of the high lift common research model (HL-CRM): a representative high lift configuration for transonic transports. https://doi.org/10.2514/6.2016-0308
5. Francioso L, De Pascali C, Pescini E, De Giorgi MG, Siciliano P (2016) Modeling, fabrication and plasma actuator coupling of flexible pressure sensors for flow separation detection and control in aeronautical applications. J Phys D Appl Phys 49(23):235201. https://doi.org/10.1088/0022-3727/49/23/235201
6. Moghaddam T, Neishabouri NB (2017) On the active and passive flow separation control techniques over airfoils. IOP Conf Ser Mater Sci Eng 248:012009. https://doi.org/10.1088/1757-899X/248/1/012009
7. Zhai X et al (2020) Lift characteristics of two tandem airfoils in the globally unstable wake of a heated cylinder. J Therm Anal Calorim. https://doi.org/10.1007/s10973-020-10358-x
8. Islam MS (2017) Improvement of aerodynamic characteristics of an airfoil by static extended trailing edge. Khulna University of Engineering and Technology
9. Moreau E (2007) Airflow control by non-thermal plasma actuators. J Phys D Appl Phys 40(3):605–636. https://doi.org/10.1088/0022-3727/40/3/S01
10. Zhang B, Li Z, He J (2017) A numerical model of acoustic wave caused by a single positive corona source. Phys Plasmas 24(10):103521. https://doi.org/10.1063/1.4998352
11. Mikropoulos PN, Zagkanas VN (2016) Negative DC corona inception in coaxial cylinders under variable atmospheric conditions: a comparison with positive corona. IEEE Trans Dielectr Electr Insul 23(3):1322–1330. https://doi.org/10.1109/TDEI.2015.005517
12. Chawdhury P, Bhargavi KVSS, Selvaraj M, Subrahmanyam C (2020) Promising catalytic activity by non-thermal plasma synthesized SBA-15-supported metal catalysts in one-step plasma-catalytic methane conversion to value-added fuels. Catal Sci Technol 10(16):5566–5578. https://doi.org/10.1039/D0CY00900H
13. Rodrigues FF, Pascoa JC, Trancossi M (2016) Analysis of innovative plasma actuator geometries for boundary layer control. https://doi.org/10.1115/IMECE2016-66495
14. Borghu F (2017) Characterization of a three-electrodes surface corona actuator. ING - Scuola di Ingegneria Industriale e dell'Informazione

15. Lee JW, Sohn DK, Ko HS (2020) Experimental study on momentum of electrohydrodynamic flow induced by corona discharge in the pin to two cylinder configuration. J Mech Sci Technol 34(9):3679–3686. https://doi.org/10.1007/s12206-020-0820-1
16. Chen S, Li K, Wang F, Sun Q, Zhong L (2019) Effect of humidity and air pressure on the discharge modes transition characteristics of negative DC corona. IET Sci Meas Technol 13(8):1212–1218. https://doi.org/10.1049/iet-smt.2019.0032
17. Li Z, Liu J, Lu X (2020) A large atmospheric pressure nonequilibrium open space air plasma based on a rotating electrode. Plasma Sources Sci Technol 29(4):045015. https://doi.org/10.1088/1361-6595/ab7aaf
18. Corke T, Post M (2005) Overview of plasma flow control: concepts, optimization, and applications. https://doi.org/10.2514/6.2005-563
19. Sound generation by turbulence and surfaces in arbitrary motion. Philos Trans R Soc Lond Ser A Math Phys Sci 264(1151):321–342 (1969). https://doi.org/10.1098/rsta.1969.0031
20. Ladson CL (1988) Effects of independent variation of Mach and Reynolds numbers on the low-speed aerodynamic characteristics of the NACA 0012 airfoil section. NASA TM 4074
21. Kostic C (2015) Review of the Spalart-Allmaras turbulence model and its modifications to three-dimensional supersonic configurations. Sci Tech Rev 65:43–49. https://doi.org/10.5937/STR1501043K
22. Kernighan BW, Ritchie D (1988) C programming language, 2nd edn. Pearson
23. Jespersen DC, Pulliam TH, Childs ML. Overflow turbulence modeling resource validation results. NASA Ames Research Center, NAS Technical Report: NAS-2016-01

Performance Comparison of Silica Gel and Natural Desiccant Material-Based Evaporative Cooling System

Sandeep Shukla, Rohit Misra, and G. D. Agrawal

Abstract The present study experimentally explores the feasibility of using natural and biodegradable material viz. Cocopeat powder as a desiccant material. The thermal performance of Cocopeat powder has been investigated to see whether it carries the potential to be used as a desiccant to improve the performance of the evaporative cooling system. Its performance has been compared with that of commercially popular chemical desiccant Silica gel. The study also examines the effect of airflow velocity on the moisture absorption and desorption capacity of the Cocopeat and Silica gel. Frictional characteristic of these materials has also been investigated in terms of pressure drop offered by them. The study reveals the relative humidity (RH) of air at the exit of the desiccant chamber having Cocopeat as a natural desiccant is minimum for most of the duration of experimentation. Up to an initial 30–40 min of operation, the RH of air at the exit of the silica gel chamber (SGC) is minimum and the temperature of air at the exit of SGC has been observed to be 1.5–2.0 °C higher up to 30–40 min of operation as compared to Silica gel. The study concludes that the Cocopeat powder may potentially be used as a desiccant material in place of costly chemical desiccants like Silica gel and others.

Keywords Silica gel · Direct evaporative cooling · Indirect evaporative cooling · Cocopeat powder · Thermal regeneration

1 Introduction

In a country like India, the majority of the population tries to make themselves comfortable using the direct evaporative cooler during the summer season which prevails from the middle of March to the end of September, of course, from July to August there is a monsoon season and frequent rains are observed. From mid

S. Shukla · G. D. Agrawal
Malaviya National Institute of Technology Jaipur, Jaipur, India

R. Misra (✉)
Government Engineering College Ajmer, Ajmer, India
e-mail: rohiteca@rediffmail.com

D. P. Mishra et al. (eds.), *Recent Trends in Thermal and Fluid Sciences*, Lecture Notes in Mechanical Engineering, https://doi.org/10.1007/978-981-19-3498-8_25

of March to mid of June, especially in the Northwestern part of India, the climate is hot and dry, however, the central part of India and Northeastern part of India observe a hot and humid climate. Undoubtedly, the passive methods of cooling the buildings cannot completely replace the mechanical vapor compression refrigeration system (MVCRS), but the detrimental effect of the same on our environment and their energy-intensive nature are something which compel every researcher, design engineer and architect to seriously look at the evaporative cooling systems as an alternative of conventional MVCRS. We know that evaporative cooling systems (ECS) have been in existence since ancient times. ECS is absolutely an eco-friendly method of cooling down hot air. ECS is really a blessing for places having hot and dry climates.

However, under the hot and humid climatic conditions, direct evaporative cooler (DEC) fails to perform and during the rainy season, they perform badly. DECs may prove to be more useful when they are integrated with other cooling systems like ground heat exchangers and when assisted with some desiccant system they may result in an increased duration for which the thermal comfort conditions can be maintained. Commercially popular solid and liquid desiccants have been used for absorbing the moisture from air, but these chemical desiccants are quite costly and their use may be justified only when the cooling is required at a larger level. Integration of a desiccant chamber and desiccant wheel employing chemical compounds is going to ruin the overall economics of DEC/IEC, because integrating such a desiccant system with DEC/IEC having a small cooling capacity may not be justified. The cost of a desiccant system may be 5–10 times the cost of the DEC/IEC system. Therefore, it is the need of the hour to explore some natural materials which are easily available all around us. Apart from that, these natural desiccant materials are biodegradable, cheap and easy to use.

Singh et al. [1] experimentally investigated the natural desiccant materials like dried cow dung, cocopeat and sawdust to be used as a desiccant in the DEC system and compared their sorption and desorption characteristics with silica gel. Moisture sorption and desorption time in cocopeat and silica gel were found to be 28–85% RH and the total amount of moisture sorption and desorption time was 11.45 gm and 12 min for silica gel and 10.8 gm and 14 min for cocopeat. Based on the results, they concluded that the cocopeat can be a suitable desiccant material for the DEC system.

Dhamneya et al. [2] experimentally evaluated the performance of DEC by using a different configuration of Aspen fibers based on input temperature of air, humidity and mass flow rate. Triangular configuration for case-I results in the highest saturation efficiency of 97% followed by case-IV 89%, case-II 88% and case-III Nil. It was reported that the saturation efficiency for triangular cooling media configuration was higher than other developed systems for case-IV.

Alasfour et al. [3] experimentally investigated a pottery (staggered and aligned) evaporative cooling system to cool inlet air (temperature varied from 40 to 48 °C and airflow rate of 300–1300 m^3/hr) with minimal change in RH and low water consumption. It was reported that the staggered case had a larger air temperature drop as compared to the aligned one.

Jain [4] experimentally investigated an evaporative cooler working on two-stage cooling used to enhance the performance of the system under low temperature and high humidity conditions. A study claimed that the temperature drops through a two-stage evaporative cooler range from 8 to 16 and by the use of a two-stage evaporative cooler, and cooling temperature could be dropped up to ambient air WBD.

Comino et al. [5] experimentally access the performance of solar desiccant ECS (SDECS) developed to maintain indoor air conditions in a research room located in Martos, Spain COP, percentage of renewable energy and thermal energy delivered by the SDECS. Seasonal COP and seasonal latent COP were claimed to be 2.1 and 0.5, respectively, whereas the total seasonal COP value was 2. It was concluded that as the outdoor temperature increases, instantaneous COP also increases and SDECS gives the best result.

Kulkarni and Rajput [6] theoretically investigated the cooling capacity and efficiency of rectangular, hexagonal and cylindrical-shaped cooling pads made up of rigid corrugated paper, cellulose, high-density polythene and aspen fiber material. The study concludes that the highest saturation efficiency (97%) is for aspen hexagonal-shaped pads followed by cylindrical (90%) and rectangular (89%) ones. There is an inverse relationship between the mass flow rate of air and saturation efficiency and a direct relationship between the mass flow rate of air and cooling capacity with a minimum value of 35,826 kJ/h for a rectangular pad with an air mass flow rate of 0.3 kg/s, and for same flow rate cylindrical pad of aspen material had the highest cooling capacity 119,015 kJ/h.

Kashif et al. [7] investigated experimentally a solar-assisted desiccant-based heating and humidification system under dry and cold weather conditions from December to February. Silica gel was used as a desiccant and the thermal COP, electric COP, heating capacity and auxiliary thermal power requirement for heating were reported to be 1.48, 2.14, 5.9 kW and 2.75 kW, respectively. In cold and dry conditions, solar-assisted heating and humidification systems performed reasonably.

Grasso [8] analyzed a fabric desiccant window dehumidifier air conditioning system used to dehumidify the air. Absorption of moisture was carried out with 16 different arrangements of fabric and desiccant. The study revealed that a combination of silica gel/rayon and silica gel/rayon/polyglycol had a maximum moisture absorption of 19.3% and 18.9%, respectively.

La et al. [9] experimentally investigated the performance of an innovative chiller with an open-cycle rotary desiccant cooling system. The objective of the study was to extend the capability of handling sensible heat of a rotary desiccant cooling system. According to the study, the innovative chiller is an excellent alternative for space chilling with a low-grade heat source. The supplied cooled water from this unit had a temperature of roughly 15–20 °C. The thermal COP was reported to be 0.3–0.6 in terms of chilled water production. The rotary desiccant cooling system was also shown to be superior to the traditional rotary desiccant cooling system in terms of handling sensible heat. When both chilled water and dry air were produced, the specific thermal COP of the innovative rotating desiccant cooling system was around 0.8–0.9.

Kumar et al. [10] studied the effect of integration of inlet air cooling steam injection and film cooling on the performance of the gas turbine cycle. For analysis, two methods viz. inlet fogging and evaporative cooling were used. Results show that the inlet fogging is better than evaporative cooling for achieving lower temperatures at the inlet of the compressor which finally enhances the overall performance of the gas turbine cycle.

Kumar et al. [11] experimentally evaluated the performance of steam-injected gas turbine-based power plants with inlet evaporative cooling. Results show 3.2% enhancement in cycle thermal efficiency due to the lowering of compressor inlet temperature from 318 to 282 K at a 5% steam to air ratio. It was also concluded that at 1850 K turbine inlet temperature and cycle pressure ratio of 24, the thermal efficiency of the gas turbine cycle increases. Steam injection in the combustion chamber increases the specific expansion work in a gas turbine that increases at the rate of 2.95% for every increase in steam to air ratio by 2%. Finally, they concluded that for obtaining more efficiency and power, the gas turbine cycle with inlet evaporative cooling (IEC) steam injection and film cooling is best.

It is evident from the literature that the feasibility and performance study of natural desiccant materials like Cocopeat powder, etc. used as a desiccant material with DEC/IEC has not been reported. Though there is a great potential for the natural materials to be used as a desiccant in order to enhance the effectiveness of DEC/IEC systems, this study is an attempt to bridge that gap and aims to explore the performance of cheap and bio-degradable natural desiccant materials (Cocopeat powder) to be used in place of commercially accepted chemical desiccants like Silica gel to reduce the moisture content of incoming air.

2 Description of Experimental Setup

The test setup was fabricated and assembled in the Heat Transfer Laboratory of Govt. Engineering College, Ajmer, Rajasthan, India. Ajmer is situated at the center of Rajasthan state, which is actually in the western part of India.

A laboratory-scale fixed-bed desiccant system integrated with an evaporative cooling system has been developed. It primarily consists of two Humidifying Chambers (HCs) and two Desiccant Chambers (DCs) (one meant for Silica gel and the other for natural desiccant) arranged in parallel and commonly connected to a heating chamber and a centrifugal blower (CB). To augment the flow of air through the PVC pipes of the test section as well as through the humidifier units, a CB has been used. Natural desiccant material viz. Cocopeat was planned to be tested for its thermal performance, moisture sorption and desorption behavior so that its performance can be compared with that of chemical desiccant, Silica gel. Figure 1 depicts the line diagram and Fig. 2 depicts a real experimental setup fabricated for carrying out the investigation.

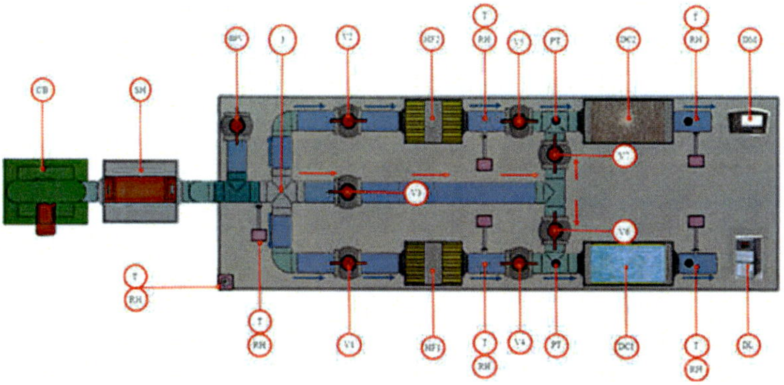

DC 1 DC2: Desiccant Chamber
HF1 HF2: Humidifier
RH: Relative Humidity
CB: Centrifugal Blower
SH: Sensible Heater
V1, V2, V3, V4, V5, V6 & V7: Flow Control Valve

DM: Digital Micro Manometer
BPV: By Pass Valve
PT: Pressure Tapping
DL: Data Logger
T: Temperature
J: Joint

Fig. 1 Plan view of experimental setup

3 Result and Discussion

Experimentation was conducted at an airflow velocity of 2.0 m/s. During the experimentation, the RH of air in the laboratory varied between 24.1 and 28.5%, and its DBT varied between 34.5 and 35.20 °C. At the start of experimentation, both the humidifiers deliver the moist air having RH of 60.7% with a DBT of 31.8 °C. Gradually, the RH of air at the exit of both the humidifiers increases, and its DBT drops. It is observed from Fig. 3 that at the start of experimentation, RH and DBT of air at the exit of both the DCs are almost equal. It is noticed that at the start of experimentation, the RH values of air at the exit of Silica gel and Cocopeat chambers are 24.2% and 24.0%, respectively.

Corresponding values of DBT are noted to be 36.3 °C and 38.7 °C, respectively. It is noticed in Fig. 3 that as the moist air stream supplied by the humidifier passes through the DC1, due to a chemical reaction taking place in Silica gel upon absorption of moisture, the exothermic process results in an increase in the DBT of air at the exit of DC1. The moist air stream supplied by the humidifier passes through the DC1, and due to a chemical reaction taking place in Silica gel upon absorption of moisture, the exothermic process results in an increase in the DBT of air at the exit of DC1.

It is further noted from Fig. 3 that the DBT of air coming out of the Silica gel chamber is 0.5–1.5 °C higher than the DBT of air exiting from the Cocopeat chamber, during the initial 10–15 min of operation. After 10 min of operation, the RH of air at the exit of DC1 and DC2 is 31.6% and 36.1%, respectively, and the corresponding DBT is 36.7 °C and 34.1 °C, respectively. Thus, due to the chemical reaction taking

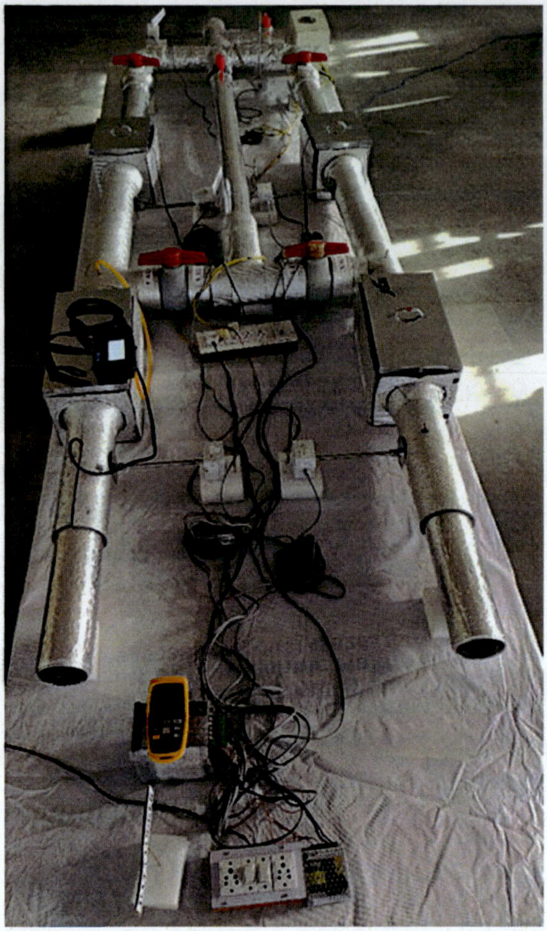

Fig. 2 Actual experimental setup

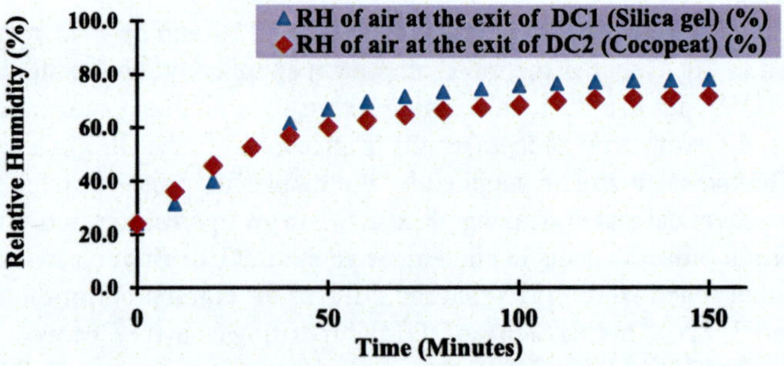

Fig. 3 Variation in relative humidity of air at the exit of DCs during absorption at 2.0 m/s flow velocity

place in the Silica gel and as an outcome of an exothermic reaction, the RH of air exiting from DC1 is lesser than the RH of air exiting from DC2. However, as the moist air continues to pass through both the DCs, both the desiccant materials keep on absorbing the moisture and the RH of air at the exit of DC1 and DC2 gradually starts increasing, revealing that due to continuous absorption of moisture by them, their moisture absorption capacity decreases with time.

After 30 min of operation, the RH of air at the exit of SGC and CPC are noted to be 53.4% and 52.5%, respectively. It points out that with time, the magnitude of exothermic reaction in SG reduces, and the continuous supply of cold and humid air from humidifiers offsets the heat again of air, as the temperature at the exit of SGC and CPC are noted to be 28.2 °C and 27.7 °C, respectively, after 30 min of continuous operation. The pressure difference during charging across the Silica Gel chamber (Pa) was 2.70 and across the Cocopeat chamber (Pa) was 3.15. Figure 4 reveals that initially when the heater is turned on, the RH and DBT of air at the exit of the heating chamber are 21.9% and 36.6 °C, respectively.

As the heating goes on, the DBT of air increases gradually and after 15 min of continuous operation, DBT varies very little and becomes constant at 51.9 °C. After 25 min of operation, it is noticed that the RH of air at the exit of the heating chamber varies from 21.9 to 11.3%. Initially, the RH of air at the exit of the Silica gel chamber and Cocopeat chamber is 78.0% and 71.7%, respectively. It is observed that Silica gel regenerates much faster than the Cocopeat and after 20 min of regeneration, the RH of air at the exit of the Silica gel chamber is 39.0% whereas, the RH of air at the exit of the Cocopeat chamber is 48.5%. Thus, within 20 min of time span, the RH of air at the exit of the Silica gel chamber drops by 12.3% and that at the exit of the Cocopeat chamber by 9.3%. After 30 min of regeneration, the RH of air at the exit of Silica gel and the Cocopeat chamber is noticed to be 24.4% and 31.2%, respectively. Since, after 30 min of regeneration with hot air, most of the moisture is taken away by hot air, a further decrease in RH of air at the exit of both the DCs reduces. It can be seen from Fig. 4 that after 50 min the average of air at the exit of the Silica gel chamber and Cocopeat chamber drops to 14.8% and 15.3%, respectively. Beyond 50 min of continuous regeneration, the release of moisture from both the DMs is

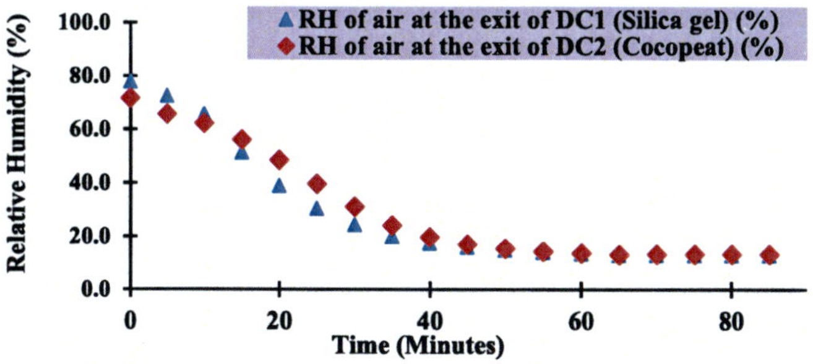

Fig. 4 Variation in relative humidity of air at the exit of DCs during regeneration at 2.0 m/s flow velocity

very sluggish and after 90 min, the RH of air is noted to be 12.9% and 13.2% at the exit of the Silica gel chamber and Cocopeat chamber, respectively.

4 Conclusion

In the present study, the thermal performance of natural desiccant material Cocopeat powder has been investigated in terms of its moisture absorption capacity and its behavior during regeneration in an experimental setup at Ajmer, India. The performance of Cocopeat powder has been compared with that of commercially popular chemical desiccant Silica gel. The following major findings of the study are

- RH of air at the exit of the desiccant chamber having Cocopeat as natural desiccant is minimum for most of the duration of experimentation, up to an initial 30–40 min of operation, because of the liberation of heat due to chemical changes in SG and transfer of heat to air flowing over the plates of SG. It actually brings down the RH.
- During the initial 30–40 min of operation, the temperature of air at the exit of SGC has been observed to be 1.5–2.0°C higher up to 30–40 min of operation as compared to natural desiccant material viz. Cocopeat.
- After 30–40 min of operation, cocopeat performs best and even better than the Silica gel.
- The study reflects that Cocopeat powder may potentially be used as a desiccant material in place of costly chemical desiccants like Silica gel and others.

References

1. Singh A, Kumar S, Dev R (2019) Studies on cocopeat, sawdust and dried cow dung as desiccant for evaporative cooling system. Renew Energy 142:295–303. https://doi.org/10.1016/j.renene.2019.04.122
2. Dhamneya AK, Rajput SPS, Singh A (2018) Thermodynamic performance analysis of direct evaporative cooling system for increased heat and mass transfer area. Ain Shams Eng J. https://doi.org/10.1016/j.asej.2017.09.008
3. Alasfour FN, Quadri MMA, Alasfour FN (2014) Pottery evaporative cooling system: a novel approach to cool inlet air with minimal change in relative humidity and low water consumption. Exp Heat Transf: J Thermal Energy Gener Transp Storage Convers 37–41. https://doi.org/10.1080/08916152.2012.757675
4. Jain D (2007) Development and testing of two-stage evaporative cooler. Build Environ 42(7):2549–2554. https://doi.org/10.1016/j.buildenv.2006.07.034
5. Comino F, Castillo González J, Navas-Martos FJ, Ruiz de Adana M (2020) Experimental energy performance assessment of a solar desiccant cooling system in Southern Europe climates. Appl Therm Eng 165. https://doi.org/10.1016/j.applthermaleng.2019.114579
6. Kulkarni RK, Rajput SPS (2013) Comparative performance analysis of evaporative cooling pads of alternative configurations and materials. Int J Adv Eng Technol 6(4):1524–1534

7. Kashif A, Ali M, Sheikh NA, Vukovic V, Shehryar M (2020) Experimental analysis of a solar assisted desiccant-based space heating and humidification system for cold and dry climates. Appl Therm Eng 175. https://doi.org/10.1016/j.applthermaleng.2020.115371
8. Grasso MM (1994) Analysis of a fabric/desiccant window cavity dehumidifier regenerative heat exchanger
9. La D, Dai YJ, Li Y, Tang ZY, Ge TS, Wang RZ (2013) An experimental investigation on the integration of two-stage dehumidification and regenerative evaporative cooling. Appl Energy 102:1218–1228. https://doi.org/10.1016/j.apenergy.2012.06.053
10. Kumar S, Singh O (2016) Thermodynamic analysis of steam injected gas turbine cycle power plant with inlet air cooling. Int J Ambient Energy 556–566. https://doi.org/10.1080/01430750.2016.1155495
11. Kumar S, Singh O (2016) Performance evaluation of steam injected gas turbine-based power plant with inlet evaporative cooling. Appl Therm Eng 102:454–464 (2016). https://doi.org/10.1016/j.applthermaleng.2016.03.136

Numerical Investigations on Performance of Triple Pipe Heat Exchanger with Helical Fins and Twisted Tape Inserts

Gudi Dheeraj, Pabbathi Deepak, and T. Srihari Vikram

Abstract In the present work, numerical simulations are carried out to investigate the heat transfer and fluid flow characteristics of a triple pipe heat exchanger with helical fins and twisted tape inserts. The effect of the mass flow rate of cold fluid, hot fluid, fin and twisted tape configurations are studied. Helical fins and twisted tape inserts provide heat transfer enhancement with an increase in pressure drop. It was found that based on the study, six helical fins and a twisted tape ratio of 9.75 were found to perform better for the conditions considered. The pumping power requirement seems very less for both hot and cold fluids ranging to be between 1 to 15 W depending on the flow rate of fluids.

Keywords Triple pipe heat exchanger · Helical fins · Twisted tape · Heat transfer · Pressure drop

1 Introduction

A heat exchanger is a system that transfers thermal energy between fluids of different temperatures with or without any direct contact between them. Heat exchangers are used in several industries such as chemical and petrochemical plants, food processing industry, etc., to recover/minimize waste heat. It helps to achieve an increased effectiveness in utilization of heat, thus increasing the efficiency of the system. The triple pipe heat exchanger is an extended version of a double pipe heat exchanger with an intermediate pipe being added to it. This additional pipe increases the heat transfer per unit length. The triple pipe heat exchangers are gaining more interest in many processing industries as they are more compact and effective when compared to double pipe heat exchangers.

Zuritz [1] developed analytical equations for estimating parameters such as bulk fluid temperature and heat transfer rate for the triple heat exchangers and verified them by a case study and found them to be in good agreement. It has been concluded

G. Dheeraj · P. Deepak · T. S. Vikram (✉)
School of Mechanical Engineering, SASTRA Deemed to be University, Thanjavur 613401, India
e-mail: harivikram@mech.sastra.edu

D. P. Mishra et al. (eds.), *Recent Trends in Thermal and Fluid Sciences*, Lecture Notes in Mechanical Engineering, https://doi.org/10.1007/978-981-19-3498-8_26

that the inclusion of an annular region within the inner pipe increases the overall heat transfer efficiency and reduces the heat exchanger length requirement by almost 25%. Valladares [2] carried out a 1D steady and transient analysis of a triple pipe heat exchanger. Saraireh [3] carried out experimental investigations to study the performance of triple pipe heat exchangers under various input conditions and validated them with numerical results. Quadir et al. [4] studied the performance of a triple heat exchanger for different fluids and conditions using FEM and carried out the parametric analysis. Pătrăşcioiu and Rădulescu [5] predicted and compared the performance of petroleum product and water in a triple pipe heat exchanger considering various parameters such as dimensions of the tube, temperature and mass flow rate of the fluids. Batmaz and Sandeep [6] proposed a methodology for estimating the overall heat transfer coefficient for double pipe and triple pipe heat exchangers by knowing the temperatures and heat capacities of the fluids and compared the performance of the two. A comparison of double and triple pipe heat exchangers for cooling a petroleum product with water as a cooling medium was carried out by Rădulescu et al. [7].

Fins, twisted tapes, swirl generators, etc., are widely used for thermal performance enhancement. Kumar et al. [8] reviewed various techniques for thermohydraulic enhancement of heat exchangers with twisted tape inserts. Eiamsa-ard et al. [9] investigated the influence of twisted tape inserts, circular rings and swirl generators on the enhancement of heat transfer, thermal efficiency and pressure drop. Bhuiya et al. [10] developed correlations to predict performance parameters of heat exchangers with twisted tape inserts based on experiments. Kumar et al. [11] studied the performance of triangular twisted tapes with v cuts for double pipe heat exchangers. El-Maakoul et al. [12] concluded that helical fins yield 3–24% better heat transfer ratio than longitudinal fins for the same geometrical parameters. The objective of the present work is the study the heat transfer and fluid flow characteristics of the triple pipe heat exchanger with and without helical fins and twisted tape inserts.

2 Modeling of Triple Pipe Heat Exchanger

The geometrical model of the triple pipe heat exchanger was developed in PTC Creo. The twisted tape inserts are placed in the inner region; helical fins are provided in the intermediate and outer regions. The side view of the triple pipe heat exchanger with six helical fins in the intermediate and outer region and a twisted tape insert in the inner region is shown in Fig. 1. The inner diameter of inner, intermediate and outer regions are 15 mm, 44 mm and 68 mm, respectively, and a tube thickness of 2 mm is considered. The width of the twisted tape insert is considered to be 10 mm and the width of the helical fin for the intermediate and outer regions are considered as 9.5 mm and 7.5 mm, respectively. The thickness of the twisted tape insert and helical fin is considered to be 1 mm and 1.5 mm, respectively. The number of helical fins is varied from 4 to 8 fins. For all the configurations of triple pipe heat exchangers, the helical turn ratio is considered as 0.5, and four different twisted tape ratios such

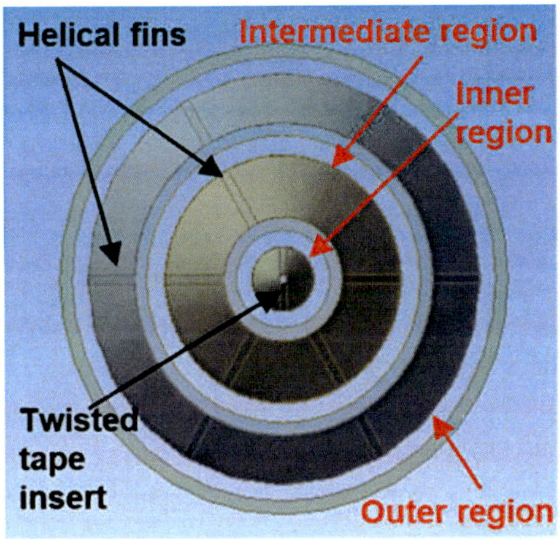

Fig. 1 Triple pipe heat exchanger with six helical fins and a twisted tape insert

as 4.85, 6.5, 9.75 and 19.5 are considered. The twisted tape ratio (TTR) is defined as the ratio between pitch length and width. The tube, fin and insert materials are considered to be steel.

2.1 *Governing Equations*

The flow is considered to be steady, uniform and turbulent with no-slip conditions at the wall. The governing equations are given by the following [13]:

Continuity equation is given by

$$\frac{\partial u_i}{\partial x_i} = 0 \tag{1}$$

Momentum equation is given by

$$\frac{\partial u_i u_j}{\partial x_i} = -\frac{1}{\rho}\frac{\partial p}{\partial x_i} + \frac{\partial}{\partial x_j}\left((\nu + \nu_t)\left(\frac{\partial u_i}{\partial x_j} + \frac{\partial u_j}{\partial x_i}\right)\right) \tag{2}$$

Energy equation is given by

$$\frac{\partial u_i \mathrm{T}}{\partial x_i} = \rho\frac{\partial}{\partial x_i}\left(\left(\frac{\nu}{Pr} + \frac{\nu_t}{Pr_t}\right)\frac{\partial \mathrm{T}}{\partial x_i}\right) \tag{3}$$

k-ε turbulence model is considered in the present work which is well suited for flows involving rotation, boundary layers effect under a strong adverse gradient of pressure and recirculation.

Turbulent kinetic energy k is given by

$$\frac{\partial u_i k}{\partial x_i} = \frac{\partial}{\partial x_i}\left(\left(\nu + \frac{\nu_t}{\sigma_k}\right)\frac{\partial k}{\partial x_i}\right) + \Gamma - \varepsilon \tag{4}$$

Turbulent energy dissipation ε is given by

$$\frac{\partial u_i \varepsilon}{\partial x_i} = \frac{\partial}{\partial x_i}\left(\left(\nu + \frac{\nu_t}{\sigma_\varepsilon}\right)\frac{\partial \varepsilon}{\partial x_i}\right) + c_1 \Gamma \varepsilon - c_2 \frac{\varepsilon^2}{k + \sqrt{\nu\varepsilon}} \tag{5}$$

where Γ represents the generation of turbulence kinetic energy k due to the mean velocity gradients. It is given by

$$\Gamma = -\overline{u_i u_j}\frac{\partial u_i}{\partial x_i} = \nu_{turb}\left(\frac{\partial u_i}{\partial x_j} + \frac{\partial u_j}{\partial x_i}\right)\frac{\partial u_i}{\partial x_i} \tag{6}$$

The turbulent kinematic viscosity is

$$\nu_t = c_\mu \frac{k^2}{\varepsilon} \tag{7}$$

The empirical constants are given by

$$c_1 = \max\ [0.43,\ \mu/(\mu_t + 5)];\ c_2 = 1.9;\ \sigma_k = 1.0;\ \sigma_\varepsilon = 1.2$$

2.2 *Boundary Conditions*

Cold fluid (water) flows in the inner and outer tube annulus and hot fluid (engine oil) flows in the intermediate tube annulus. The inlet temperature of cold fluid is considered as 283 K and the inlet temperature of hot fluid is considered as 354 K. Mass flow inlet and pressure outlet is set as inlet and outlet boundary conditions, respectively. The different flow rates of the cold fluid and hot fluid were considered. The flow rate of hot fluid is changed from 0.03 to 0.13 kg/s and the flow rate of cold fluid is changed from 0.3 to 0.7 kg/s. The numerical simulations for parallel and counter flow conditions are carried out and observed that for counter flow, more reduction in outlet temperature of hot fluid is observed. Hence, counter flow is considered for all simulations. The properties of tube (steel) and working fluids are taken from Kothandaraman and Subramaniam [14].

2.3 Numerical Procedure and Grid Independent Study

3D numerical simulations are carried out to study the thermal and fluid flow characteristics in the triple pipe heat exchanger. k-ε turbulence model with enhanced wall function is considered for turbulence modeling with default values of constants. The governing equations were solved iteratively using ANSYS Fluent with SIMPLEC algorithm. The second-order upwind scheme is considered for pressure, momentum and energy, and first-order upwind is used for turbulence kinetic energy. Inlet temperatures are kept constant for cold and hot fluid in all the configurations. Convergence criteria for continuity, momentum, energy, k and ε are taken as 1×10^{-6}. Grid independent study is carried out considering three different mesh sizes with 3,77,521, 4,10,024 and 4,55,649 cells. The heat transfer rate is compared for three meshes and found that a mesh with 4,10,024 cells is sufficient for further analysis.

2.4 Model Validation

The validation of the present numerical model is carried out with Saraireh [3]. The numerical simulations are carried out and the variation of temperature along the length of the tube for different sections is presented in Table 1. It is observed that the maximum deviation is 0.5 K at 0.8 m length and the minimum deviation is 0 K at 0.4 m length of the inner tube. It is observed that temperature increases along with the length and the maximum deviation is 0.65 K at 0.4 m length and the minimum deviation is 0 K at 0.8 m length of intermediate tube. It is observed that temperature increases along with the length and the maximum deviation is 0.65 K obtained at 0.4 m length and the minimum deviation is 0 K obtained at 0 m, 0.2 m length of the outer tube.

Table 1 Comparison of temperature along the tube length

Tube length, m	Temperature of fluid, K					
	Inner tube		Intermediate tube		Outer tube	
	Present model	Saraireh [3]	Present model	Saraireh [3]	Present model	Saraireh [3]
0	287.5	287.2	321.7	321.3	287.0	286.9
0.2	289.0	289.3	323.0	323.2	289.0	289.0
0.4	291.0	291.0	325.0	324.7	291.0	290.7
0.6	292.6	292.3	327.0	327.3	292.0	292.2
0.8	294.5	294.0	329.1	329.0	293.6	293.3
1.0	297.9	297.7	331.1	330.6	294.4	294.2

3 Heat Transfer and Fluid Flow in Triple Pipe Heat Exchanger

The heat transfer rates for the three different regions are calculated as follows [15]:

Heat transfer rate of cold fluid in the inner and outer annulus is given by,

$$Q_{c1} = \dot{m}_{c1} c_{pc} (T_{c1o} - T_{c1i}) \tag{8}$$

$$Q_{c2} = \dot{m}_{c2} c_{pc} (T_{c2o} - T_{c2i}) \tag{9}$$

Heat transfer rate of hot fluid in the intermediate annulus is given by,

$$Q_h = \dot{m}_h c_{ph} (T_{ho} - T_{hi}) \tag{10}$$

$$\text{Flow velocity is given by, } v = \frac{\dot{m}}{\rho A} \tag{11}$$

$$\text{Reynolds number is given by, } \mathrm{Re} = \frac{\rho v D_h}{\mu} \tag{12}$$

$$\text{Nusselt number is given by, } \mathrm{Nu} = \frac{h D_h}{k} \tag{13}$$

where D_h is the hydraulic diameter.

4 Results and Discussion

The thermal and fluid flow characteristics of a triple pipe heat exchanger with and without twisted tape inserts and helical fins were studied using numerical simulations. The effect of various parameters such as cold and hot fluid mass flow rate, twisted tape ratio (TTR) on the Nusselt number and friction factor are investigated.

4.1 Effect of Twisted Tape Ratio (TTR) on Heat Transfer and Pressure Drop

The variation of heat transfer and pressure drop for the flow rate of cold fluid in the inner tube for different tape twist ratios is shown in Fig. 2. It is observed that for the larger tape twist ratio (TTR), the heat transfer rate is better. It can be observed that heat transfer for all the twist ratio tapes increases with an increase in the flow

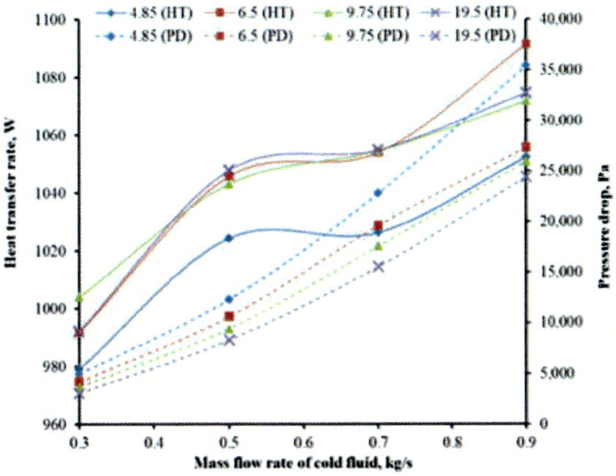

Fig. 2 Variation of heat transfer rate and pressure drop with the flow rate of cold fluid in the inner tube

rate of cold fluid. It is observed that pressure drop for all the TTR increases with an increase in the flow rate of the inner fluid. From the analysis, it is seen that the twisted tape ratio of 9.75 has better performance when compared to other twisted tape ratios. Hence, twisted tape with a TTR of 9.75 is considered.

4.2 *Effect of Mass Flow Rate of Hot Fluid on Heat Transfer and Pressure Drop*

The variation of heat transfer rate and pressure drop with a mass flow rate of hot fluid for all the configurations is shown in Fig. 3. Heat transfer rate is found to be higher

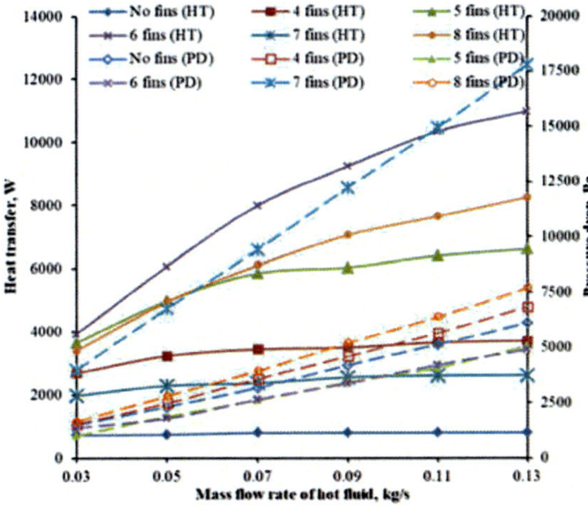

Fig. 3 Variation of heat transfer and pressure drop with a mass flow rate of engine oil

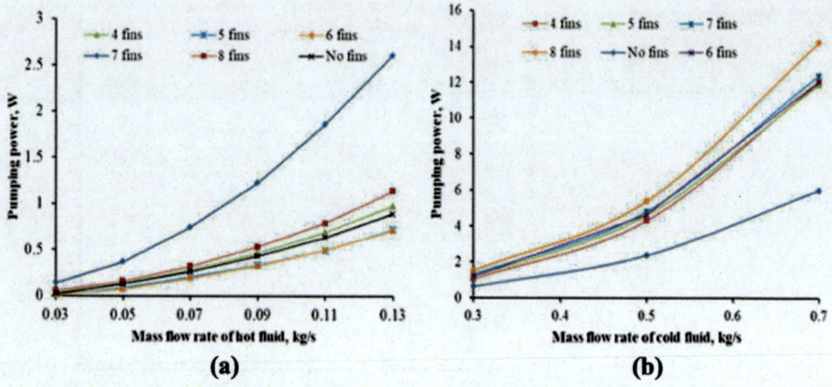

Fig. 4 Variation of pumping power with flow rate of **a** hot fluid and **b** cold fluid

for six and eight fins case and increases with flow rate and lower being for no fins case. It is well known that pressure drop increases with an increase in flow rate, and pressure drop is found to be high for seven fins case and lower being for five and six fins for different flow rates.

4.3 Variation of Pumping Power with Mass Flow Rate of Hot and Cold Fluid

The pumping power required for hot fluid and cold fluid is depicted in Fig. 4a and b, respectively. Pumping power is directly proportional to mass flow rate. Pumping power varies between 1–3 W for hot fluid and 1–15 W for cold fluid depending on the flow rate of the water. It is noticed that six fins configuration requires less pumping power whereas eight fins configuration requires more pumping power.

5 Conclusions

The numerical simulations are carried out for different configurations of triple pipe heat exchangers to estimate the performance with helical fins and twisted tape inserts. The effect of the mass flow rate of cold fluid and hot fluid, the number of fins and twisted tape ratios is considered and found that six helical fins configuration with a 9.75 twisted tape ratio performs better based on configurations considered. The pumping power required for the hot and cold fluids ranges between 1 to 15 W for different flow rates.

References

1. Zuritz C (2007) On the design of triple concentric-tube heat exchangers. J Food Process Eng 12:113–130
2. Valladares GO (2004) Numerical simulation of triple concentric-tube heat exchangers. Int J Therm Sci 43:979–991
3. Saraireh M, Experimental and numerical thermal study of triple pipe heat exchanger. Int J Simul Syst Sci Technol 19:13.1–13.4
4. Quadir G, Badruddin I, Salman NJ (2014) Numerical investigation of the performance of a triple concentric pipe heat exchanger. Int J Heat Mass Transf 75:165–172
5. Pătrăşcioiu C, Rădulescu S (2015) Prediction of the outlet temperatures in triple concentric—tube heat exchangers in laminar flow regime: case study. Heat Mass Transfer 51:59–66
6. Batmaz E, Sandeep KP (2005) Calculation of overall heat transfer coefficients in triple tube heat exchangers. Heat Mass Transfer 41:271–279
7. Radulescu S, Negoiţă L, Onutu I (2016) Analysis of the heat transfer in double and triple concentric tube heat exchangers. IOP Conf Ser Mater Sci Eng 147:012148
8. Kumar B, Srivastava G, Kumar M, Patil A (2017) A review of heat transfer and fluid flow mechanism in heat exchanger tube with inserts. Chem Eng Process 123:126–137
9. Eiamsa-Ard S Kongkaitpaiboon V, Nanan K (2013) Thermo hydraulics of turbulent flow through heat exchanger tubes fitted with circular-rings and twisted tapes. Chinese J Chem Eng 21:585–593
10. Bhuiya MMK, Chowdhury MSU, Shahabuddin M, Saha M, Memon LA (2013) Thermal characteristics in a heat exchanger tube fitted with triple twisted tape inserts. Int Commun Heat Mass Transfer 48:124–132
11. Kumar R, Nandan G, Dwivedi G, Shukla AK, Shrivastava R (2021) Modeling of triangular perforated twisted tape with V-Cuts in double pipe heat exchanger. Mater Today Proc 46:5389–5395
12. El-Maakoul A, El-Metoui M, Abdellah BA, Saadeddine S, Meziane M (2017) Numerical investigation of thermohydraulic performance of air to water double-pipe heat exchanger with helical fins. Appl Therm Eng 127:127–139
13. Zhang C, Wang D, Xiang S, Han Y, Peng X (2017) Numerical investigation of heat transfer and pressure drop in helically coiled tube with spherical corrugation. Int J Heat Mass Transfer 113:332–341
14. Kothandaraman CP, Subramanyan S (2008) Heat and mass transfer data book. New Age International
15. Incropera FP and Dewitt (2011) Fundamentals of heat and mass transfer, 6th edn. Wiley

Energy Efficient Coatings for Improved Wear and Tear Performance of a Circulating Water Pump Employed in Thermal Power Plant—A Case Study

S. Narendhiran and M. Velan

Abstract Circulating water pumps employed in large installations like thermal power plants impose challenging issues to be solved on the energy efficiencies of fluid transfer, achieving consistent performance under varied wear and tear conditions of transfer operations. In the present study, a similar deployment of coating in hydraulic passages of circulating water pumps using polymer was carried out to achieve enhanced pump efficiency in a thermal power plant. The scope of the study can be applied in various other sectors with ameliorated coat-able materials involving similar transfer applications working with decreased energy requirements. Based on the research and assessment carried out, after the coating, the increase in pump efficiency was observed from 84.17 to 88.42%, corresponding to the flow rate from 15,862 to 16,662 m^3/hr resulting in an enhanced pump's efficiency of 4.25% to handle an increased flow rate of 800 m^3/hr which accounts for an overall saving of power for the installed motor capacity of 1255 KW power generation from NLCIL plant supplying southern parts of India.

Keywords Coatings · Wear and tear · Energy efficient · Circulating pump

1 Introduction

A thermal power plant makes use of fossil fuels to produce electricity. Every major advancement in the centuries of human evolution arises from an undisputable source which is electricity. In thermal power plants, circulating water pumps supply the need for condensers. The cold circulating water stored is supplied to the circulating water pumps from suction pits. The discharge pipe from the pump carries cold circulating water to the condenser. Lee and Lim [1] conducted research stating that the reduction

S. Narendhiran (✉)
Mechanical Engineering, Indian Institute of Technology Patna, Bihta, Bihar, India
e-mail: narendhiran2000@gmail.com

M. Velan
NLCIL, Chennai, Tamil Nadu, India

D. P. Mishra et al. (eds.), *Recent Trends in Thermal and Fluid Sciences*, Lecture Notes in Mechanical Engineering, https://doi.org/10.1007/978-981-19-3498-8_27

in pumping efficiency of water inside the cooling system is due to the surface erosion-corrosion phenomena. Investigation [2] was carried out on corrosion, scaling, and biofouling, as they were core problems noticed in the power plant cooling system. In addition, the influence of inhibitors and biocides was seen in the control of these problems for copper-nickel and mild steel alloys in the power plant's cooling water. Corrosion avoidance measures are essential in water distribution systems because of modified water quality [3]. Therefore, the study analyzed the corrosion rankings for old water pipelines on their composition and structure.

The latest advancement in improving the pump's efficiency by coating epoxy material on the impeller and other hydrophobic passages of the pump having a roughness value of 4.39 μm Ra resulted in increased pump efficiency up to 4.75% reported for water pump operation [4]. It is evident from the study [5] that wear-resistant coatings were utilized to increase the efficiency of the production process to meet the surplus demand. The above study found that determining a complementary microstructural, thermo-physical, and mechanical fingerprint of wear-resistant hard coatings allows us to picture the complex structure–property associations in these materials and enhance their performance. The development of innovative adsorbent coating techniques [6] in the heat exchanger has increased the thermal plant's overall efficiency. It is also reported that the deposit of light coatings on the surface of the heat exchanger could lower the heat transfer resistance found on the adsorbent material side, therefore, improving the whole adsorption dynamics. Practical plant engineers acknowledged the effect of pump surface roughness motivated to lead a forefront in the efficiency and operation of the pumping system. The wear and tear of the impeller and feeding lines are often subjected to electrochemical corrosion and mechanical erosion through the turbulent profiles trajected with multiple solids composition and transit of fluid regime. The electrochemical and galvanic decays and pitting are projected to cause physical deformations and removal of materials both from the casings outer cover and surface of impeller blades. Different pumping schedules requiring highly inflexible flow rates over periodic intervals cause a rapid reduction in head loss and hence need more electrical power [7]. The recent advancements in the coatings have resulted in reduced wear and tear, reduced corrosion, and increased overall efficiency of the thermal power plant.

This case study brings out the exercise and attempt on the lining technologies to stabilize and improve the efficiency of pumps, minimizing the effect of surface wear and tear due to erosion and corrosion.

2 Circulating Water Pump

In the case study of the work, in Neyveli Lignite Corporation India Ltd. (NLCIL, TN), Thermal Power plant, there are five stage vertical mixed flow pumps. Each pump is of capacity 16,500 m^3/hr at a delivery head of 24 m of the water column and is coupled to a motor of power 1450 KW and running at 497 rpm.

Two circulating water pumps can keep up with the condenser requirements of 1 unit, while one pump is assigned as standby for both units. The cold circulating water stored in the forebay is circulated from suction pits to the circulating water pumps. The delivery of each pump is distributed via a steel pipe consisting of a butterfly discharge valve. Both discharge pipes from the two circulating water pumps of a unit join a cold-water line, and this pipe also can be fed from the standup pump. The condenser receives this cold-water pipe carrying the cold circulating water (CW), and it is supplied by two intake pipes attached just before entry to the condenser. In addition, two intake pipes merge with the front water boxes, located at their bottom.

The circulating water mixing with the water box flows in two different passes inside the condenser. The reversing chamber at the back merges both water passes. From the topside of the front water boxes, two outlet pipes connect a standard hot water pipe for carrying hot circulating water to the cooling tower. Make-up to the circulating water system is withdrawn from an artificial lake, which procures its water supply from the Mines. Considerable water losses in the circulating water system are passed on to the Ash disposal pump house, and evaporation loss in the cooling tower and drift loss in the cooling tower are also observed. The make-up water is drawn from the lake to the ore bay through two lines with level control valves.

3 Materials and Methods

The procedures followed for coating begin by dismantling the pump impeller, bowl, column shaft, and suction bell, then surface cleaning and drying were carried. The surface is degreased, and the bowl is heavily pitted. During the case study, the bowl cone was observed heavily damaged (Fig. 1). Bowl drying and cleaning (Fig. 2)

Fig. 1 Bowl damaged

Fig. 2 Cleaned impeller

took place at the workshop and again the use of sprays to degrease the surface. Grit blasting was done as per SA 2.5 on the pump bowl, and the impeller surface was roughened (Fig. 3) ideally by grit blasting.

In continuation, surface grinding (Fig. 4) was carried out, which creates enhanced surface region and "edges" to observe and is necessary for booming applications. In order to achieve proper adhesion to the substrate, the repair area is heated up before

Fig. 3 Impeller damaged

Fig. 4 Surface grinding

using brush-able ceramic to dry-off wetness, contamination, or solvents, which helps the ceramic system to achieve the highest adhesion.

The primer application (Fig. 5) is the applied coat tolerating possible rust, wet, UV, acid, high temperature, alkali and oil, and filling dented surfaces (Fig. 6) with ceramic wear-resistant putty. Finally, coating and curing (Fig. 7) were carried out by applying a coat on the pump bowl, impeller, and bell for repairing and protecting surfaces subject to erosion, wear, and corrosion. The applied brush-able ceramic will reach a tack-free finish in 2–3 h. The cure (Fig. 8) may be accelerated by heating after the coating has been fully let to solidify under ambient states. Each coat should be 0.5–1.0 mm in thickness per coat (Fig. 9). Two coats provide a pinhole-free lining.

Fig. 5 Application of primer

Fig. 6 Filling dented surfaces

Fig. 7 Before coating

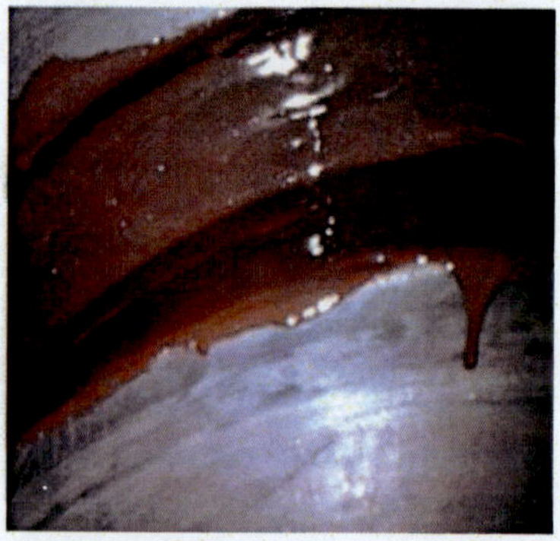

Fig. 8 Coating

Fig. 9 After coating

4 Result and Discussion

4.1 Effect of Impeller Surface Roughness—Theory and Field Practical Issues

The natural surface and coating surfaces are greatly affected by flow rate variations in pumping and during switching on and off repeatability that exits in practice over a specified pumping schedule.

Material selection plays an effective solution in the erosion of surfaces and impeller's wear and tear resistance during varied flow rates and transfer operations [8]. Theoretically, the properties of fluid handled can also influence erosion leaving enormous scope for the composition of multiphase fluid systems. The external environment accelerates the deterioration and pitting due to induced roughness reduction perpetuating sulfate/chloride dissolution. This effect is visualized theoretically to detach the surface and fluid by a contact interface, namely lining/coating [9]. Practically, the issues faced are as follows: (i) ineffective fit for cost-effective frame, (ii) cost of lining is expensive and its ready availability limitation, (iii) dissolution and its physical deformation strength reduction, (iv) prolonged times of repair and rehabilitation and sophisticated equipment requirements, and (v) decolorization of coatings [10].

The present case study's intersection with both material and fluid phenomenon properties is often affected by the effect of hydrophobic properties of surfaces in the flow part of centrifugal pumps and their resulting operational performance [11]. A challenge faced in the case study was to unveil how the heating surface under continuous operation affects the coating. It was interesting to note several reasons to support the performance change due to analysis for finding critical wall roughness variations on the centrifugal pump at design and off-design conditions as investigated by Deshmukh and Samad [12]. The case study reveals that it enhances the smooth transition and reduces the formation of pitting and corrosion due to decreased retention of solids at the impeller peripheries at different flow regimes, as depicted by the results and testing methods [13, 14].

4.2 Case Study and Modernization—Combinatorial Properties of Fluid and Casing Materials

A close comparison has been made with the hydrophobic properties of the flow regime before and after coating. A direct combination of choosing a polymeric coating did not support the wear and tear of the intervening surface. A complex blend of fluid and casing material properties advances the understanding and implementation of the trials for a converged solution requiring less cost and labor. A similar line of the study investigated by Wang et al. [15] realized the effect of coating thickness on a solid–liquid two-phase flow operated with a centrifugal pump underwater solid suspension.

The efforts of operational procedure were in conformation with internationally recognized methodologies in place of provisions by ISO, NACE International, and ASTM. Tested following ASTM D45419 and ISO 4624.10 [16–18].

4.3 Theoretical Assessments of the Case Study

The roughness of the interior surface of a circulating water pump affects the flow characteristics. In the boundary layer, micro and macro turbulent eddies are formed due to a rough surface, causing an enhanced velocity gradient. This following velocity gradient develops resistance in the course of fluid flow. With the advantages of the coatings, surface roughness is ablated by multiple folds, thereby enhancing the fluid flow by reducing the flow turbulence [7, 15]. Due to this phenomenon, overall performance, i.e., head and flow characteristics, is enhanced. Explaining further, the surface roughness of a cast iron substrate is Ra 18.8 μm. However, the surface roughness of a flu glide coating is as low as 0.4–0.08 μm (Fig. 10), which depicts the lowered coefficient of friction between the fluid and coating applied on the hydrophobic nature of the layer, which reduces the pump's power consumption, which results in enhanced pump efficiency in agreement with the findings of the work reported on the pump performance enchantment of centrifugal pump by minimizing casing losses [19].

The application of the coating in the hydrophobic passages of circulating water pumps using polymer coating resulted in an advantage to seal and protect new equipment exposed to erosion and corrosion, protecting pump impeller blades, casings, and water boxes that provide increased efficiency pumps (Table 1). In addition, based

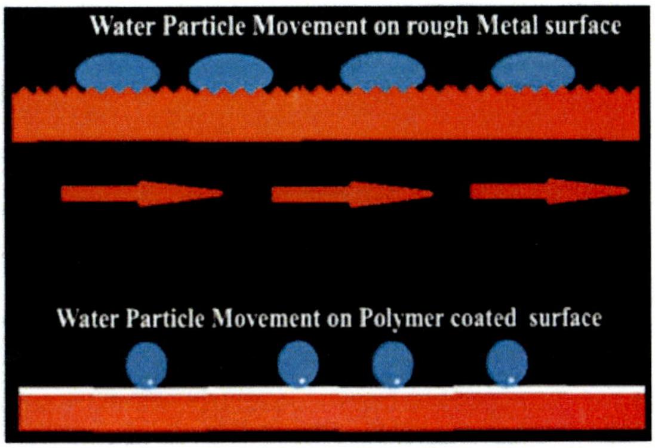

Fig. 10 A Hypothetical water movement in different types of surfaces to visualize

Table 1 CWP performance test results

S. No.	Parameters	Before coating (Fig. 7)	After coating (Fig. 9)
1	Flow in m^3/hr	15,862	16,662
2	Observed head in meter	22	22
3	Power consumed in KW	1255.17	1255.17
4	Pump efficiency in %	84.17	88.42

on the analysis and assessment of the performance obtained before and after the coating, the flow rate enhances along with the pump's efficiency increase (Table 1) and is comparable to the investigation reported [10].

5 Conclusions

In the present study conducted in a thermal power plant, there is an increase in flow rate of about 800 m^3/hr, which will increase the heat transfer in the condenser, thereby increasing the overall efficiency of the thermal power plant. After the coating, the increase in pump efficiency was observed from 84.17 to 88.42%, corresponding to the flow rate of 15,862 to 16,662 m^3/hr resulting in an enhanced pump's efficiency of 4.25% to handle an increased flow rate of 800 m^3/hr which accounts for an overall saving of power for the installed motor capacity of 1255 KW power generation from NLCIL plant supplying southern parts of India. The mentioned coating methods offer sealed and protected equipment exposed to erosion and corrosion in better long-term durability, and also the pump casings, impeller blades, and water boxes are protected. The scope of the study can be applied in various other sectors like water supply pumping mains and distribution systems and industries handling corrosive chemicals and fluids for transfer operations such as petrochemical industries, textile manufacturing industries, paper mills, and leather pharmaceuticals industries. Adopting improved coat-able materials in the above-mentioned industrial pumping lines involving similar transfer applications will decrease energy requirements for similar working demands.

References

1. Lee DI, Lim HC (2009) Erosion-corrosion of water-pump impeller. Int J Automot Technol 10(5):629–634
2. Rahmani Kh, Jadidian R, Haghtalab S (2015) Evaluation of inhibitors and biocides on the corrosion, scaling and biofouling control of carbon steel and copper–nickel alloys in a power plant cooling water system. Desalination 393(1):174–185
3. Chawala V, Gurbuxani PG, Bhagure G (2012) Corrosion of water pipes : a comprehensive study of deposits. J Miner Mater Charact Eng 11(05)
4. Ulaganathan K, Karuppuswamy P (2018) Study the effects of surface coating on impeller and volute casing to improve the efficiency of the water pump. Desalin Water Treat 132:79–88
5. Michael T, Nina S, Rostislav D, Jozef K, Christoph C, Christian M (2016) Advanced characterization methods for wear resistant hard coatings: a review on recent. Surf Coat Technol 285:31–46
6. Angela C, Andrea F, Luigi C (2020) Recent developments in coating technologies for adsorption heat pumps: a review. Coatings 10(9):855
7. Kouhi S, Ghamari B, Yeganeh R (2018) The effect of nano particle coating on anticorrosion performance of centrifugal pump blades. Jordan J Mech Ind Eng 12:117–122
8. Oharriz O (2016) Internal corrosion management of process vessels in the oil and gas industry. In: 2016 NACE corrosion risk management conference. NACE International, Houston, TX
9. Postlethwaite J, Nešić S (2000) Erosion-corrosion in single and multiphase flow. In: Revie W (ed) Uhlig's corrosion handbook, 2nd edn. Wiley, Hoboken, NJ
10. Kocaaslam O, Ozgoren M, Aksoy MH, Babayigit O (2016) Experimental and numerical investigation of coating effect on pump impeller and volute. J Appl Fluid Mech 9:2475–2487
11. Volkov AV, Khovanov GP, Fialova S (2015) Analysis of the effect of hydrophobic properties of surfaces in the flow part of centrifugal pumps on their operational performance. Therm Eng 62:817–824
12. Deshmukh D (2019) Samad A CFD based analysis for finding critical wall roughness on centrifugal pump at design and off design conditions. J Braz Soc Mech Sci Eng 41:1–18
13. ASTM D4541-17 (2017) Standard test method for pull-off strength of coatings using portable adhesion testers. ASTM, West Conshohocken, PA
14. ISO 4624:2016 (2016) Paints and varnishes—pull-off test for adhesion. ISO, Geneva, Switzerland
15. Luo K, Wang Y, Liu H, Chan J (2019) Effect of coating thickness on a solid–liquid two phase flow centrifugal pump under water medium. J Mech Eng 65:251–261
16. ISO 2812-1:2017 (2017) Paints and varnishes—determination of resistance to liquids—Part 1: Immersion in liquids other than water. ISO, Geneva, Switzerland

17. NACE No. 2/SSPC-SP 10. Near-white metal blast cleaning. NACE, Houston, TX
18. ISO 8501-1:2007 (2007) Preparation of steel substrates before application of paints and related products—visual assessment of surface cleanliness—Part 1: Rust grades and preparation grades of uncoated steel substrates and of steel substrates after overall removal of previous coatings. ISO, Geneva, Switzerland
19. Deshmukh PA, Deshmukh KD, Mandhare NA (2020) Performance enhancement of centrifugal pump by minimizing casing losses using coating. SN Appl Sci 2:252

Experimental Investigation of Modified Staggered V-Ribs Type Artificially Roughened Solar Air Heater

Prabudh Morya, Ghanshyam Das Agrawal, and Rohit Misra

Abstract In the proposed research, the heat transfer and friction attributes of an artificially roughened solar air heater (SAH) having modified staggered V-ribs as a turbulence inducer are presented. The roughness parameters taken into account are relative staggered rib pitch (P′/P) of 0.4 to 0.7, relative roughness pitch (P/e) of 10, relative staggered width position (w/g) of 0.4–0.7, angle of attack (a) of 60°, relative gap width (g/e) of 4, number of gaps (Ng) of 3, relative roughness height (e/D) of 0.043, and the Reynolds number (Re) ranging from 6000 to 18,000. The effect of variation of parameters, namely relative staggered rib pitch (P′/P) and relative staggered width position (w/g), on the performance of artificially roughened solar air heater is examined. The maximum elevation in the values of Nusselt number (Nu) and friction factor (f) for modified staggered V-ribs surface was noticed to be 3.94 and 3.92 times that of smooth surface, respectively, in case of varying P′/P while in case of w/g variation, the values of Nu and f were 3.87 and 3.92 times that of the surface without roughness, respectively. The utmost value of thermohydraulic performance parameter was observed to be 3.22 at P′/P = 0.6 and w/g = 0.5 for Re = 12,000.

Keywords Solar air heater · Roughness parameters · Nusselt number · Friction factor

P. Morya · G. Das Agrawal
Department of Mechanical Engineering, Malaviya National Institute of Technology, Jaipur, India
e-mail: 2019rme9513@mnit.ac.in

G. Das Agrawal
e-mail: rohiteca@ecajmer.ac.in

R. Misra (✉)
Department of Mechanical Engineering, Govt. Engineering College, Ajmer, India
e-mail: gdagarwal.mech@mnit.ac.in

D. P. Mishra et al. (eds.), *Recent Trends in Thermal and Fluid Sciences*, Lecture Notes in Mechanical Engineering, https://doi.org/10.1007/978-981-19-3498-8_28

1 Introduction

The declining available energy assets, increased pollution, and fast degradation of conventional energy sources have made an immediate scenario to concentrate on the exertion of inexhaustible energy sources such as hydro, solar, and wind. One of the cleanest, green, accessible, and readily available sources of energy is solar energy. A solar air heater (SAH) is an uncomplicated device that utilizes solar radiation to heat the working medium, which is air. The solar heating of air has wide applications in many areas, especially in colder countries, for the warming of houses, preventing the ice formation and rusting of equipment, medical purposes, drying of crops, agricultural activities, etc. The critical components of a solar air heater are one or more transparent covers, an absorber plate, passage for the movement of air, and insulation around the sides and bottom of the outer structure of the device. Air acts as a working fluid, and a blower is used for supplying the air. However, the major drawback of this device is the poor heat transfer rate and low efficiency due to radiative and convective losses. Currently, researchers try to impart roughness on the absorber part, which not only enhances the efficiency but also optimizes the power consumption of the blower. An experimental and numerical study of a v-rib roughened solar air heater was examined by Patel and Lanjewar [1]. The roughness parameters taken into account were the pitch of rib to height ratio (p/e), the rib height to hydraulic diameter (e/Dh) ratio, the gap width to rib height (g/e) ratio, the angle of attack (α), the staggered rib pitch to pitch (p′/p) ratio, the ratio of staggered rib size to rib height (r/e), the relative gap width of the additional gap (g′/e), number of gaps (Ng), and Reynolds number (Re). The thermohydraulic performance parameter value at $p/e = 10$ was observed to be highest and varying in the range of 1.14–1.59. Gill et al. [2] studied an experimental work to evaluate the efficiency of broken arc rib in a SAH for values of Reynold number 2000–16,000 having the position of relative gap variation of 0.2–0.8 and realized that at 65% relative gap, the values of Nu and f were highest, but declination on either side of this gap position was noticed. Maithani and Saini [3], through an experimental estimation, evaluated the heat transfer and friction factor aspects of V-ribs roughened type SAH with symmetrical gaps. The range of Reynolds number considered was 4000–18,000, and the highest augmentation in the values of f and Nu was discovered to be 3.67 and 3.6 times that of surface without roughness, respectively. Heydari and Mesgarpour [4] performed both experimental and numerical analyses to observe the reverberation of helical-shaped channeling on a SAH. The results obtained reveal that there was enough increase in the average thermal efficiency (14.7%) of the proposed setup when compared to the simple duct. Poongavanam et al. [5] conducted an analysis to study the V-corrugated absorber surface type SAH. The shot blasting process was used to roughen the absorber surface so that turbulence could be created and the dead regions inside SAH could be reduced. The main conclusion drawn from the study was that the friction factor, as well as the Nusselt number for the absorber surface having V-corrugation, is higher than the usual. The maximum augmentation in thermohydraulic performance factor was observed to be in the range of 1.35–1.56 than that of the standard absorber plate type solar air heater. Aharwal et al. [6] presented an experimental study on discrete square inclined ribs roughened solar air heaters and observed that maximum augmentation in Nu (2.83) and f (3.60) as juxtaposed to the surface without roughness was noticed. Chamoli and Thakur [7] opted analytical method to study a SAH having V-down perforation, and exergy methodology had been adapted for optimizing roughness parameters. The exergetic efficiency of such type of SAH was noticed to be 76% higher than the one having a smooth absorber plate.

This work covers the experimental analysis of variation of two parameters, namely P′/P and w/g, to evaluate the performance of the proposed SAH, along with the investigation of thermohydraulic performance parameters.

2 Details of the Experimental Setup

2.1 *Experimental Setup*

A full-scale indoor setup of the proposed SAH was designed and installed at the energy lab of the Mechanical engineering department of MNIT Jaipur. The setup (as shown in Figs. 1 and 2) comprises a rectangular duct (wooden) with the dimension of 1900 mm × 300 mm × 30 mm. The rectangular duct was divided into sections such as an inlet section (entry) of length 500 mm, a test section of length 1000 mm, and an exit (outlet) section of length 400 mm. As per ASHRAE standard (93–1986) [8], the dimension of the entry was considered as $5\sqrt{W*H}$ and the dimension of the exit section as $2.5\sqrt{W*H}$. The top horizontal side absorber plate of the rectangular duct was fabricated using galvanized iron (GI) of dimensions 1000 mm × 300 mm, and it served the purpose of a test section. The absorber surface was colored black so that the incident solar radiations could be enhanced.

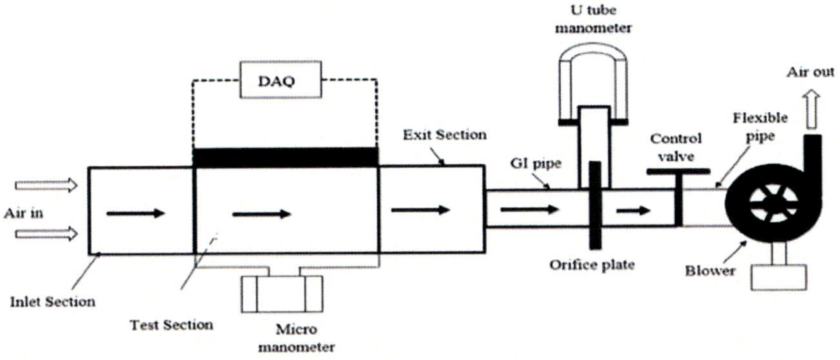

Fig. 1 Schematic representation of the experimental setup

Fig. 2 Photographic view of the experimental setup

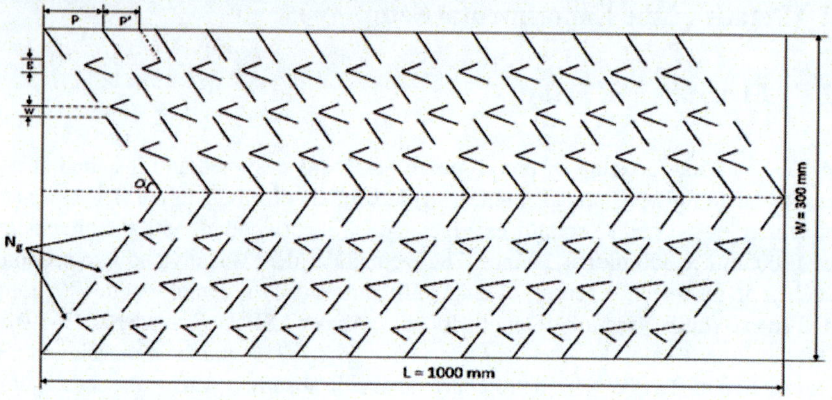

Fig. 3 Representation of artificial geometry on absorber plate

Artificial roughness on the absorber plate was created by attaching GI wires (2.3 mm diameter) in a broken V-rib pattern (pitch of 23 mm) combined with modified staggered ribs as shown in Fig. 3 (2D representation). The GI wires were glued to the absorber plate with the help of adhesive (Araldite) after performing proper marking on the absorber plate.

2.2 *Experimental Procedure*

To assess the thermal performance of the proposed SAH having modified staggered V-shaped ribs pasted on the absorber plate, first, the experimentation is performed on the SAH of having absorber plate without roughness by directing heat flux of intensity 1000 W/m^2 using a solar simulator. Then, the air was circulated inside the duct by a blower (centrifugal and induced type) placed at the outlet of the SAH duct and was allowed to flow through the duct, and the flow was regulated using a regulating valve. The air introduced inside the duct was at atmospheric temperature. The temperature values were measured with the help of thermocouples attached to the absorber plate (fifteen thermocouples), one thermocouple at the entry part of the test section and five thermocouples at the exit were wired to the data acquisition system for data display and storage. With time, the temperature of the absorber surface increases due to incident radiations, and then the accumulated heat from the smooth plate is transferred to the flowing air inside the duct. A steady state was assumed to be achieved when no variation in temperature was noticed for the next fifteen minutes.

For pressure drop measurement occurring across the rectangular duct, a digital manometer was used. Thereafter, variation of Re varying from 6000 to 18,000 was taken into consideration. From the obtained experimental values of temperature and pressure drop, values of Nu and f in relation to the Reynolds number were calculated and were compared with the standard values of Nu and f obtained for the surface without roughness.

Further, it is noticed that the experimentally obtained values for smooth absorber plate are more or less close to the theoretical standard values with a maximum deviation of 5.85 and 6%, respectively, which argues the practicality and preciseness of the experimental work. As the obtained results validate experimentation for the surface without roughness, the same procedure is adopted for evaluating the thermohydraulic performance parameter of the SAH with modified staggered V-ribs.

3 Results and Discussion

The mentioned work was conducted between February and March, and data are recorded accordingly. The thermal attributes of the SAH having V-shaped ribs combined with V-shaped staggered elements, i.e. modified staggered ribs, have been thoroughly explored based on data obtained from the experimental study. Experimental research has been conducted for distinct roughness parameters with Re varying from 6000 to 18,000.

3.1 Effect of Relative Staggered Rib Pitch

The change in values of Nu and f with the relative staggered rib pitch at distinct values of Re is depicted in Fig. 4. As the P′/P values elevated from 0.4 to 0.6, the Nusselt number improves and reaches the maximum value of 211.16 at 0.6 for Re = 18,000. The reason for such behavior is the formation of secondary flow and flow diversion by staggered ribs and having more acceleration of flow through multiple gaps. The flow separation and reattachment points are a function of the placement of the staggered rib element, so it is evident that the V-shaped staggered piece is placed

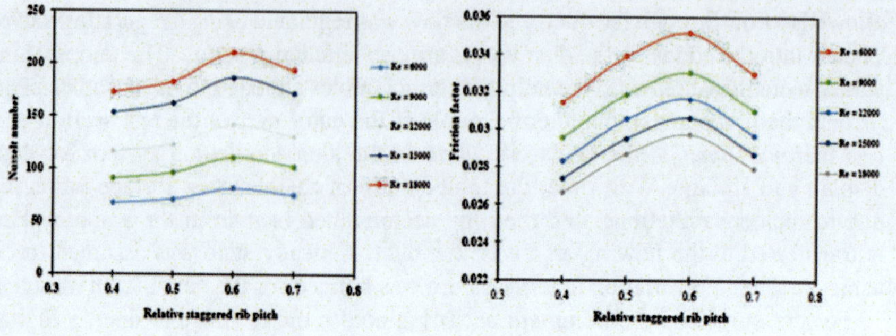

Fig. 4 Change in Nu and f values with the relative staggered rib pitch

in the midway of two inclined ribs, which helps in separating the flow and extracting maximum energy, thus resulting in more significant heat movement.

Up to the relative staggered rib pitch of 0.6, these factors are dominant and promote heat exchange. In contrast, it is well known that heat exchange enhancement using roughness elements also increases friction factor, which massively boosts up power requirements.

3.2 *Effect of the Relative Staggered Width Position*

The change in values of Nu and f with the relative staggered width position (w/g) at distinct values of Re is depicted in Fig. 5. As relative staggered width position elevated from 0.4 to 0.7, the Nusselt number improves and reaches to maximum value of 211.16 at w/g = 0.5 for Re = 18,000. For w/g = 0.4–0.7, the Nusselt number enhancement ratio is noticed to be 3.12–3.43, 3.12–3.9, 3.30–3.6, and 3.1–3.51, respectively for Re values of 6000–18,000. The maximum augmentation in the Nu is observed to be 3.9 as compared to the surface without roughness at w/g = 0.6 and Re = 12,000. In contrast, as the relative staggered width position increases from 0.4 to 0.7, the friction faction factor begins to elevate and reaches to maximum

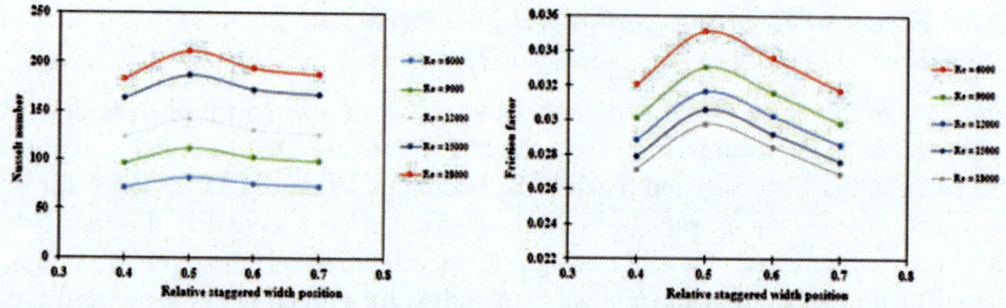

Fig. 5 Change in the values of Nu and f with the staggered width position

value of 0.034 at w/g = 0.6 for Re = 6000. For the relative staggered width position of 0.4–0.7, the value of the friction factor enhancement ratio is noticed to be 3.38–3.58, 3.67–3.92, 3.54–3.67, and 3.32–3.54, respectively, for all values of Re. The maximum augmentation in f is found to be 3.92 at w/g = 0.5 concerning the smooth duct.

3.3 Thermohydraulic Performance

Thermohydraulic performance computation is a crucial basis for concluding the thermal as well as pressure drop occurring inside SAH. To justify the utilization of artificial roughness pasted or soldered on the absorber plate of the SAH, the thermo-hydraulic performance computation is performed. This parameter, as represented in Eq. 1, is the ratio of the Nu and f for roughened element to that of the smooth element. The greater value of the parameter indicates that the proposed solar air heater will have efficient performance. This parameter, as presented by Webb and Eckert [9], can be expressed as

$$THPP(\eta) = \left(\frac{Nu_r}{Nu_s}\right) \Big/ \left(\frac{f_r}{f_s}\right)^{1/3} \quad (1)$$

In the case of varying relative staggered rib pitch, a similar trend in thermohydraulic performance is noticed for all P'/P values for the Reynolds number values varying from 6000 to 18,000. Firstly, a fall in the thermohydraulic performance values is realized for Re values of 6000–9000, which might be because of the governing effect of the enhancement in both Nu and f values for the available range of Re. According to observations, the augmentation in the thermohydraulic performance is the steepest at Re equals to 12,000 for all the relative staggered rib pitch values as the ratio of the roughened friction factor and the smooth friction factor is analyzed to be the lowest. The domination of the friction factor comes into play, and the thermohydraulic performance parameter begins to decrease after Re = 12,000 for all P′/P values. The maximum value of the thermohydraulic performance parameter is noticed to be 3.22 at Re = 12,000 for P′/P = 0.6, as represented in Fig. 6. For P′/P equal to 0.4, 0.5, 0.6, and 0.7, the maximum thermohydraulic performance parameters are 2.95, 2.90, 3.22, and 3.15, respectively, for the value of Re varying from 6000 to 18,000. In the case of varying the relative staggered width position (w/g), a similar trend in thermohydraulic performance is noticed for w/g = 0.4–0.7 at the Reynolds number values varying from 6000 to 18,000. The value of the thermohydraulic performance parameter is noticed to be 3.22 (maximum) at Re = 12,000 for w/g = 0.5, as shown in Fig. 5. For w/g equal to 0.4, 0.5, 0.6, and 0.7, the maximum thermohydraulic performance parameters are 3.07, 3.22, 3.10, and 3.17, respectively for Re = 6000 to 18,000.

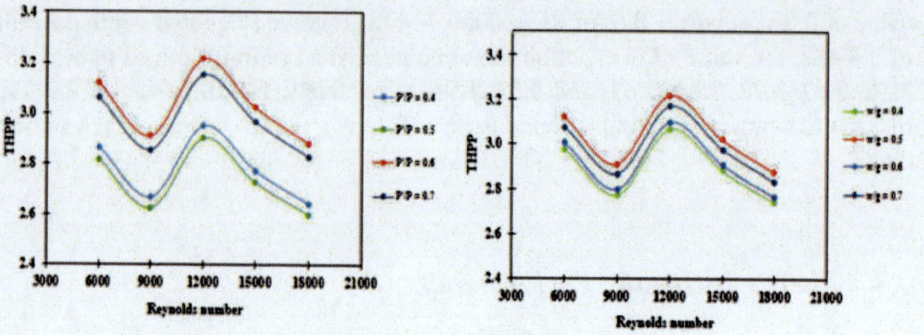

Fig. 6 Graphical representation of thermohydraulic performance parameter

4 Conclusion

The heat transfer and friction characteristics of the proposed solar air heater possessing modified staggered V-shaped ribs on the absorber surface were experimentally evaluated, and the following notable findings are perceived:

i. A relevant augmentation in the heat relocation between absorber surface and working medium and friction factor is spotted by inserting a staggering V-shaped piece.
ii. The maximum lift in the heat transfer was noticed at P′/P of 0.6 for all available Re values, and the Nusselt number enhancement value is observed to be 3.9 at Reynolds number = 15,000.
iii. The friction factor for the present study was observed to be 3.9 times higher than the conventional surface at P′/P of 0.6 and Re = 18,000.
iv. In case of varying w/g, the maximum Nusselt and friction factor enhancement values were noticed to be 3.9 at w/g = 0.6 (Re = 12,000) and 3.92 at w/g = 0.5, respectively.
v. The maximum thermohydraulic performance parameter was found as 3.22 at P′/P = 0.6 and w/g = 0.5 for Re = 12,000.

5 The Research Scope for Future Work

Based on the present research, the following points that can be implemented are indicated below:

(a) Effect of variation of other roughness parameters like angle of attack, rib roughness height, etc., can be used to investigate the performance of modified staggered V-shaped roughened solar air heater.
(b) The study can be carried out in an actual environment, and a comparison can be made with the present study.

(c) Exergy analysis gives an insight into useful energy, so such analysis can be performed.
(d) Real-life implementation of SAH in industries can be studied.

References

1. Singh Patel S, Lanjewar A (2019) Experimental and numerical investigation of solar air heater with novel V-rib geometry. J Energy Storage 21:750–764. https://doi.org/10.1016/j.est.2019.01.016
2. Gill RS, Hans VS, Singh S (2017) Investigations on thermo-hydraulic performance of broken arc rib in a rectangular duct of solar air heater. Int Commun Heat Mass Transf 88:20–27. https://doi.org/10.1016/j.icheatmasstransfer.2017.07.024
3. Maithani R, Saini JS (2017) Heat transfer and fluid flow behaviour of a rectangular duct roughened with V-ribs with symmetrical gaps. Int J Ambient Energy 38:347–355. https://doi.org/10.1080/01430750.2015.1100681
4. Heydari A, Mesgarpour M (2018) Experimental analysis and numerical modeling of solar air heater with helical flow path. Sol Energy 162:278–288. https://doi.org/10.1016/j.solener.2018.01.030
5. Poongavanam GK, Panchabikesan K, Leo AJD, Ramalingam V (2018) Experimental investigation on heat transfer augmentation of solar air heater using shot blasted V-corrugated absorber plate. Renew Energy 127:213–229. https://doi.org/10.1016/j.renene.2018.04.056
6. Aharwal KR, Gandhi BK, Saini JS (2009) Heat transfer and friction characteristics of solar air heater ducts having integral inclined discrete ribs on absorber plate. Int J Heat Mass Transf 52:5970–5977. https://doi.org/10.1016/j.ijheatmasstransfer.2009.05.032
7. Chamoli S, Thakur NS (2014) Exergetic performance evaluation of solar air heater having V-down perforated baffles on the absorber plate. J Therm Anal Calorim 117:909–923. https://doi.org/10.1007/s10973-014-3765-8
8. LPTS_ASHRAE_Standard_93-1986_solar_collectors.pdf
9. Webb RL, Eckert ERG (1972) Application of rough surfaces to heat exchnager design. Int J Heat Mass Transf 15:1647–1658